The Quest to HarmonyOS App Development

李宁 ◉ 编著

人 民 邮 电 出 版 社
北 京

图书在版编目（CIP）数据

鸿蒙征途 : App开发实战 / 李宁编著. -- 北京 : 人民邮电出版社, 2021.8
ISBN 978-7-115-56386-6

Ⅰ. ①鸿… Ⅱ. ①李… Ⅲ. ①移动终端－应用程序－程序设计 Ⅳ. ①TN929.53

中国版本图书馆CIP数据核字(2021)第068652号

内 容 提 要

本书是一本基于 Java 的鸿蒙操作系统（HarmonyOS）App 开发指南。全书系统全面、由浅入深地介绍了 HarmonyOS App 开发的必备知识、相关经验和技巧。本书以理论与实战相结合的方式向读者呈现 HarmonyOS App 开发的整个过程。全书分为两个部分共 12 章，第一部分（第 1 章～第 10 章）详细讲解 HarmonyOS App 开发所需的知识，并提供大量的真实案例代码供读者练习；第二部分（第 11 章和第 12 章）给出两个实战项目，分别为应用类 App 项目（在线电子词典）和游戏类 App 项目（俄罗斯方块），通过这两个项目，读者可以很好地回顾和总结前面 10 章讲解的知识点，并付诸实践。

本书内容通俗易懂，循序渐进，既是 HarmonyOS 初学者的入门图书，也是 HarmonyOS 开发人员的进阶读物。

◆ 编　　著　李　宁
　责任编辑　张　涛
　责任印制　王　郁　焦志炜

◆ 人民邮电出版社出版发行　　北京市丰台区成寿寺路 11 号
　邮编　100164　　电子邮件　315@ptpress.com.cn
　网址　https://www.ptpress.com.cn
　北京市艺辉印刷有限公司印刷

◆ 开本：800×1000　1/16
　印张：18.5
　字数：418 千字　　　　2021 年 8 月第 1 版
　印数：1 – 3 000 册　　　2021 年 8 月北京第 1 次印刷

定价：89.80 元

读者服务热线：(010)81055410　印装质量热线：(010)81055316
反盗版热线：(010)81055315
广告经营许可证：京东市监广登字 20170147 号

前　言

作为华为认证的“首批 HarmonyOS 系统课程开发者”，我在 2020 年 8 月应邀到华为总部学习，有幸在鸿蒙操作系统（HarmonyOS）宣布升级至 2.0 版本（2020 年 9 月 10 日）前一个多月接触到 HarmonyOS，HarmonyOS 的理念和技术让我感到非常震撼！

HarmonyOS 与 Android 和 iOS 一样，是独立的操作系统，支持多种硬件设备，包括智能手机、平板电脑、个人计算机、电视机、智能手表等。但 HarmonyOS 又与 Android 和 iOS 不同，HarmonyOS 的侧重点是物联网（Internet of Thing，IoT），因为 21 世纪将是物联网的世纪，也可以称为“物联网元世纪”。随着 5G 时代以及以后的 6G 时代、7G 时代的来临，大量的物联网设备将得到网络支持，不再是独立的存在，而是像连入全球物联网大脑的一个神经元。数以万亿计的物联网设备产生的数据将在这颗全球物联网大脑中流动，产生难以置信的“力量”。华为公司正是看准了这个爆发点，才顺应时代的潮流推出了面向物联网的 HarmonyOS。

作为首批 HarmonyOS 系统课程开发者，我承担着支持 HarmonyOS 生态发展的职责。而 HarmonyOS 生态发展的首要任务就是让广大的程序员了解 HarmonyOS 并为 HarmonyOS 开发 App。所以我在哔哩哔哩网站我的主页（https://exl.ptpress.cn:8442/ex/l/db0f514e）上传了多套免费的 HarmonyOS 视频课程。但有很多学员提出，视频用来演示开发过程比较好，想查询 HarmonyOS 的某项技术并不方便。因此，为了满足大量学员的需求，我决定将数月的 HarmonyOS App 开发的实践经验，以及遇到的各种问题写成一本书，以方便广大学员配合我上传在哔哩哔哩网站上的免费视频课程学习 HarmonyOS App 开发。

在华为正式推出安装了 HarmonyOS 的手机（后文中简称为“HarmonyOS 手机”）或大范围开放手机安装 HarmonyOS 之前，大多数人只能使用 HarmonyOS 模拟器来体验 HarmonyOS App 开发。不过由于和华为有合作的关系，我提前得到了几部用于测试的 HarmonyOS 手机，其实就是安装了 HarmonyOS 的 P40。使用 HarmonyOS 模拟器可以测试大部分功能，但涉及硬件（如摄像头、传感器、蓝牙等）的功能就无法测试了，所以要完整测试 HarmonyOS 的各种功能，就必须使用 HarmonyOS 手机或其他安装了 HarmonyOS 的设备。本书的内容正是基于这些 HarmonyOS 手机完成的。不管正在阅读本书的读者是否拥有 HarmonyOS 手机，都可以通过本书体验使用 HarmonyOS 手机的感觉。

经过几个月来不断在网上输出关于 HarmonyOS 的内容（主要是文章和视频），我还收集了网友反馈的很多关于 HarmonyOS 的问题，其中被问得最多的一个问题就是 HarmonyOS 使用什么语言做开发，是否容易学习。关于这个问题，读者不用担心，HarmonyOS 的核心开发语言是 Java，而目前 Java 语言在程序员群体中非常流行，并且各种学习资料极其丰富。

HarmonyOS 除了开发语言是 Java，开发方式也与 Android 类似。这样设计主要是为了兼容 Android，所以，如果读者对 Android 开发有一定的了解，学习 HarmonyOS 开发会更容易。

写作本书时，我使用的 IDE 和 HarmonyOS 的版本，都是最新的。华为在 2020 年 12 月发布了 HarmonyOS 2.0 手机开发者 Beta 版本，本书中的所有代码都是基于这个版本编写的。

本书的读者将有幸成为国内第一批从事 HarmonyOS App 开发的开发人员，同时，本书是一本基于 Java 的 HarmonyOS App 开发教程，而且是基于 HarmonyOS 2.0 编写的。我真诚地希望读者可以用心阅读这本书，多掌握一项技能，为自己未来的求职和工作增加更多的筹码，让自己获得更多的机会和自信。让我们开启 HarmonyOS App 开发之旅吧！

建议的阅读方式

本书内容通俗易懂，由浅入深，不仅适合初学者入门，还适合专业开发人员阅读。在学习本书内容之前，读者需要有 Java 基础，但并不需要有 Android 基础。如果读者以前没有接触过 Java，可以到哔哩哔哩网站我的主页学习免费的 Java 视频课程。尽管 Android 并不是阅读本书的必备技能，但了解 Android 有助于学习 HarmonyOS App 开发。如果读者想对比 Android 一起学习，也可以到哔哩哔哩网站我的主页学习免费的 Android 视频课程。

阅读本书时，读者可以根据自身的情况来决定如何阅读。初学者可以从第 1 章开始由浅入深地学习本书的内容，这样做可以不费劲儿地理解所学的内容。已经有一定的 Java 开发经验，并且从事过或正在从事 Android App 开发的开发人员，可以采用跳跃式阅读的方式，挑自己感兴趣的内容阅读。无论如何阅读都请记住，本书中的每一章内容、每一行代码都非常精彩！

本书内容

本书的内容是非常系统化的，其中介绍了 HarmonyOS 目前支持的大多数核心技术，尤其是全面讲解了 HarmonyOS 的分布式特性，包括 Feature Ability 与 Feature Ability、Feature Ability 与 Particle Ability 之间的交互。全书一共 12 章，其中前 10 章是基础知识，主要包括 HarmonyOS 开发环境的配置和调试、Page Ability 和 AbilitySlice、布局、UI 组件、对话框、数据管理、Data Ability、Service Ability、多媒体、传感器、定位、蓝牙等。

第 11 章给出一个应用类 App 项目——在线电子词典。这个项目演示了如何将多种技术综合在一起完成一个复杂的 App，其中涉及数据库、资源文件、文件存储、网络爬虫、使用第三方库等知识，而且这个项目除了使用了 Java，还使用了 Python 生成离线词库，如果读者不熟悉 Python，可以到哔哩哔哩网站我的主页学习免费的 Python 视频课程。

第 12 章给出一个游戏类 App 项目——俄罗斯方块。这个项目的核心是通过 Canvas 在窗口中绘制各种游戏元素，如游戏背景、方块等，并通过复杂的算法完成各种动作，如方块的移动、快速移动、旋转等，同时支持游戏积分机制。这款游戏类 App 演示了如何利用 HarmonyOS 中的绘图功能实现可交互的图形界面。游戏本身就是一类复杂的、可交互的绘图程序，因此也是展示 HarmonyOS App 开发的一个理想的例子。

本书各章的内容相对独立，因此，读者除了可以循序渐进完成对本书的学习，还可以将本书作为参考手册，随时查阅。

资源下载

为了方便读者学习，本书提供了书中所有的源代码（包括项目的源代码）。但是，建议读者不要急着看源代码，读者最好先理解书中的内容，然后自己实现一遍，最后再参考源代码，这样学习效果会更好。要下载本书的源代码和其他资源，请扫描下面二维码关注微信公众号“极客起源”，并输入 174496 获取相关资源的下载地址。

公众号：极客起源

勘误

尽管我和孙喆思、张涛编辑以及其他为本书付出努力的人员已经尽力对本书的内容进行了反复核对，但书中难免会存在一些未被发现的错误。读者可以在人民邮电出版社异步社区上查看所有已发现的错误，也可以通过在我的公众号中输入 174496 获取最新的勘误表。如果读者在阅读本书时发现了还未被确认的错误，也欢迎直接在异步社区的本书页面上提交勘误，或者通过公众号“极客起源”以及哔哩哔哩网站我的主页联系我，我会及时对读者反馈的错误进行确认。

目　录

第1章 开启鸿蒙之旅

2020年9月10日，华为在东莞松山湖举办华为开发者大会，并在会上宣布鸿蒙操作系统升级至2.0版本（HarmonyOS 2.0）。HarmonyOS是鸿蒙的英文名字，为了方便，在本书后面的章节，基本上都会用HarmonyOS来指代鸿蒙，这也是华为推荐的名字。HarmonyOS 2.0是HarmonyOS的第2个版本，这一版本不仅支持海量的物联网（IoT）设备，还支持包括手机、平板电脑、个人计算机、电视机、智能手表等富设备。而对广大开发人员来说，他们又多了一个研究方向，也许这是一次难得的机会。所以本章将由浅入深地带领大家从HarmonyOS开发环境开始，学习HarmonyOS App的开发。

通过阅读本章，读者可以掌握：

- HarmonyOS的定位与核心技术；
- 如何搭建HarmonyOS开发环境；
- 如何创建HarmonyOS工程；
- 如何使用HarmonyOS虚拟机；
- 如何运行HarmonyOS App；
- 如何对HarmonyOS App签名；
- 如何在真机上调试HarmonyOS App；
- HarmonyOS工程的结构；
- config.json配置文件的基本结构；
- 如何开发跨设备的HarmonyOS App；
- 如何通过设置断点和输出日志来调试代码；
- 如何解决gradle下载慢的问题。

1.1 了解HarmonyOS

HarmonyOS其实在几年前就已经发布了，只不过当时是1.0版本，而且只能运行在是IoT设备[①]上。而这次开发者大会上发布的HarmonyOS 2.0，可以运行在任何设备上，这些设备包括只有

---

① IoT设备的硬件配置通常不高，这里指的硬件配置主要是内存和CPU。内存通常只有几兆字节到几十兆字节。所以无法运行像Android这样的系统。而HarmonyOS由于拥有LiteOS内核，所以可以很容易地运行在底端的IoT设备上。

几百 KB 的 IoT 设备，也包括拥有数百 GB 的高端设备。为了让广大读者了解 HarmonyOS 的前世今生，本节将从宏观上介绍 HarmonyOS 系统。

1.1.1 HarmonyOS 的由来

听到“鸿蒙”这个词，可能有很多读者会感到好奇，这套 OS 为什么叫“鸿蒙”呢？而且它的英文名为什么叫 Harmony 呢？本节将对“鸿蒙”和 Harmony 的由来做一个详细的解释。

“鸿蒙”这个名字原来是华为内部一个研究操作系统内核的技术项目的代号。几年前，华为手机的销售量飞速增长，但这引起了华为的担心。如果手机的销售量不再增长了，那么华为该怎么办呢？所以华为认为需要有自己的操作系统，而且这个操作系统一定是面向未来 IoT 的。那面向未来的 IoT 需要什么样的操作系统呢？这个系统和现有的操作系统肯定是不一样的。在万物互联的时代，孤立的设备是没有价值的，所以这个面向未来的 IoT 的操作系统一定要可以弹性部署到大大小小的设备上，让这些设备像人一样用同样的语言交流，让设备的连接变得非常容易。

另外，仅仅连接这些设备还不够，不同类型设备之间的能力差异是千差万别的，华为希望能够有机融合这些能力，这就需要让系统将这些设备看作同一个设备，也就是虚拟设备。

华为确定了操作系统的思路后，发现“鸿蒙”这个名字和华为希望做的操作系统的定位是很匹配的：“鸿蒙”取自《山海经》，有开天辟地的意思，表示华为希望开拓一个新的纪元；鸿蒙的英文名字 HarmonyOS 是后来取的，Harmony 是和谐的意思，表示 HarmonyOS 能让多设备互联互通，同时又表示产业合作、和谐共处。

1.1.2 HarmonyOS 的技术定位和目标

HarmonyOS 的技术定位是一款面向未来、面向全场景（移动办公、运动健康、社交通信、媒体娱乐等）的分布式操作系统。在传统的单设备系统能力的基础上，HarmonyOS 提出了基于同一套系统能力、适配多种终端形态的分布式理念，以支持多种终端设备。

- 对消费者而言，HarmonyOS 能够对生活场景中的各类终端进行能力整合，可以实现不同的终端设备之间的快速连接、能力互助、资源共享，匹配合适的设备，提供流畅的全场景体验。
- 对应用开发者而言，HarmonyOS 采用了多种分布式技术，使应用程序的开发实现与不同终端设备的形态差异无关。这能够让开发者聚焦上层业务逻辑，更加便捷、高效地开发应用。
- 对设备开发者而言，HarmonyOS 采用了组件化的设计方案，可以根据设备的资源能力和业务特征进行灵活裁剪，满足不同形态的终端设备对操作系统的要求。

根据 HarmonyOS 的市场定位，HarmonyOS 需要运行在各种各样的设备上，例如，我们最常用的智能手机、平板电脑、个人计算机等。还有很多物联网设备，如打印机、投影仪等。华为将这些设备称为“1 + 8 + N”，如图 1-1 所示。这里的 1 是指智能手机，8 是指强计算设备，目前可分为如下 8 类设备[①]：

---

① 强计算设备未来可能不只有 8 类，可能会更多，这里的 8 可以理解为所有的强计算设备。

- 个人计算机；
- 平板电脑；
- 电视机；
- 音响；
- 眼镜；
- 智能手表；
- 车载电脑；
- 耳机。

而这里的 N 是指泛 IoT 设备，如打印机、投影仪、门铃、电冰箱、智能手环，甚至是茶杯、闹钟、床等。也就是说，凡是带芯片[①]的设备，都应该属于泛 IoT 设备。也就是说，HarmonyOS 的最终目标是将一切有计算能力和存储能力的设备连接起来，然后再加入人工智能技术，这就是 AIoT。如果将全世界数以百亿计的 IoT 设备连接起来，并能有效地利用这些 IoT 设备的各种能力，让这些 IoT 设备高效互相协作，那么我们将会实现梦寐以求的智能城市，智能社会，甚至智能星球！

图 1-1 HarmonyOS 的“1 + 8 + N”战略

1.1.3 HarmonyOS 的系统架构

HarmonyOS 在系统架构上与 Android 类似，采用了分层设计方案，从下向上依次为内核层、系统服务层、框架层和应用层。系统功能按照“系统→子系统→功能/模块”逐级展开，在多设备部署场景下，支持根据实际需求裁剪某些非必要的子系统或功能/模块。HarmonyOS 系统架构如图 1-2 所示。

---

① 这里的“芯片”只是泛化概念，包括目前广泛流行的硅基芯片，以及未来的碳基芯片、量子芯片、生物芯片等，只要拥有计算能力和存储能力，都应该属于芯片范畴。

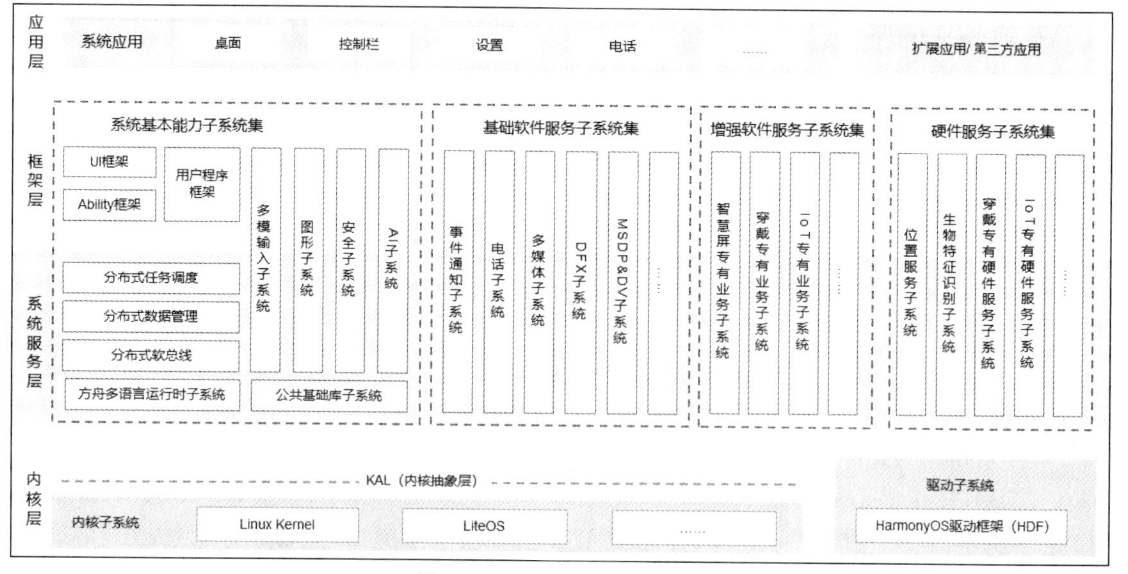

图 1-2 HarmonyOS 系统架构

下面分别对这 4 层进行解释。

1. 内核层

内核层分为内核子系统和驱动子系统。

- 内核子系统：HarmonyOS 采用多内核设计，支持针对不同资源受限的设备选用适合的 OS 内核。内核抽象层（Kernel Abstract Layer，KAL）通过屏蔽多内核差异，为上层提供基础的内核能力，包括进程/线程管理、内存管理、文件系统、网络管理和外设管理等。
- 驱动子系统：HarmonyOS 驱动框架（HDF）是 HarmonyOS 硬件生态开放的基础，它提供了统一外设访问能力以及驱动开发和管理框架。

2. 系统服务层

系统服务层是 HarmonyOS 的核心能力集合，通过框架层为应用提供服务。该层包含以下几个部分。

- 系统基本能力子系统集：为分布式应用在 HarmonyOS 多设备上的运行、调度、迁移等操作提供了基本能力，由分布式软总线、分布式数据管理、分布式任务调度、方舟多语言运行时子系统、公共基础库子系统、多模输入子系统、图形子系统、安全子系统、AI 子系统等组成。其中，方舟多语言运行时子系统提供了 C/C++/JavaScript 多语言运行时和基础的系统类库，也为使用方舟编译器静态化的 Java 程序（即应用程序或框架层中使用 Java 语言开发的部分）提供运行时服务。
- 基础软件服务子系统集：为 HarmonyOS 提供公共的、通用的软件服务，由事件通知子系统、电话子系统、多媒体子系统、DFX（Design For X）子系统 、MSDP&DV 子系统等组成。

- 增强软件服务子系统集：为 HarmonyOS 提供针对不同设备的、差异化的能力增强型软件服务，由智慧屏专有业务子系统、穿戴专有业务子系统、IoT 专有业务子系统等组成。
- 硬件服务子系统集：为 HarmonyOS 提供硬件服务，由位置服务子系统、生物特征识别子系统、穿戴专有硬件服务子系统、IoT 专有硬件服务子系统等子系统组成。

根据不同设备的部署环境，基础软件服务子系统集、增强软件服务子系统集、硬件服务子系统集内部可以按子系统粒度裁剪，每个子系统内部又可以按功能粒度裁剪。

3. 框架层

框架层为 HarmonyOS 的应用提供了 Java/C/C++/JavaScript 等多语言的用户程序框架和 Ability 框架，以及各种软硬件服务对外开放的多语言框架 API；同时为采用 HarmonyOS 的设备提供了 C/C++/JavaScript 等多语言的框架 API，不同设备支持的 API 与系统的组件化裁剪程度相关。

4. 应用层

应用层包括系统应用和扩展应用/第三方应用。HarmonyOS 的应用由一个或多个 FA（Feature Ability）或 PA（Particle Ability）组成。其中，FA 有 UI，提供与用户交互的能力；而 PA 无 UI，提供后台运行任务的能力以及统一的数据访问抽象。基于 FA/PA 开发的应用，能够实现特定的业务功能，支持跨设备任务调度与分发，为用户提供一致、高效的应用体验。

1.2 搭建 HarmonyOS 开发环境

HarmonyOS 拥有自己的集成开发环境（Integrated Development Environment，IDE）——DevEco Studio。这款 IDE 是基于 IntelliJ IDEA 社区版二次开发的，所以使用方式与 IntelliJ IDEA 非常类似。目前有很多 IDE 都是基于 IntelliJ IDEA 社区版二次开发的，例如，用于开发 Android App 的 Android Studio 就是一个非常典型的例子。DevEco Studio 之所以选择基于 IntelliJ IDEA 社区版二次开发，主要是为了让尽可能多的开发人员尽快上手，因为目前使用 IntelliJ IDEA 及其衍生 IDE 的开发人员非常多。

DevEco Studio 是跨平台的，目前 DevEco Studio 有 Windows 版和 macOS 版。读者可以到下面的页面下载 DevEco Studio 的最新版。

https://developer.harmonyos.com/cn/develop/deveco-studio

进入下载页面，找到图 1-3 所示的下载列表，单击右侧的下载链接即可下载。注意，下载 DevEco Studio 需要有一个华为账号，没有账号的读者需要注册一个。

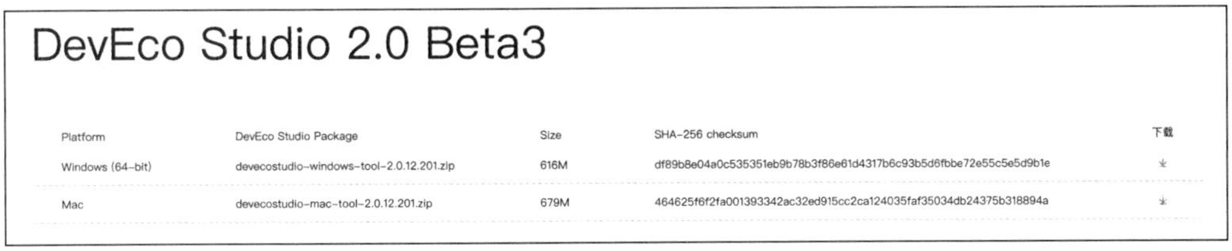

DevEco Studio 2.0 Beta3

Platform	DevEco Studio Package	Size	SHA-256 checksum	下载
Windows (64-bit)	devecostudio-windows-tool-2.0.12.201.zip	616M	df89b8e04a0c535351eb9b78b3f86e61d4317b6c93b5d6fbbe72e55c5e5d9b1e	⤓
Mac	devecostudio-mac-tool-2.0.12.201.zip	679M	464625f6f2fa001393342ac32ed915cc2ca124035faf35034db24375b318894a	⤓

图 1-3 下载 DevEco Studio

下载的 DevEco Studio 是一个 zip 压缩文件，将其解压，如果使用的是 Windows 版本，里面是一个 exe 安装文件，直接双击安装即可。如果使用的是 macOS 版本，里面是一个 dmg 文件，同样双击安装即可。安装完成后，Windows 版会在桌面生成一个 DevEco Studio 图标，macOS 版会安装在应用程序（Applications）中，双击 DevEco Studio 图标即可运行 DevEco Studio。

运行 DevEco Studio 后，会看到图 1-4 所示的欢迎界面。如果是第一次使用 DevEco Studio，右侧的 Recent Project 列表为空；如果以前创建过 HarmonyOS 工程，那么在 Recent Project 列表中会看到曾经创建过的 HarmonyOS 工程，单击某个 HarmonyOS 工程，就会打开该工程。

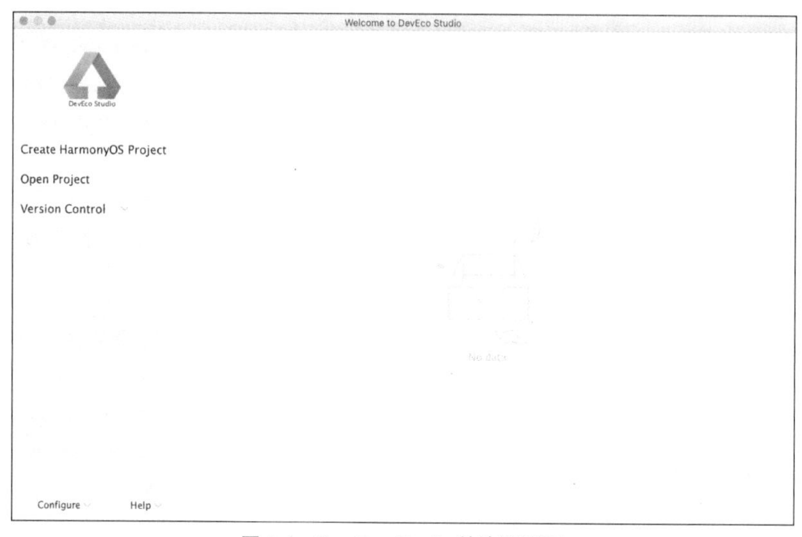

图 1-4　DevEco Studio 的欢迎界面

读者可以单击左侧的 Create HarmonyOS Project 链接创建新的 HarmonyOS 工程，或者单击 Open Project 链接打开一个 HarmonyOS 工程。下一节将会详细介绍如何使用 DevEco Studio 来开发我们的第一个 HarmonyOS App。

1.3 创建第一个 HarmonyOS App

本节会详细讲解使用 DevEco Studio 开发 HarmonyOS App 的完整过程。

1.3.1　创建 HelloWorld 项目工程

在 DevEco Studio 的欢迎界面单击 Create HarmonyOS Project 链接，或在开发主界面菜单栏选择图 1-5 所示的 File→New→New Project 选项打开 Create HarmonyOS Project 窗口。

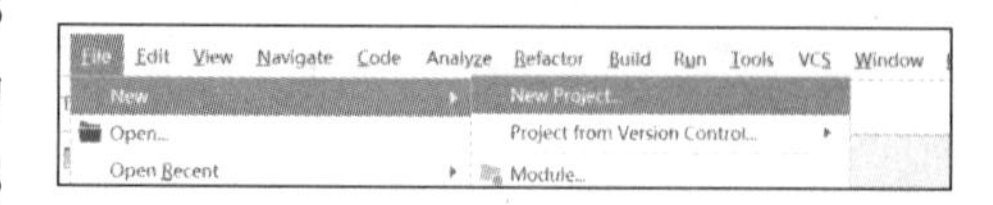

图 1-5　New Project 菜单项

打开的 Create HarmonyOS Project 窗口如图 1-6 所示。

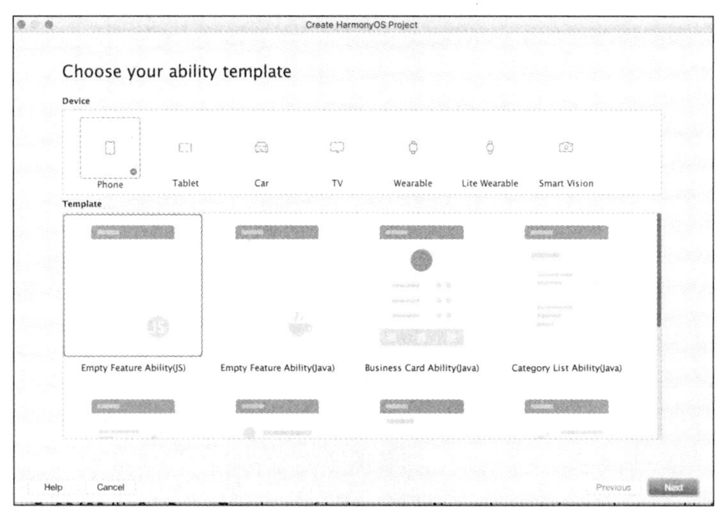

图 1-6 Create HarmonyOS Project 窗口

在创建 HarmonyOS 工程之前，先来看一下 Create HarmonyOS Project 窗口中的内容。这个窗口分为两个区域——Device 和 Template。其中，Device 区域允许用户选择设备。因为不同设备的屏幕尺寸、分辨率、API 都有所不同，所以创建 HarmonyOS 工程时需要针对不同的设备。目前 HarmonyOS 支持如下 7 种设备：

- Phone（智能手机）；
- Tablet（平板电脑）；
- Car（车载电脑）；
- TV（电视机），即华为智慧屏以及任何安装了 HarmonyOS 的智能电视；
- Wearable（高端智能手表）；
- Lite Wearable（低端智能手表）；
- Smart Vision（带摄像头的设备）。

在这 7 种设备中，因为 Lite Wearable 和 Smart Vision 的硬件配置较低，所以不支持用 Java 开发 App，只支持用 JavaScript 开发 App；另外 5 种设备同时支持用 Java 和 JavaScript 开发 App。

Java 开发的方式与 Android 类似，很多概念也与 Android 类似，这主要是为了让 Android 开发人员更容易上手开发 HarmonyOS App。JavaScript 开发的方式类似于微信小程序的开发技术。也就是说，JavaScript 开发方式采用的是 Web 栈技术。这主要是针对微信小程序开发人员和 Web 开发人员的。这就意味着，你只要是 Android 开发人员、微信小程序开发人员或 Web 开发人员，就非常容易上手 HarmonyOS App 开发，即使你不是上述开发人员，这些技术比起 C++来说也容易得多，所以对于任何领域的开发人员来说，上手 HarmonyOS App 开发都是相当容易的，关键是如何起步，而本书的目的就是让这些想进入 HarmonyOS App 开发领域的开发人员有一个好的开始。

为了让 HarmonyOS 开发人员有一个更好的开始，Create HarmonyOS Project 窗口的 Template 区域提供了若干个 HarmonyOS 工程模板来生成一些必要的代码，相当于 HarmonyOS App 的 Hello World。为不同设备提供的模板是不同的。这里以 TV 为例，为 TV 设备提供的主要模板及描述如下。

- Empty Feature Ability(JS)：一个空的 JavaScript 版 HarmonyOS 工程，包含了一个 Hello World Demo，可以在屏幕上输出 Hello World。
- Empty Feature Ability(Java)：一个空的 Java 版 HarmonyOS 工程，包含了一个 Hello World Demo，可以在屏幕上输出 Hello World。
- List Container Ability(Java)：带有列表功能的 Java 版 HarmonyOS 工程。目前很多 App 的主窗口是一个列表，如微博、微信等，所以 HarmonyOS 提供了这类模板。
- List Feature Ability(JS)：带有列表功能的 JavaScript 版 HarmonyOS 工程。
- Split Panel Ability(Java)：带分隔面板的 Java 版 HarmonyOS 工程。
- Tab Feature Ability(JS)：带标签页的 JavaScript 版 HarmonyOS 工程。

其中，Ability 是指 Feature Ability，即带 UI 的 Ability。还有一种不带 UI 的 Ability，即 PA，后面的章节会详细介绍这些内容。在本章中，读者不需要了解 FA 有哪些具体特性，以及相关的 API，只要知道 FA 类似于 Android 中的 Activity 即可。下一章会详细介绍 FA，以及 Ability Slice①的特性和相关 API。

本书主要讲解如何用 Java 开发 HarmonyOS App，所以忽略 JavaScript 版的模板，关于 JavaScript 版的内容，请大家关注我后期出版的 HarmonyOS App 开发系列图书。

本章为了从头演示创建 HarmonyOS 工程的过程，使用了 Empty Feature Ability(Java)模板。选择 TV 设备和 Empty Feature Ability(Java)模板后，单击 Next 按钮，进入 Configure your project 窗口，如图 1-7 所示，在该窗口中我们可以配置 HarmonyOS 工程。

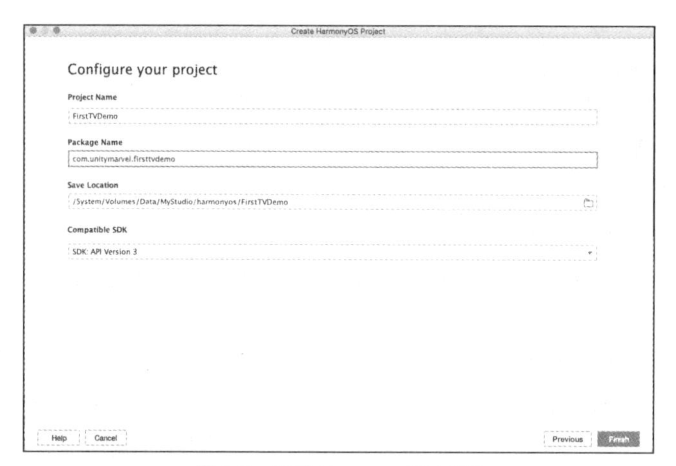

图 1-7 配置 HarmonyOS 工程

配置 HarmonyOS 工程需要指定以下 4 个信息。

- Project Name：HarmonyOS 工程名。
- Package Name：HarmonyOS 工程的包名，与 Android 中包的含义相同，用于在 HarmonyOS 设备中唯一标识 HarmonyOS App。

① Ability Slice 类似 Android 中的 Fragment。

- ❑ Save Location：指定 HarmonyOS 工程保存的目录，注意，这个目录应该是空的，DevEco Studio 会直接将与工程相关的目录和文件都放到这个保存目录中。
- ❑ Compatible SDK：HarmonyOS SDK 的 API 版本，大多数设备都有 3 和 4 两个版本，读者可以选择其中一个。

完成配置后，单击 Finish 按钮创建新的 HarmonyOS 工程，HarmonyOS 工程的结构如图 1-8 所示。

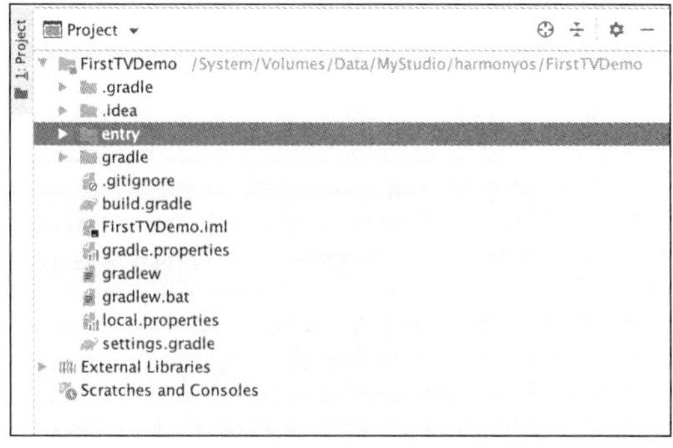

图 1-8 HarmonyOS 工程的结构

1.3.2 启动模拟器

在没有真机的情况下，只能在模拟器中运行 HarmonyOS App。不过，因为目前 HarmonyOS 还没有 x86 版的模拟器，所以我们通过 ARM 服务器模拟 HarmonyOS 的各种硬件设备，只不过这些 ARM 服务器需要通过华为账号申请。申请成功后，会在 IDE 中显示一个预览界面，可与 ARM 服务器交互。申请的过程如下。

首先在 DevEco Studio 中选择 Tools→HVD Manager 菜单项，如果是第一次选择该菜单项，会自动弹出一个华为账号登录窗口，如图 1-9 所示。

如果没有华为账号，可以单击“注册”链接进行注册，如果有华为账号，直接登录即可。

成功登录后，会弹出图 1-10 所示的对话框，提示“HUAWEI DevEco Studio 想要访问您的华为帐号”，单击“允许”按钮。

图 1-9 华为账号登录

图 1-10 授权 DevEco Studio 访问华为账号

成功授权后，回到 DevEco Studio，会弹出 Virtual Device Manager 对话框，并显示两个设备（TV 和 Wearable），如图 1-11 所示。

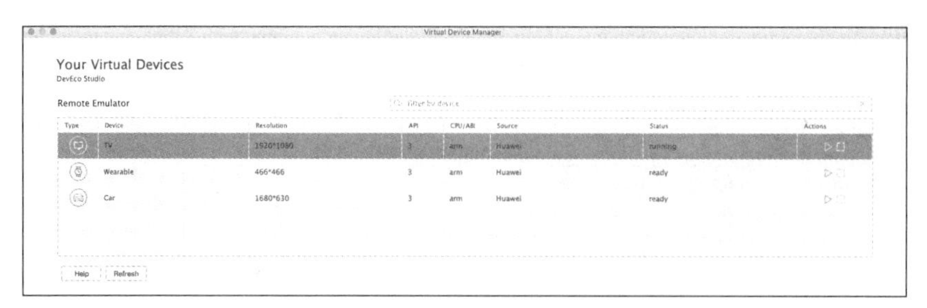

图 1-11　HarmonyOS 预览器列表

其中，TV 是华为智慧屏，Wearable 是华为智能手表。因为我们创建的工程是 TV 类型，所以双击 TV 或单击 Actions 列的三角形按钮申请 TV 预览器[1]。

成功申请 TV 模拟器后，会显示一个视图，里面是 TV 预览器，如图 1-12 所示。该预览器支持鼠标点击、滑动等操作。在 TV 预览器中右侧的按钮用于控制预览器，例如，最上面的叉号，用来释放预览器，最下面的三角形按钮是回退键。

图 1-12　TV 预览器

注意，每次申请，使用时间只有 1 小时（为了防止用户长时间占用 TV 预览器），如果过期了，再按前面的步骤重新申请即可。

1.3.3　运行 HelloWorld 工程

现在已经准备就绪，单击 IDE 上方工具栏中图 1-13 所示的三角形按钮运行 HarmonyOS App。

图 1-13　三角形按钮

单击三角形按钮后，会弹出图 1-14 所示的设备选择对话框。不管是真机，

① 这里还不能称为模拟器，因为不是在本地运行，而是在服务端通过 ARM 服务器模拟出了若干个 HarmonyOS 设备。在运行时，按一定的时间频率，将屏幕显示的数据流发送给前端（预览器），然后预览器会将鼠标和键盘的动作返给服务端，接下来服务端将响应后的屏幕结果返回给预览器。所以，预览器相当于一个外接显示器，并没有处理任务的能力。

还是模拟器，只要成功连接，都会在设备选择列表中显示。但要注意，该列表列出的不只是 HarmonyOS 设备，如果本机连接了 Android 设备，也同样会列出。因为 HarmonyOS App 是不能运行在 Android 设备上的，所以应在列表中选择 HarmonyOS 设备来运行 HarmonyOS App。

选择要运行 HarmonyOS App 的设备后，单击 OK 按钮，会在选中的设备中运行当前的 HarmonyOS App，运行效果如图 1-15 所示。

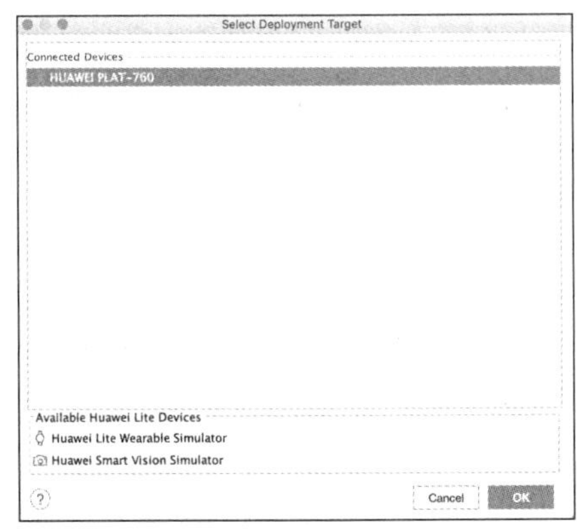

图 1-14 设备选择对话框

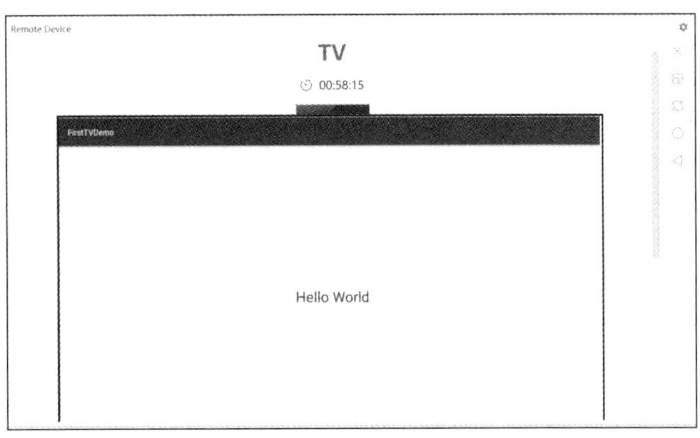

图 1-15 运行效果

1.4 分析第一个 HarmonyOS 工程

本节将分析 HarmonyOS 工程的结构，主要包括 HarmonyOS 工程的目录结构和配置文件 config.json 的核心选项。

1.4.1 HarmonyOS 工程的目录结构

HarmonyOS 工程的目录结构如图 1-16 所示。

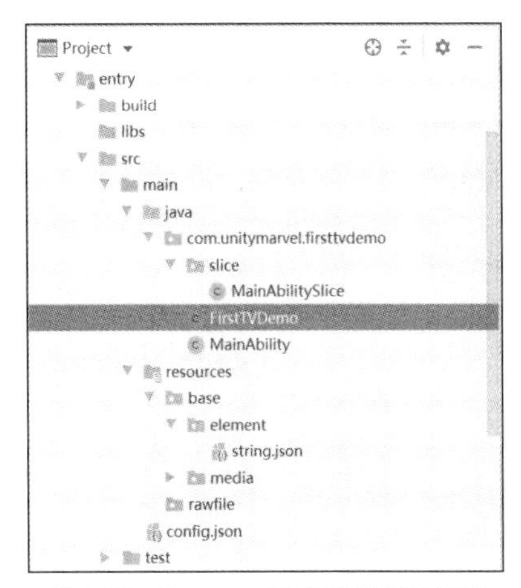

图 1-16 HarmonyOS 工程的目录结构

尽管 HarmonyOS 工程的目录结构比较复杂，但并不是所有的目录和文件都经常被使用。本节将介绍两个比较常用的目录——java 目录和 resources 目录。这两个目录的位置如下。

- java 目录：<HarmonyOS 工程根目录>/entry/src/main/java。
- resources 目录：<HarmonyOS 工程根目录>/entry/src/main/resources。

java 目录用来保存 HarmonyOS App 中的 Java 代码文件，resources 目录用来保存 HarmonyOS App 中的各种资源，如图像、字符串、数据库等。

我们打开 java 目录，会看到该目录中会有若干个 java 包，如果是新创建的 HarmonyOS 工程，里面只有一个 java 包，如本例的 com.unitymarvel.firsttvdemo。在包目录中有若干个 Java 文件，主要的编码工作会在该目录中完成。

如果 HarmonyOS App 中要使用资源，而且这些资源要被嵌入到 hap 文件中，则需要将资源文件放到 resources 目录下，后面的章节会详细描述具体规则。

1.4.2 HarmonyOS 工程的主配置文件——config.json

绝大多数工程都会有一个主配置文件，HarmonyOS 工程也不例外，这个主配置文件就是 config.json。很明显，config.json 是 JSON 格式的配置文件。

config.json 文件由 3 个顶层属性组成，这 3 个属性如表 1-1 所示。

表 1-1 config.json 的顶层属性

属性名称	含义	数据类型	是否必须设置
app	包含 App 的全局配置信息。如版本号、bundleName、vendor 等。同一个 App 的不同 HAP 包的 app 属性必须保持一致。	对象	否
deviceConfig	包含与具体设备相关的配置信息，如车载电脑（Car）、华为智慧屏（TV）、智能手表（Wearable）等。	对象	否
module	包含 HAP 包的配置信息，如 Ability 相关的信息。该标签下的配置只对当前 HAP 包生效。	对象	否

在这 3 个属性中，app 和 module 的使用比较频繁，尤其是 module 属性，修改的次数会比较多。因为 HarmonyOS App 的重要组成部分——Ability，必须在 module 中进行配置后才能使用。

DevEco Studio 提供了自动创建 Ability 的功能，在 HarmonyOS 工程中，单击鼠标右键弹出菜单，选择 New→Ability 菜单项，会弹出图 1-17 所示的创建各种类型的 Ability 的菜单项。在创建某一个类型的 Ability 后，config.json 文件中将自动添加相关的信息，通常不需要开发人员手动添加 Ability 信息，当然，开发人员可以在自动添加的 Ability 信息的基础上进行修改。

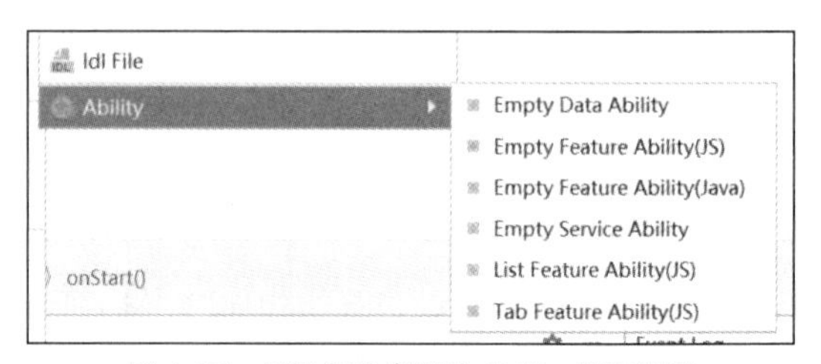

图 1-17 创建各种类型的 Ability 的菜单项

假设要创建一个名为 MyAbility 的 FA，在 config.json 中将自动生成如下配置代码：

```
{
    "orientation": "landscape",
    "formEnabled": false,
    "name": "com.unitymarvel.firsttvdemo.MyAbility",
    "icon": "$media:icon",
    "description": "$string:myability_description",
    "label": "entry",
    "type": "page",
    "launchType": "standard"
}
```

这段配置代码涉及的属性比较多，例如，orientation 用来指定默认的方向（横屏或竖屏），name 用来指定当前 Ability 的类名（包含包名），label 用来指定 Ability 的标题文本等。关于 config.json 文件中配置的详细描述，读者可以参阅官方的文档，链接如下：

https://developer.harmonyos.com/cn/docs/documentation/doc-guides/basic-config-file-elements-0000000000034463#ZH-CN_TOPIC_0000001050708780__table34298422421

1.5 开发跨设备的 HarmonyOS App

现在的智能设备种类越来越多，而且这些智能设备的屏幕尺寸、分辨率都不同。例如，比较常见的智能设备有手机、平板电脑、车载电脑、电视机、智能手表等，尽管这些设备都有屏幕，但它们的屏幕千差万别，有的屏幕尺寸小，有的屏幕尺寸大；有的是纵向的屏幕，有的是横向的屏幕；有的带触摸功能，有的不带触摸功能；甚至有的设备的屏幕是圆形的（如智能手表），这给开发 App 带来了麻烦。现在几乎每一个智能设备厂商，如 Apple、华为都面临这个问题。这就要求我们开发的 App 尽可能适合更多的智能设备。

当然，最简单直接的方式是为每一类智能设备单独开发 App。例如，为手机开发一款 App，为智能电视开发一款 App，为智能手表开发一款 App。这么做尽管从技术上这是可行的，但这些不同设备的 App，尽管在 UI 展现上不同，但大多数逻辑代码是相同的。如果为不同的设备单独开发 App，会造成大量的代码冗余。所以推荐的方案是让一个 App 同时适用于不同的智能设备。基本的原理是在 App 运行时自动检测当前设备，然后运行与特定设备相关的代码，使用与特定设备相关的布局和资源。

这个方案的关键点是检测当前的设备类型。在创建 HarmonyOS 工程时，要么创建 TV（华为智慧屏）工程，要么创建 Wearable（智能手表）工程，要么创建其他设备的工程。也就是说，使用 IDE 创建的 HarmonyOS 工程只能在一种特定的设备上运行。要想让 App 在多种设备上运行，需要设置 config.json 文件的 deviceType 属性，该属性指定了当前工程可以运行的设备类型，如果创建的是 TV 工程，deviceType 属性的值如下：

```
"deviceType": [
     "tv"
]
```

如果创建的是 Wearable 工程，deviceType 属性的值如下：

```
"deviceType": [
     "wearable"
]
```

如果 deviceType 属性的值是 tv，当前工程是不能在智能手表上运行的，反之亦然。要想让当前工程同时在 TV 和 Wearable 上运行，需要同时指定 tv 和 wearable（要手动修改 config.json 文件），配置代码如下：

```
"deviceType": [
     "tv",
     "wearable"
]
```

当完成 deviceType 属性的设置后，当前工程就可以同时在 TV 和 Wearable 上运行了。不过因为 TV 和 Wearable 的屏幕尺寸相差太大，所以布局通常会采用完全不同的样式。在 HarmonyOS 中，可以使用 Java 语言动态创建组件的布局方式，也可以使用布局文件。关于布局文件的使用，后面的章节会详细介绍。本节主要讨论使用 Java 语言动态创建组件的布局方式。

在创建的 HarmonyOS 工程中会自动生成一段样例代码，这些代码主要集中在 MainAbilitySlice.java 文件的 onStart 方法中，代码如下：

```
public void onStart(Intent intent) {

        super.onStart(intent);

        LayoutConfig config = new LayoutConfig(LayoutConfig.MATCH_PARENT,
                                               LayoutConfig.MATCH_PARENT);
        myLayout.setLayoutConfig(config);
        ShapeElement element = new ShapeElement();
        element.setRgbColor(new RgbColor(255, 255, 255));
        myLayout.setBackground(element);
        Text text = new Text(this);
        text.setLayoutConfig(config);
        text.setText("Hello World");
        text.setTextColor(new Color(0xFF000000));
        text.setTextSize(50);
        text.setTextAlignment(TextAlignment.CENTER);
        myLayout.addComponent(text);
        super.setUIContent(myLayout);
}
```

读者并不需要对这段代码的每一行都了解，只需要知道这段代码将背景颜色设为白色，并且创建了一个用于显示文本的 Text 组件，该组件会显示文本 Hello World。如果在 TV 设备上运行，效果如图 1-15 所示。不过运行代码，可以发现，在所有设备中 UI 都一样，但我们的目的是让不同的设备显示不同的 UI，所以就需要通过下面的代码判断当前设备的类型：

```
if(DeviceInfo.getDeviceType().equals("tv")) {
   ...
} else if(DeviceInfo.getDeviceType().equals("wearable")) {
   ...
}
```

其中，getDeviceType 方法返回的值就是 App 当前运行设备的类型。如果运行在 TV 上，值为 tv；如果运行在智能手表上，值为 wearable。所以，可以用下面的代码来替换 onStart 方法中的代码：

```
public void onStart(Intent intent) {
        super.onStart(intent);
        LayoutConfig config = new LayoutConfig(LayoutConfig.MATCH_PARENT,
                                               LayoutConfig.MATCH_PARENT);
        myLayout.setLayoutConfig(config);
        ShapeElement element = new ShapeElement();
        element.setRgbColor(new RgbColor(255, 255, 255));
        myLayout.setBackground(element);
        Text text = new Text(this);
        text.setLayoutConfig(config);
        if(DeviceInfo.getDeviceType().equals("tv")) {
```

```
            // 运行在 TV 上的代码
            text.setText("华为智慧屏");
            text.setTextColor(new Color(0xFFFF0000));
            text.setTextSize(200);
        } else if(DeviceInfo.getDeviceType().equals("wearable")) {
            // 运行在 Wearable 上的代码
            text.setText("华为智能手表");
            text.setTextColor(new Color(0xFF0000FF));
            text.setTextSize(50);
        }
        text.setTextAlignment(TextAlignment.CENTER);
        myLayout.addComponent(text);
        super.setUIContent(myLayout);
    }
```

在这段代码中，将 TV 和 Wearable 上显示的文本内容、文本尺寸和文本颜色做了改变，所以在 TV 和 Wearable 上显示的文本是不同的。在 TV 上显示的效果如图 1-18 所示，在 Wearable 上显示的效果如图 1-19 所示。

图 1-18 在 TV 上显示的效果

图 1-19 在 Wearable 上显示的效果

1.6 调试代码

调试是开发 App 的必备技能，毕竟任何一个比较复杂的 App 都不可能一次编码成功。如果 App 的效果与自己期望的不同，或由于某些原因导致 App 异常中断，就需要查找原因，这些都需要调试代码。

调试代码有多种方法，比较常用的两种方式是设置断点和输出日志，本节将详细介绍这两种调试 App 的方式。

1.6.1 设置断点

如果认为需要跟踪的代码或 bug 在某行代码的附近，可以单击该行代码前面的部分为该行设置断点，这时在这行代码前面会出现一个红点，如图 1-20 所示。

然后单击上方工具栏中图 1-21 所示的调试运行按钮。

```
62          }
63      });
64      System.out.println("hello world");
65
66      text.setTextAlignment(TextAlignment.CENTER);
```

图 1-20 设置断点

图 1-21 调试运行按钮

如果未遇到断点，单击该按钮会正常运行程序，一旦遇到断点，程序运行就会在断点处停止运行。直到按 F7 或 F8 键才继续一行一行运行程序，以便监控程序中相关变量的状态。F7 键表示 Step Into，Step Into 表示会跟踪到方法内部，例如，当前正在运行一个方法，按 F7 键会继续跟踪到方法内部，然后继续一行一行代码运行。F8 键表示 Step Over，Step Over 会将方法当作一行代码运行，不会跟踪到方法内部。所以，如果认为某一个方法肯定没有问题，就按 F8 键，如果认为该方法可能会有一些问题，就按 F7 键。

调试代码的过程如图 1-22 所示。我们可以看到，在下方的 Variables 视图中会显示相关变量的当前值。

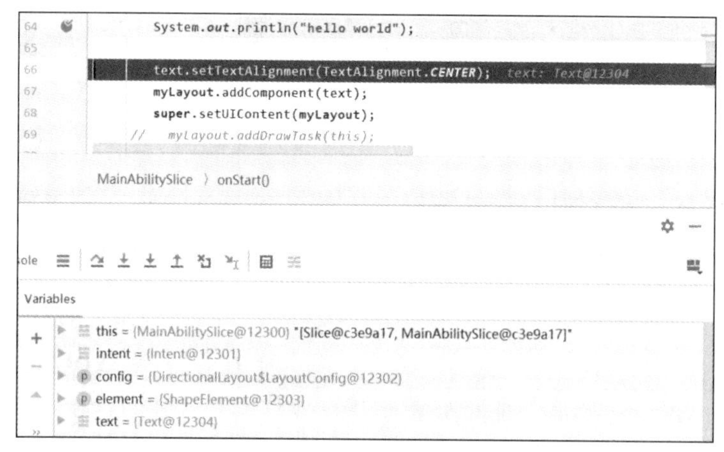

图 1-22 调试代码的过程

1.6.2 输出日志

另外一种调试代码的方式是输出日志，可以直接使用 System.out.println()将信息输出到 Logcat 视图中的方式，例如，运行下面的代码，会在 Logcat 视图中看到图 1-23 所示的信息。

```
System.out.println("hello world");
```

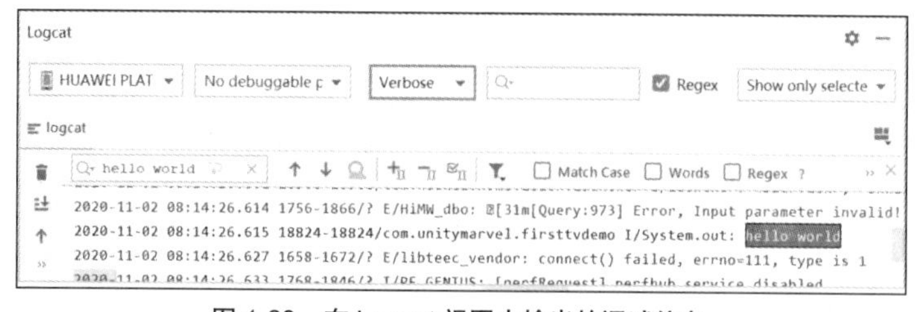

图 1-23 在 Logcat 视图中输出的调试信息

HarmonyOS 还提供了另外一种输出日志的方式，这就是 HiLog 类，该类提供了多个静态方法，用于输出不同级别的日志信息。这些静态方法及功能描述如表 1-2 所示。

表 1-2 HiLog 类中用于输出日志信息的方法

方法名	功能
debug	用于输出调试日志信息
info	用于输出普通日志信息
warn	用于输出警告日志信息
error	用于输出普通错误日志信息
fatal	用于输出致命错误日志信息

这 5 个方法的参数完全相同，例如，error 方法的原型如下：

```
public static int error(HiLogLabel label, String format, Object... args);
```

其中，label 参数表示输出日志的相关信息，类型是 HiLogLabel 对象；format 参数表示要输出的日志文本（可以包括格式化占位符）；args 参数表示格式化的参数值，是可选参数。

使用输出日志方法的案例如下：

```
HiLogLabel label = new HiLogLabel(HiLog.LOG_APP ,223, "MY_TAG");
HiLog.error(label, "这是一行错误信息，原因:%{private}s","Url 不可访问");
HiLog.warn(label,"这是一个警告，原因是：%{public}s", "变量的值可能是负数");
```

其中，HiLogLabel 构造方法的第 1 个参数表示日志类型，目前只能设置为 HiLog.LOG_APP，后续会开放更多的日志类型；第 2 个参数是 domain，就是一个整数类型；第 3 个参数是日志标签。domain 和日志标签都会以某种形式显示在日志信息上。

在输出日志信息时，运行格式化日志信息，也就是为日志信息指定占位符。占位符需要在%和符号（如 s、d）之间加{private}或{public}，如果加{private}，输出的信息就是<private>，也就是说会隐藏占位符对应的信息；如果加{public}，则输出原始的占位符信息。domain 在输出时被转换为十六进制数放到日志标签的前面，中间用斜杠（/）分隔。

运行这段代码，会在 HiLog 视图中输出图 1-24 所示的日志信息。

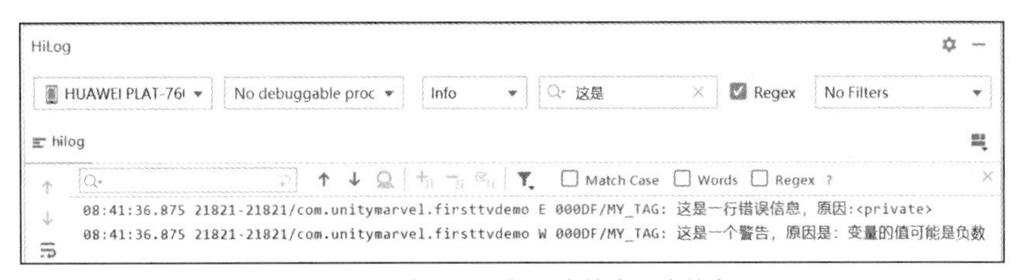

图 1-24 在 HiLog 视图中输出日志信息

使用 error 方法会输出深红色的日志信息，日志的内容与其他方法输出的日志内容相同。不过要注意，使用 HiLog 的相关方法输出的日志分为 5 个级别，分别是 DEBUG（调试）、INFO（信息）、WARN（警告）、ERROR（错误）、FATAL（致命错误）。这 5 个级别分别用 5 个整数表示，这些整数都在 HiLog 类中定义，代码如下：

```
public final class HiLog {
    public static final int DEBUG = 3;
    public static final int INFO = 4;
    public static final int WARN = 5;
    public static final int ERROR = 6;
    public static final int FATAL = 7;
    ...
}
```

如果要在 HiLog 视图中过滤这些级别的信息，只有小于或等于当前级别的信息才会显示。例如，要过滤 WARN 信息，只有 DEBUG、INFO 和 WARN 这 3 个级别的信息才会被显示，因为 ERROR 和 FATAL 的级别值都比 WARN 大，所以这两个级别的信息不会被显示。

1.7 在真机上运行 HarmonyOS App

到现在为止，相信大家已经基本了解了开发一款 HarmonyOS App 的流程，但我们一直是在 HarmonyOS 模拟器上运行 App。在实际开发过程中，需要在真机上测试后才能对外发布，而且有很多功能（如蓝牙、NFC、传感器等）在 HarmonyOS 模拟器上是无法测试的。所以本节会讲解如何在手机上安装并测试 HarmonyOS App。

1.7.1 生成签名文件

在真机上运行 App 与在模拟器上运行 App 不同。在真机上不管是调试还是发布，都需要对 hap 文件进行签名。签名一个 hap 文件需要 4 类文件，即 p12、csr、cer 和 p7b 文件。其中 p12 和 csr 文件可以自助生成，cer 和 p7b 文件需要到 App Grallery Connect 去申请，然后下载。

生成这 4 类文件，需要 7 个步骤，下面就具体介绍一下这 7 个步骤。

1. 生成 p12 文件

执行下面的命令，会在当前目录下生成一个名为 demo.p12 的文件。

```
keytool -genkeypair -alias "myalias" -keyalg EC -sigalg SHA256withECDSA -dname "C=CN,O=Huawei CBG,OU=HOS Development IDE Team,CN=ide_demo_pk Debug" -keystore demo.p12 -storetype pkcs12 -storepass 1234abcd
```

在执行上面的命令之前，需要将如下 3 个信息修改成自己的内容（其他内容不需要改）：

- -alias 命令行参数后面的别名，本例是 myalias；
- -keystore 命令行参数后面的文件名，本例是 demo.p12；
- -storepass 命令行参数后面的密码，本例是 1234abcd。

2. 生成 csr 文件

执行下面的命令，会在当前目录下生成一个名为 demo.csr 的文件。

```
keytool -certreq -alias "myalias" -keystore demo.p12 -file demo.csr
```

其中，-alias 命令行参数指定的别名要与生成 demo.p12 文件时指定的别名相同，demo.csr 文件可以改成其他文件名。

注意，在执行上面的命令之前，要先生成 demo.p12 文件。在生成 demo.csr 文件的过程中，会要求输入生成 demo.p12 文件时指定的密码，本例是 1234abcd。一定要记住这个密码，因为在后面配置 App 签名时还要用到。

3. 创建项目

通过下面的 URL 打开页面，并单击“我的项目”按钮：https://developer.huawei.com/consumer/cn/service/josp/agc/index.html 进入创建项目页面，如图 1-25 所示。可以在该页面创建一个或多个项目，每一个项目可以包含 0 到多个安装包。

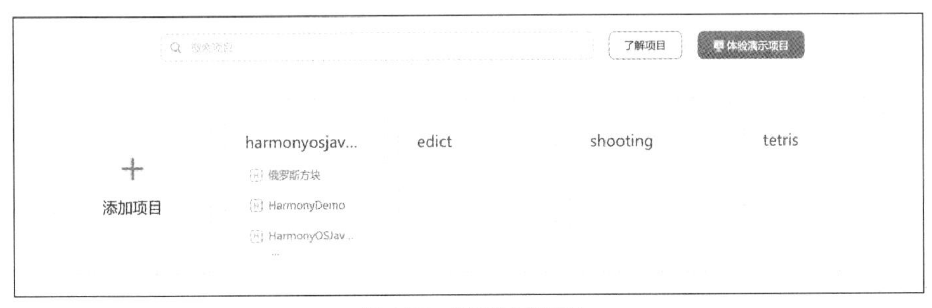

图 1-25 创建项目页面

单击左侧的“添加项目”链接，会弹出图 1-26 所示的页面，输入项目名，单击“确定”按钮就可以创建新项目。

4. 通过 csr 文件在线申请 cer 文件（调试证书文件）

通过下面的 URL 打开页面，并单击“用户与访问”按钮：

https://developer.huawei.com/consumer/cn/service/josp/agc/index.html

进入该页面后，在左侧导航栏选择“证书管理”选项，进入证书管理页面，单击“新增证书”按钮弹出新增证书窗口，并按要求输入相应的内容，如图 1-27 所示。

图 1-26 创建新项目

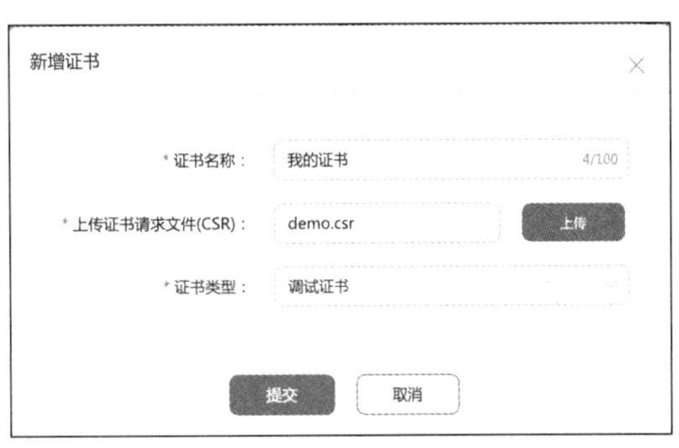

图 1-27 创建证书

创建完证书后，会在证书列表中看到新创建的证书，如图 1-28 所示。单击右侧的“下载”链接，下载对应的 cer 文件，本例是“我的证书.cer”文件。

图 1-28 下载 cer 文件

5. 添加设备

通过下面的 URL 打开页面，并单击“用户与访问”按钮：

https://developer.huawei.com/consumer/cn/service/josp/agc/index.html

进入该页面后，在左侧导航栏选择“设备管理”，进入设备管理页面。然后单击右侧的“添加设备”按钮，会弹出图 1-29 所示的窗口。这里需要选择设备类型，输入设备名称和 UDID，然后单击“提交”按钮添加设备。只有在这里添加的设备才能使用前面申请的证书安装 HarmonyOS App。

图 1-29 添加设备

要想获得设备的 UDID，需要用 USB 数据线连接 HarmonyOS 设备，然后执行下面的命令：

```
adb shell dumpsys DdmpDeviceMonitorService
```

成功执行命令后，在输出结果中的 Local device info 部分找到 dev_udid，后面的字符串就是该设备的 UDID。

如果有多部设备连接到了计算机上，需要使用-s 命令行参数指定设备标识。

6. 创建鸿蒙应用

通过下面的 URL 打开页面，并单击“我的项目”按钮：

https://developer.huawei.com/consumer/cn/service/josp/agc/index.html

进入该页面后，其中会列出在该项目下创建的应用，如图 1-30 所示，读者可以单击“添加应用”链接创建新的应用。

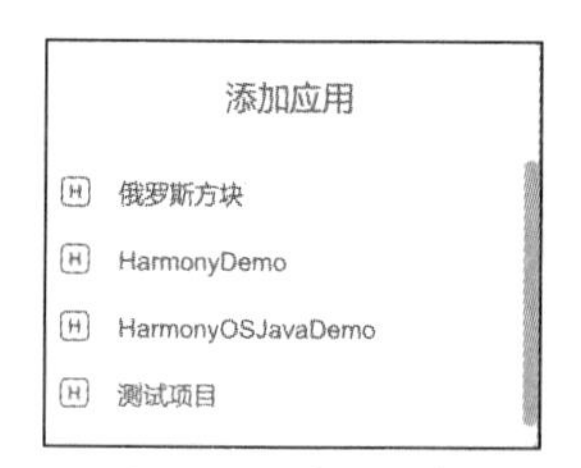

图 1-30 应用列表

添加应用页面如图 1-31 所示。读者需要在该页面按图 1-31 所示填写必要的信息（需要按自己 App 的情况填写相应的内容），其中“支持设备”保持默认值就可以在 HarmonyOS 手机上运行。

注意，如果 App 的包变了，需要重新按这一步添加应用，否则 App 将无法部署在 HarmonyOS 手机上。

7. 获得 p7b 文件

进入上一步创建的应用的页面，单击右上角的“添加”按钮，添加一个 HAP Provision Profile。这一步是一个总绑定，将 App 的包、证书和设备绑定在一起。也就是说，只有特定包的 App，使用特定的证书，才能在特定的设备上部署。进入页面后，可以仿照图 1-32 所示填写这个页面的内容。

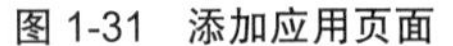

图 1-31 添加应用页面

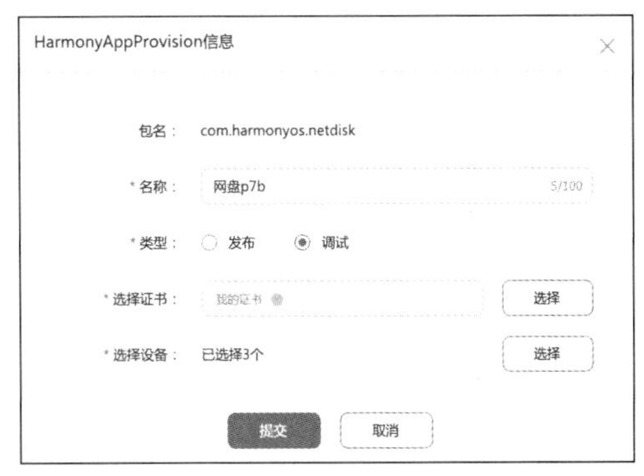

图 1-32 添加 HAP Provision Profile

成功添加 HAP Provision Profile 后，会在列表中显示刚才创建的 HAP Provision Profile，如图 1-33 所示。单击右侧“下载”链接，就会下载一个名为“网盘 p7bDebug.p7b”的文件。

图 1-33 下载 p7b 文件

到目前为止，所有必要的文件都已经备齐，通过这 7 个步骤获得了图 1-34 所示的 4 个文件，其中 demo.csr 文件只是一个中间文件，在签名时并不需要它。

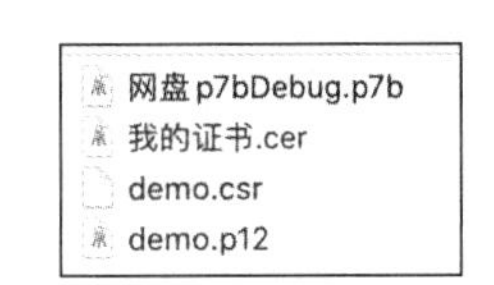

图 1-34 用于签名的文件

注意，有的读者可能会感到生成证书比较麻烦，需要这么多步骤。其实只有第一次签名要这么多步骤，再创建新的应用时，只需要做第 6 步和第 7 步就可以了。也就是说，如果设备不变，在签名时只需要更换 p7b 文件就可以了。

1.7.2 签名 HarmonyOS App

将 HarmonyOS App 在手机上部署，需要对 hap 文件签名。首先通过 Create HarmonyOS Project 窗口创建一个 Phone 工程，如图 1-35 所示。这里选择 Empty Feature Ability(Java)模板。

在配置工程时，工程名可以任意输入，但 Package Name 必须输入创建应用时指定的包的名称，即 com.harmonyos.netdisk，如图 1-36 所示。最后单击 Finish 按钮创建工程。

创建完工程后，选择 File→Project Structure 菜单项，打开 Project Structure 窗口，单击左侧的 Modules，选择 Signing Configs 选项卡，进入配置页面，并按图 1-37 所示对签名文件进行配置。其中，Store Password 和 Key Password 使用 1.7.1 节中生成 demo.p12 文件时使用的密码，本例是 1234abcd。配置完后，单击 OK 按钮保存配置。

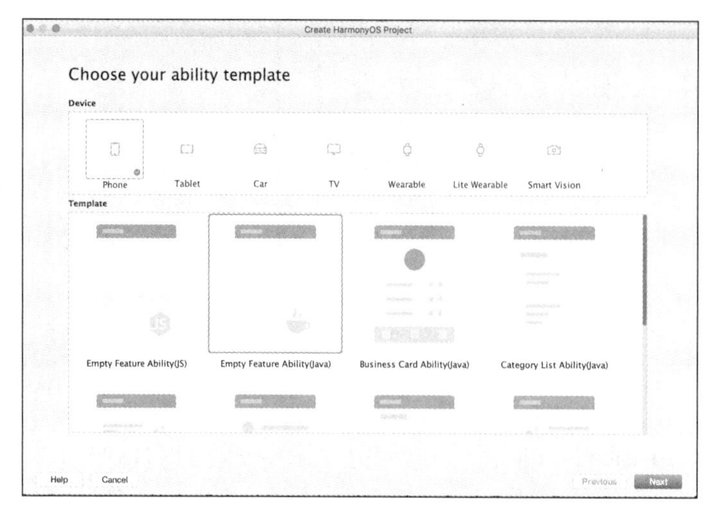

图 1-35 创建 Phone 工程

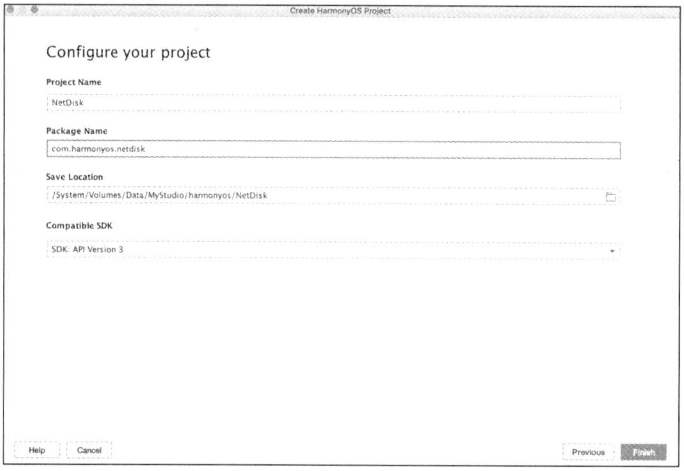

Create HarmonyOS Project

Configure your project

Project Name

NetDisk

Package Name

com.harmonyos.netdisk

Save Location

/System/Volumes/Data/MyStudio/harmonyos/NetDisk

Compatible SDK

SDK: API Version 3

Help Cancel Previous Finish

图 1-36 配置工程

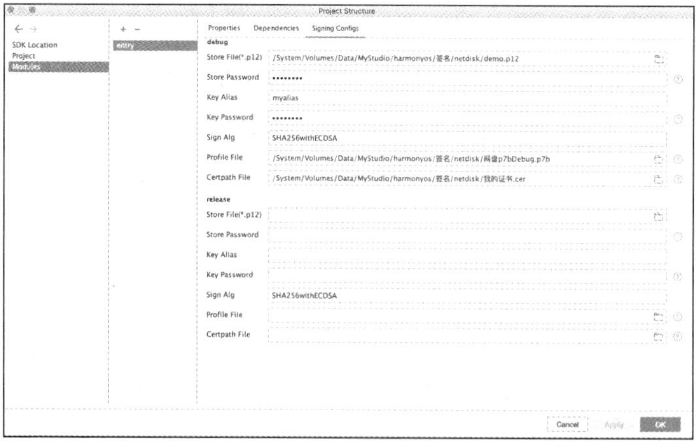

图 1-37 配置签名文件

到这里，签名的工作就已经完成了，现在运行程序，会弹出一个图 1-38 所示的选择设备对话框。列表中显示连接了两部 HarmonyOS 手机（通过 USB 数据线连接），可以选择其中一部用于运行程序。

运行效果如图 1-39 所示。

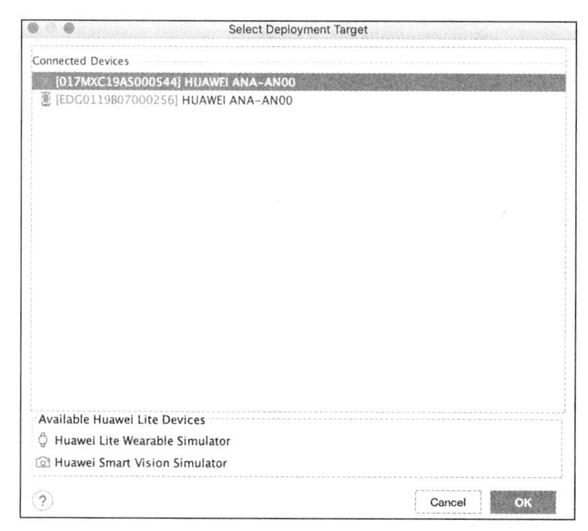

图 1-38 选择设备对话框

图 1-39 App 在手机上的运行效果

1.7.3 网络部署 App

通常在真机上运行 HarmonyOS App 时要用 USB 数据线将手机和计算机连接起来，这种方式比较复杂，而且如果使用的是 USB 接口并不富余的计算机，如苹果的 Mac，还需要连接 USB Hub 等设备，所以本节介绍另外一种部署 App 的方式——网络部署。

其实网络部署与通过 USB 数据线部署在本质上是一样的，只是这种方式不需要通过 USB 数据线将手机与计算机进行连接，只需要手机与计算机在同一个网段即可。

在默认情况下，手机的网络调试功能是关闭的，所以需要使用 adb 或 hdc 命令打开手机的网络调试功能。adb 是 Android 的命令，hdc 是 HarmonyOS 的命令，这两个命令的功能类似。这两个命令的使用方法如下（执行其中一个即可）：

```
adb tcpip 5555
hdc tmode port 5555
```

注意，在执行上面的命令之前，手机必须先用 USB 数据线与计算机相连，等打开手机的网络调试服务后，就可以断开 USB 数据线的连接了。如果计算机连接了多部手机，使用 adb 命令需要添加-s 命令行参数指定特定手机的标识，使用 hdc 命令需要添加-t 命令行参数指定特定手机的标识，命令如下：

```
adb -s 017MXC19AS000544 tcpip 5555
hdc -t -t 017MXC19AS000544 tmode port 5555
```

其中，5555 是端口号，可以指定任何端口号。计算机需要通过这个端口号连接手机。如果成功打开手机的网络调试服务，可以使用下面的命令连接 HarmonyOS 手机。

- adb 方式：

```
adb connect 192.168.31.220:5555
```

- hdc 方式：

```
hdc tconn 192.168.31.220:5555
```

成功连接 HarmonyOS 手机后，就可以运行程序了。这时会在图 1-38 所示的列表中看到通过网络连接的 HarmonyOS 设备，选择设备并单击 OK 按钮运行即可，效果与通过 USB 数据线部署 App 是一样的。

1.8 解决 gradle 下载太慢的问题

DevEco Studio 使用 gradle 作为构建工具，在默认配置的情况下，创建 HarmonyOS 工程后，在初始化的过程中可能会出现超时（timeout）异常，也就是访问 gradle 服务端超时异常。

网上解决这个问题的方法很多，比较常用的方法是自己先下载 gradle 的 zip 包（大概 100 MB），例如，下载的文件是 gradle-6.3-all.zip，然后将该文件放在特定的目录下。目录的位置与操作系统有关。

- Windows 下的位置：C:\Users\用户名\.gradle。
- macOS 下的位置：/Users/用户名/.gradle。
- Linux 下的位置：/home/用户名/.gradle。

读者根据自己使用的操作系统进入不同位置的.gradle 目录，然后再进入该目录下的 wrapper/dists 子目录，会看到名称为“gradle-版本号-all”[①]的目录，例如，图 1-40 展示的是我的 macOS 上的 gradle 集合。

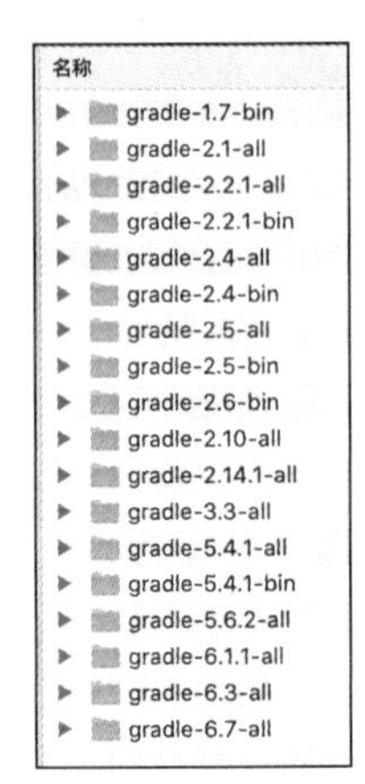

图 1-40 macOS 上的 gradle 集合

进入 gradle-6.3-all 目录（最新的 DevEco Studio 使用的是 gradle 6.3），会看到一个名称由字母和数字组成的目录，如 b4awcolw9l59x95tu1obfh9i8，这个目录的名称与 gradle 版本和路径有关。所以不同版本和路径的 gradle，目录名不同。现在将下载的 gradle-6.3-all.zip 文件复制到该目录。用 DevEco Studio 重新打开 HarmonyOS 工程，就会非常快地同步 gradle。这是因为 DevEco Studio 会优先在特定目录下搜索 gradle 的 zip 文件是否存在，如果存在，就直接使用，如果不存在，才会从网上下载。

不过有时特定的 gradle 目录还没有生成，我们不知道要将 gradle-6.3-all.zip 或类似的文件放到哪里，所以可以采用另外一种方法，换个下载速度更快的地址。

---

① 有的开发工具生成的目录是以 bin 结尾的，如 gradle-6.3-bin，这个无关紧要，因为下载 gradle-6.3-all.zip 和下载 gradle-6.3-bin.zip 是一样的，它们都包含必要的构建工具。

打开 HarmonyOS 工程，找到<工程根目录>/gradle/wrapper/gradle-wrapper.properties 文件，会看到一个 distributionUrl 属性，该属性默认指向 gradle 的官方网站，可以将其换成国内的地址，或干脆用 apache、nginx 等服务器自己搭建一个本机下载服务，如我将其换成了如下的地址：

```
http://127.0.0.1/gradle-6.3-all.zip
```

现在再重新打开 HarmonyOS 工程，就会非常快速地下载和同步 gradle 及其相关配置了。不过这里还有一个问题，如果创建新的 HarmonyOS 工程，还要改一遍这个文件。所以为了一劳永逸，我们可以修改 DevEco Studio 的模板。

首先进入<DevEco Studio 根目录>/plugins/harmony 目录，这是 DevEco Studio 的插件目录。在该目录下找到 lib/templates/gradle/wrapper/gradle/wrapper/gradle-wrapper.properties 文件，该文件就是每次创建 HarmonyOS 工程时的 gradle-wrapper.properties 文件。打开该文件，将 distributionUrl 属性的值改成 http://127.0.0.1/gradle-6.3-all.zip 即可。然后重新启动 DevEco Studio。与创建新的 HarmonyOS 工程时，DevEco Studio 就会自动使用新的 gradle 下载地址了。由于此前已经下载了 gradle，所以即使不启动本机的 http 服务器，仍然能快速使用 gradle。

1.9 总结与回顾

本章介绍了 HarmonyOS 的基础内容，目的是带领读者开启 HarmonyOS App 的开发之旅。首先让读者了解 HarmonyOS 和 HarmonyOS 的技术定位；然后一步一步教读者如何搭建 HarmonyOS 开发环境、创建第一个 HarmonyOS 工程；为了让读者了解 HarmonyOS 工程的结构，在本章中，我们分析了这个工程结构，以便让读者知其然，知其所以然；最后，讲解了如何在装有 HarmonyOS 的真机上运行 App，让读者初步具有开发 App 的能力。当然，要想成为一名优秀的 HarmonyOS App 开发者，读者需要付出非常大的努力，需要掌握更多的知识，需要通过大量的项目提升自己的实战经验。最重要的是，需要继续阅读本书后面的内容，因为开发 HarmonyOS App 需要的知识基本上本书都包含了。

第2章 Page Ability

HarmonyOS 的核心之一是 Ability。目前 Ability 分为两大类——Feature Ability（FA）和 Particle Ability（PA）。每一大类 Ability 中又分为一个或多个小类别。目前 Feature Ability 只支持 Page Ability，Page Ability 支持 Service Ability（见第 8 章）和 Data Ability（见第 7 章）。本章主要讨论 Page Ability。那么 Page Ability 到底如何使用，以及 Page Ability 有哪些特性呢？请继续看本章的内容。

通过阅读本章，读者可以掌握：

- Page Ability 的基本用法；
- Page Ability 之间的交互；
- Page Ability 的生命周期；
- Page Ability 的启动模式；
- AbilitySlice 的应用；
- AbilitySlice 之间的交互；
- Page Ability 的启动类型；
- 如何获取设备 ID 列表；
- 如何跨设备迁移 Page Ability。

2.1 Page Ability 概述

Page Ability 是 Feature Ability 唯一支持的 Ability，它本质上是一个窗口，类似于 Android 中的 Activity，用于提供与用户交互的能力。另外，HarmonyOS 还提供了 AbilitySlice，AbilitySlice 的功能与 Page Ability 类似，只是在切换时可以在同一个 Page Ability 内完成。从 Page Ability 切换到 Page Ability，相当于 Web 页面用新窗口导航到另一个页面。从 AbilitySlice 切换到 AbilitySlice，相当于 Web 页面用同一个窗口导航到另一个页面。

一个 Page Ability 可以不使用 AbilitySlice，也可以使用一个或多个 AbilitySlice。在创建 HarmonyOS 工程时，生成的 Demo 代码中包含了一个默认的 AbilitySlice（MainAbilitySlice.java）。图 2-1 是 Page Ability 和 AbilitySlice 的关系。

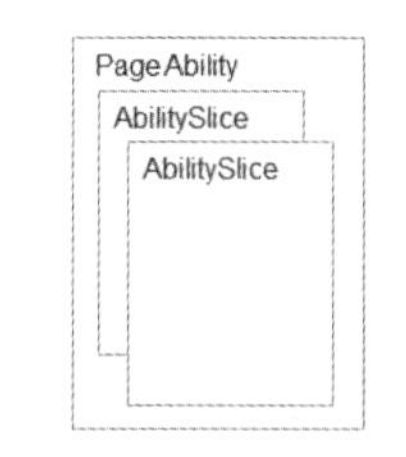

图 2-1 Page Ability 和 AbilitySlice 的关系

如果在 Page Ability 中使用多个 AbilitySlice，那么这些 AbilitySlice 提供

的业务能力应高度相关。

例如，在线视频播放器的主界面可以用一个 Page Ability 来实现，在主界面中包含了两个 AbilitySlice，一个用于展示视频列表，另一个用于播放视频。

在 DevEco Studio 中创建 HarmonyOS 工程时，IDE 会提供一些 Ability 模板，如图 2-2 所示。通过这些 Ability 模板，可以生成 HarmonyOS 工程的代码框架，其中会包含一些简单的例子，相当于 HelloWorld 工程。

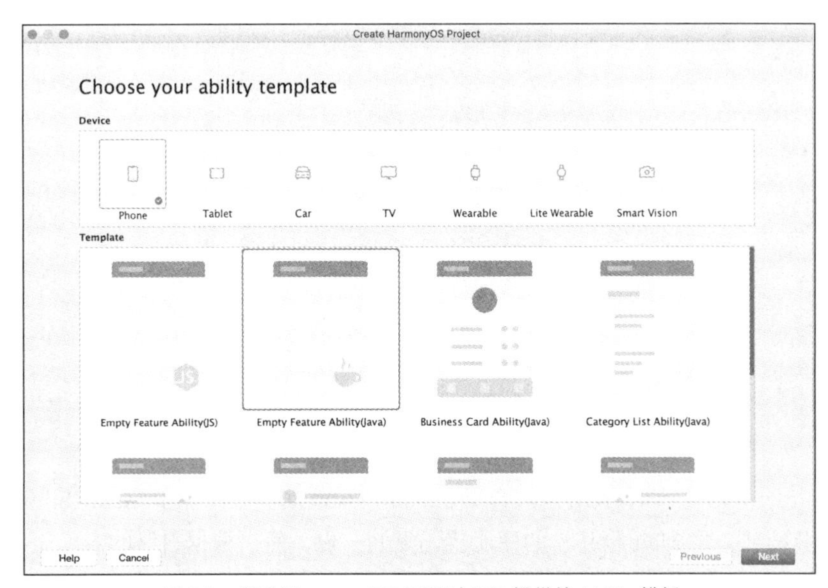

图 2-2 创建 HarmonyOS 工程时 IDE 提供的 Ability 模板

2.2 Page Ability 的基本用法

DevEco Studio 提供了自动创建 Page Ability 的功能，在创建的过程中会自动向 config.json 文件中添加相应的配置信息。不过为了更深入理解 Page Ability 的创建和使用过程，本节将手动创建一个 Page Ability。

2.2.1 手动创建 Page Ability 类

Page Ability 类是一个普通的 Java 类，所以我们首先应该创建一个 Java 类，本例创建的 Java 类是 FirstAbility 类。任何一个 Page Ability 类都必须从 Ability 类继承，该类属于 ohos.aafwk.ability 包，所以 FirstAbility 类的代码如下：

```
package com.unitymarvel.demo;
import ohos.aafwk.ability.Ability;
public class FirstAbility extends Ability {
}
```

2.2.2 在 config.json 文件中注册 Page Ability

在 HarmonyOS App 中，任何一个可用的 Page Ability 在使用前都必须在 config.json 文件中注册。Page Ability 需要在 config.json 文件中的 abilities 部分注册。abilities 是一个对象数组，其中的每一个元素是一个对象，表示一个 Ability（包括 Page Ability、Data Ability 和 Service Ability）。FirstAbility 的注册代码如下：

```
{
   "skills": [
      {
        "actions": [
           "com.unitymarvel.demo.first"
        ]
      }
   ],
   "orientation": "landscape",
   "formEnabled": false,
   "name": "com.unitymarvel.demo.FirstAbility",
   "icon": "$media:icon",
   "description": "$string:sqlite_description",
   "label": "第一个 Ability",
   "type": "page",
   "launchType": "standard"
}
```

在创建 HarmonyOS 工程时，IDE 已经自动在 config.json 文件中添加了 MainAbility 的注册信息，读者可以照葫芦画瓢，将 MainAbility 的注册代码复制一份，然后根据自己的需求略作修改。

FirstAbility 的注册代码中主要修改了 actions、name 和 label 属性。actions 用于指定一个或多个与 Page Ability 关联的动作，可以通过这些 action 来调用当前的 Page Ability。name 用于指定 Page Ability 类的全名，在本例中为 com.unitymarvel.demo.FirstAbility。label 用于指定 Page Ability 的标题文本。

2.2.3 创建布局文件

HarmonyOS App 既可以使用 Java 代码动态创建组件的方式进行布局，也可以使用布局文件进行布局。本例使用布局文件进行布局，这也是我推荐的布局方式。因为使用 Java 代码创建复杂的布局非常困难，而且代码量比较大，代码难以维护。

HarmonyOS 工程的所有布局文件都放在 resources/base/layout 目录下，现在创建一个 first_layout.xml 文件，并输入下面的代码：

```
<?xml version="1.0" encoding="utf-8"?>
<DirectionalLayout xmlns:ohos="http://schemas.huawei.com/res/ohos"
                ohos:width="match_parent"
                ohos:height="match_parent"
                ohos:orientation="vertical"
                ohos:padding="32"
                ohos:alignment="horizontal_center">
 <Button
```

```
        ohos:id="$+id:button1"
        ohos:width="300vp"
        ohos:height="50vp"
        ohos:text="按钮 1"
        ohos:bottom_margin="10vp"
        ohos:background_element="#00FFFF"/>
    <Text
        ohos:id="$+id:text"
        ohos:width="300vp"
        ohos:height="50vp"
        ohos:bottom_margin="10vp"
        ohos:text="第一个 Ability"
        ohos:background_element="#FF0000"/>
    <Button
        ohos:id="$+id:button2"
        ohos:width="300vp"
        ohos:height="50vp"
        ohos:text="按钮 2"
        ohos:background_element="#00FFFF"/>
</DirectionalLayout>
```

关于布局的更多内容在后面的章节会详细讨论，这里读者只要知道，first_layout.xml 文件中使用了方向布局，而且是垂直方向。我们在垂直方向上从上到下放置了 3 个组件，分别是 button1、text 和 button2。

2.2.4 装载布局文件

创建完布局文件后，需要与 Page Ability 关联才能显示布局中的组件。通常需要在 Page Ability 启动时装载布局文件，这就要使用 Page Ability 的生命周期方法 onStart。该方法会在 Page Ability 开始时调用，通常用于做一些初始化的工作，例如，为组件指定事件监听器。

现在需要在 FirstAbility 类中添加一个 onStart 方法，并使用 super.setUIContent 方法装载布局文件，代码如下：

```
package com.unitymarvel.demo;
import ohos.aafwk.ability.Ability;
import ohos.aafwk.content.Intent;
public class FirstAbility extends Ability {
   @Override
   public void onStart(Intent intent) {
       // 必须调用父类的 onStart 方法
       super.onStart(intent);
       super.setUIContent(com.unitymarvel.demo.ResourceTable.Layout_first_layout);
   }
}
```

在 HarmonyOS App 中，任何形式的资源都会与一个 int 类型的值绑定，以便通过该值引用资源，这些值都在 ResourceTable 类中以常量（其实是 static final 形式的变量）形式定义。这些值都是自动生成的，通常以资源文件名作为变量名，前面加上前缀。布局文件生成的 ID 需加上的前缀是 Layout，如本例的布局文件是 first_layout.xml，所以生成的 ID 是 Layout_first_layout。根据这个生成规则，还要求资源文件的命名必须符合 Java 标识符的命名规则，否则无法在 ResourceTable 类中生成 ID。

2.2.5 显示 Page Ability

到现在为止，一个最小的、完整的 Page Ability 已经完成了，最后一步就是显示这个 Page Ability。如果想让 FirstAbility 作为主 Ability 显示（HarmonyOS App 运行后显示的第 1 个 Page Ability），可以修改 FirstAbility 配置信息中的 skills 部分，将其改成如下形式：

```
"skills": [
  {
   "entities": [
     "entity.system.home"
   ],
   "actions": [
     "action.system.home"
   ]
  }
]
```

注意，可能在 config.json 文件中还有其他 Page Ability 的 actions 也设为 action.system.home。而 HarmonyOS 只会显示在 config.json 文件中遇到的第 1 个主 Ability。所以要将 FirstAbility 的注册信息调整为 abilities 的第 1 个元素，或者注释掉其他的 action.system.home。

如果想在其他的 Page Ability 中显示 FirstAbility，需要使用下面的代码：

```
Intent intent = new Intent();
intent.setAction("com.unitymarvel.demo.first");
startAbility(intent);
```

不管使用哪种方式，FirstAbility 的显示效果都会如图 2-3 所示。

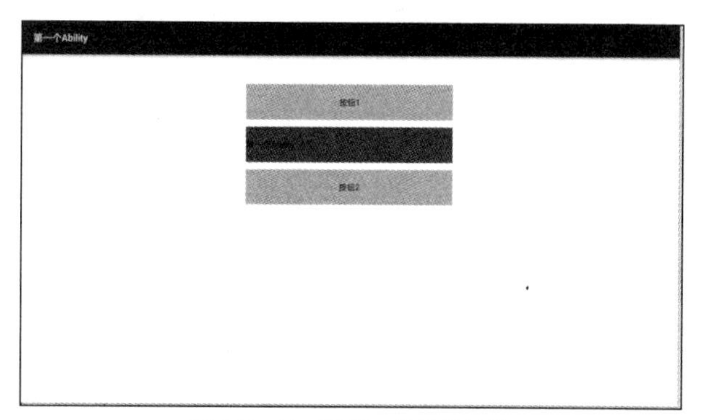

图 2-3 FirstAbility 的显示效果

2.2.6 销毁 Page Ability

在 Page Ability 用完后，我们需要关闭 Page Ability，或称为销毁 Page Ability，只需要调用下面的代码即可销毁 Page Ability：

```
terminateAbility();
```

该方法属于 Ability 类，如果在 AbilitySlice（在后面介绍）中调用该方法，需要先获得包含 AbilitySlice 的 Ability 对象。

2.3 Page Ability 之间的交互

本节将详细介绍如何在不同 Page Ability 之间进行交互，例如，两个 Page Ability 之间传递数据、通过显式和隐式的方式在一个 Page Ability 中显示另一个 Page Ability 等。

2.3.1 显式使用 Intent

显示 Page Ability 有两种方式——显式显示和隐式显示。所谓显式显示就是直接指定 Page Ability 的类名，隐式显示就是通过与 Page Ability 对应的 action 来显示 Page Ability。本节将主要介绍用显式方式显示 Page Ability。

不管用显式方式还是隐式方式显示 Page Ability，都需要使用 Intent 对象。这个对象相当于 Page Ability 之间的信使，用于传递数据和相关信息。

现在我们来做一个实验，首先创建一个名为 ExplicitIntentAbility 的 Page Ability，然后在 config.json 文件中使用下面的配置信息注册这个 Page Ability。

```
{
  "orientation": "landscape",
  "formEnabled": false,
  "name": "com.unitymarvel.demo.ability.ExplicitIntentAbility",
  "icon": "$media:icon",
  "label": "显式使用 Intent",
  "type": "page",
  "launchType": "standard"
}
```

在这段配置信息中并没有为 ExplicitIntentAbility 指定任何的 action，因为本例并不需要使用 action，所以可以不设置 action。

显式显示 Page Ability 的基本步骤如下：

（1）创建 Intent 对象；

（2）创建 Operation 对象，并设置 BundleName 和 AbilityName；

（3）将 Operation 对象与 Intent 对象关联；

（4）使用 startAbility 方法显示 Page Ability。

按照这 4 个步骤，可以用下面的代码显示 ExplicitIntentAbility。

```
// 第(1)步
Intent intent = new Intent();
// 第(2)步
Operation operation = new Intent.OperationBuilder()
        .withBundleName("com.unitymarvel.demo")
```

```
        .withAbilityName("com.unitymarvel.demo.ability.ExplicitIntentAbility")
        .build();

// 第(3)步
intent.setOperation(operation);
// 第(4)步
startAbility(intent);
```

其中，Operation 对象用于指定 Intent 对象使用的相关信息，例如，BundleName、AbilityName 等。BundleName 是 HarmonyOS App 的包名，在 config.json 文件中 bundleName 通过指定。AbilityName 是 Page Ability 的类全名（包名+类名）。最后需要调用 build 方法，因为该方法返回了 Operation 对象。

执行这段代码，会显示 ExplicitIntentAbility，效果如图 2-4 所示。

图 2-4 ExplicitIntentAbility 的显示效果

2.3.2 隐式使用 Intent

隐式使用 Intent 就是不明确指定具体的 Ability 类，而是通过与 Ability 对应的 action 来操作 Ability，具体的方法是使用 Intent.setAction 方法指定 action。

现在举一个例子，首先创建一个名为 ImplicitIntentAbility 的 Page Ability，然后在 config.json 中使用下面的代码配置 ImplicitIntentAbility。

```
{
    "skills": [
      {
        "actions": [
          "action.harmonyos.demo.intentaction"
        ]
      }
    ],
    "orientation": "landscape",
    "formEnabled": false,
    "name": "com.unitymarvel.demo.ability.ImplicitIntentAbility",
    "icon": "$media:icon",
    "label": "第一个 Ability",
    "type": "page",
    "launchType": "standard"
}
```

我们可以看到，在这段配置信息中为 ImplicitIntentAbility 指定的 action 是 action.harmonyos.demo.intentaction，所以可以使用下面的代码通过 action 显示 ImplicitIntentAbility。

```
Intent intent = new Intent();
intent.setAction("action.harmonyos.demo.intentaction");
startAbility(intent);
```

如果没有找到 action，那么代码执行后不显示 Page Ability；如果找到了 action，就会显示 action 对应的 Page Ability。

很多读者可能会发现一个问题，如果有多个 Page Ability 指定了同一个 action，那么会发生什么呢？其实这也是显式和隐式显示 Page Ability 的重要区别之一。如果使用显式方式显示 Page Ability，因为指定了确定的 Page Ability 类，所以一定会显示指定的 Page Ability。而隐式显示 Page Ability 的方式因为通过 action 确定要显示的 Page Ability，所以就可能会出现一个 action 对应多个 Page Ability 的情况。如果出现这种情况，HarmonyOS 会弹出一个列表，让用户选择到底显示哪一个 Page Ability。

现在举一个例子。首先创建两个 Page Ability——ImplicitIntentAbility1 和 ImplicitIntentAbility2。在 config.json 文件中的配置信息如下：

```
{
    "skills": [
      {
        "actions": [
          "action.harmonyos.demo.intentaction"
        ]
      }
    ],
    "orientation": "landscape",
    "formEnabled": false,
    "name": "com.unitymarvel.demo.ability.ImplicitIntentAbility1",
    "icon": "$media:icon",
    "label": "第一个 Ability",
    "type": "page",
    "launchType": "standard"
   },
   {
    "skills": [
      {
        "actions": [
          "action.harmonyos.demo.intentaction"
        ]
      }
    ],
    "orientation": "landscape",
    "formEnabled": false,
    "name": "com.unitymarvel.demo.ability.ImplicitIntentAbility2",
    "icon": "$media:icon",
    "label": "第二个 Ability",
    "type": "page",
    "launchType": "standard"
}
```

很明显，这两个 Page Ability 指定了同一个 action。现在使用下面的代码显示 Page Ability。

```
Intent intent = new Intent();
intent.setAction("action.harmonyos.demo.intentaction");
startAbility(intent);
```

执行这段代码，并不会马上显示 Page Ability，而是会弹出图 2-5 所示的窗口，让用户选择显示哪一个 Page Ability。

用户做出选择后，就会显示指定的 Page Ability。

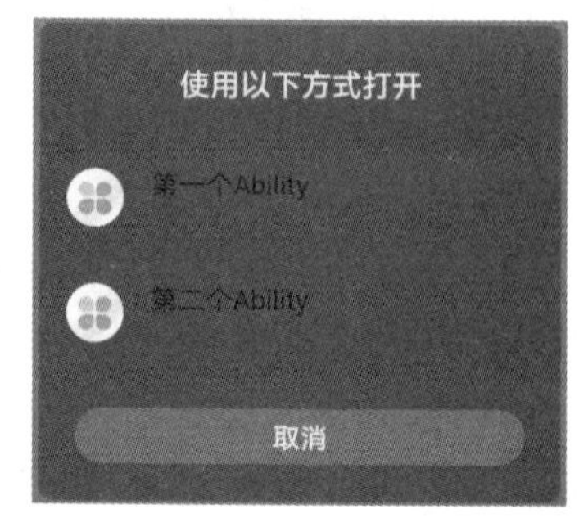

图 2-5 让用户选择显示哪一个 Page Ability

2.3.3 Page Ability 之间的交互

在很多场景下，两个 Page Ability 之间是需要交互的。所谓交互，是指两个 Page Ability 之间互相传递数据。例如，Page Ability1 给 Page Ability2 传递一个字符串类型的数据，然后 Page Ability2 给 Page Ability1 返回一个整型的值，这就是标准的交互过程。

如果传递数据的过程是单向的，也就是说，数据只是从 Page Ability1 传递给 Page Ability2，那么直接使用 startAbility 方法即可。如果需要从 Page Ability2 返回数据，就必须使用 startAbilityForResult 方法。

如果在 Page Ability1 中使用 startAbilityForResult 方法显示 Page Ability2，当 Page Ability2 关闭后，会自动调用 Page Ability1 中的 onAbilityResult 方法返回数据，该方法的原型如下：

```
protected void onAbilityResult(int requestCode, int resultCode, Intent resultData);
```

onAbilityResult 方法的 3 个参数的含义如下。

- requestCode：请求码，通过 startAbilityForResult 方法的第 2 个参数指定。
- resultCode：响应码，通过 setResult 方法指定。
- resultData：以 Intent 对象形式返回的数据。

这里重点介绍一下 requestCode 和 resultCode。因为在 Page Ability 中可能会通过 startAbilityForResult 方法显示多个 Page Ability，所以 onAbilityResult 方法可能是多个 Page Ability 共享的，这就要求在 onAbilityResult 方法中区别是哪一个 Page Ability 返回的结果。我们可以通过 requestCode 或 resultCode 单独区分不同的 Page Ability，也可以使用 requestCode 和 resultCode 共同区分不同的 Page Ability。如果在 Page Ability1 中通过 startAbilityForResult 方法显示 Page Ability2，那么 requestCode 应该在 Page Ability1 中指定，而 resultCode 应该在 Page Ability2 中指定。

【例 2.1】 本例中演示了两个 Page Ability 之间如何交互。在一个 Page Ability 中通过 startAbility ForResult 方法显示了另一个 Page Ability，并传递了几个值，然后在另一个 Page Ability 中获取这些传递过来的值，并显示在 Text 组件中，最后单击“返回”按钮，会将一些值返回给请求 Page Ability。

首先创建一个名为 target_layout.xml 的布局文件，用于待显示 Page Ability （Target Ability）的布局，该布局文件的内容如下：

```
<?xml version="1.0" encoding="utf-8"?>
<DirectionalLayout xmlns:ohos="http://schemas.huawei.com/res/ohos"
```

```
                ohos:width="match_parent"
                ohos:height="match_parent"
                ohos:orientation="vertical"
                ohos:padding="32">
    <Button
        ohos:id="$+id:button_close"
        ohos:width="match_parent"
        ohos:height="60vp"
        ohos:text_alignment="center"
        ohos:background_element="#00FFFF"
        ohos:text_size="50vp"
        ohos:text="关闭 Page Ability，并返回值"
        />
     <Text
        ohos:id="$+id:text"
        ohos:width="match_parent"
        ohos:height="match_parent"
        ohos:text_alignment="center"
        ohos:text_size="50vp"
        ></Text>
</DirectionalLayout>
```

在 target_layout.xml 文件中放置了两个组件——Button 和 Text。TargetAbility 会将传递过来的信息显示在 Text 组件上，单击这个 Button，会返回相应的信息。

接下来创建一个名为 TargetAbility 的 Page Ability 类。

代码位置：src\main\java\com\unitymarvel\demo\ability\TargetAbility.java

```
package com.unitymarvel.demo.ability;

import com.unitymarvel.demo.ResourceTable;
import ohos.aafwk.ability.Ability;
import ohos.aafwk.content.Intent;
import ohos.agp.components.Button;
import ohos.agp.components.Component;
import ohos.agp.components.Text;

public class TargetAbility extends Ability {
   @Override
   public void onStart(Intent intent) {
       super.onStart(intent);
       // 装载布局文件
       super.setUIContent(ResourceTable.Layout_target_layout);

       Text text = (Text)findComponentById(ResourceTable.Id_text);
       if(text != null) {
          // 开始获取传递过来的信息
          String name = intent.getStringParam("name");
          int age = intent.getIntParam("age", -1);
          int[] data = intent.getIntArrayParam("data");
          String dataStr = "";
          // 将 int 类型数组转换为字符串形式
          for(int value : data) {
              dataStr += value + " ";
```

```
            }
            text.setMultipleLine(true);
            text.setText(String.format("name:%s\r\nage:%d\r\ndata:%s", name,age,dataStr));
        }

        Button buttonClose = (Button)findComponentById(ResourceTable.Id_button_close);
        if(buttonClose != null) {
            buttonClose.setClickedListener(new Component.ClickedListener() {
                @Override
                public void onClick(Component component) {
                    Intent resultIntent = new Intent();
                    // 指定要返回的值
                    resultIntent.setParam("result", "OK");
                    // 指定 resultCode，其中 100 就是 resultCode
                    setResult(100, resultIntent);
                    // 关闭当前的 Page Ability
                    terminateAbility();
                }
            });
        }


    }
}
```

阅读这段代码，读者应该了解如下几点。

- onStart 方法的 intent 参数值为通过 startAbilityForResult 方法的第 1 个参数指定的 Intent 对象，所有传递过来的信息都可以从这个 Intent 对象中获得。通过 Intent 对象的 getXxxParam 方法，可以获取各种类型的值。其中 Xxx 表示 String、Int 等。
- Text 组件默认只能显示单行文本，如果要显示多行文本，需要使用 text.setMultipleLine(true) 进行设置。
- 返回结果需要使用 setResult 方法设置，该方法的第 1 个参数是 resultCode，第 2 个参数是要返回的 Intent 对象。返回的 Intent 对象可以通过 onAbilityResult 方法的第 3 个参数获得。

最后在 config.json 文件中配置 TargetAbility。

```
{
  "orientation": "landscape",
  "formEnabled": false,
  "name": "com.unitymarvel.demo.ability.TargetAbility",
  "icon": "$media:icon",
  "label": "Page Ability 之间的交互",
  "type": "page",
  "launchType": "standard"
}
```

现在待显示的 TargetAbility 已经准备完成，接下来在另外一个 Page Ability 中通过 startAbilityForResult 方法显示 TargetAbility，并接收 TargetAbility 返回的结果。

代码位置：src\main\java\com\unitymarvel\demo\slice\ability\AbilityAndSliceSlice.java

```
Intent intent = new Intent();
Operation operation = new Intent.OperationBuilder()
```

```
        .withBundleName("com.unitymarvel.demo")
        .withAbilityName("com.unitymarvel.demo.ability.TargetAbility")
        .build();

intent.setOperation(operation);
// 设置要给 TargetAbility 传递的值
intent.setParam("name","Bill");                    // 传递一个字符串
intent.setParam("age",20);                         // 传递一个整数
intent.setParam("data", new int[]{1,2,3,4,5});    // 传递一个 int 类型的数组
// 显示 TargetAbility，并指定 requestCode 为 99
startAbilityForResult(intent,99);
```

在这段代码中使用了显式方式显示 TargetAbility。最后使用下面的代码获取 TargetAbility 返回的结果。

代码位置： src\main\java\com\unitymarvel\demo\slice\ability\AbilityAndSliceSlice.java

```
@Override
protected void onAbilityResult(int requestCode, int resultCode, Intent resultData) {
   switch (requestCode) {
       case 99:                    // 请求码是 99
          switch (resultCode) {
             case 100:             // 响应码是 100
                String result = resultData.getStringParam("result");
                HiLogLabel label = new HiLogLabel(HiLog.LOG_APP ,200,
                                                       "Ability_Result_Tag");
                HiLog.info(label, result,"");
                break;
          }
          break;
   }
}
```

在这段代码中，同时对 requestCode 和 resultCode 进行判断，如果满足条件，会在 HiLog 视图中输出 TargetAbility 的返回值。

运行程序，会看到图 2-6 所示 TargetAbility 的显示效果。

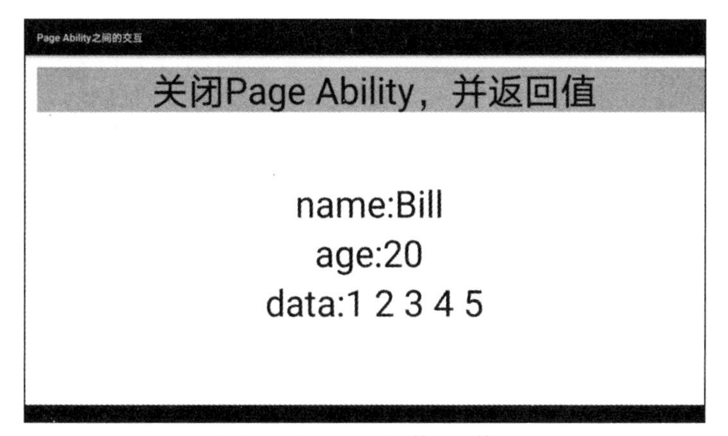

图 2-6 TargetAbility 的显示效果

单击窗口上方的“关闭 Page Ability，并返回值”按钮，会关闭 TargetAbility，然后返回结果，这时会在 HiLog 视图中输出图 2-7 所示的返回结果。

```
11-15 15:45:38.809 1830-1830/com.unitymarvel.demo D 01100/AbilityShell: AbilityShellPageComponent::onPageComponentResult called
11-15 15:45:38.809 1830-1830/com.unitymarvel.demo E 01100/AbilityShell: IntentConverter::createElementName param invalid
11-15 15:45:38.810 1830-1830/com.unitymarvel.demo D 01310/Ability: send result to ability developer, request code :65635, result code :100
11-15 15:45:38.810 1830-1830/com.unitymarvel.demo D 01310/AbilitySliceManager: remove sliceindex 1 from map
11-15 15:45:38.811 1830-1830/com.unitymarvel.demo I 000C8/Ability_Result_Tag: OK
11-15 15:45:38.812 1830-1830/com.unitymarvel.demo D 01100/AbilityShell: AbilityShellPageComponent::onRestart called
```

图 2-7　在 HiLog 视图中输出 TargetAbility 的返回结果

2.4 Page Ability 的启动类型

在 Page Ability 的配置信息中有一个启动类型属性（即 launchType 属性），在前面所有的例子中，该属性的值都是 standard，这是 launchType 属性的默认值（当不设置 launchType 时，launchType 属性值为默认值 standard）。launchType 属性值还可以被设置为 singleton。这两个属性值的区别如下：

- standard 表示在任何情况下，无论 Page Ability 被显示多少次，每次被显示都会创建一个新的 Page Ability 实例；
- singleton 表示如果要显示的 Page Ability 在栈顶，那么再次显示这个 Page Ability 时，不会再创建新的 Page Ability 实例，而是直接使用这个 Page Ability 实例。如果 Page Ability 上面有其他的 Page Ability，那么首先弹出这些 Page Ability，然后再复用这个 Page Ability。总之，拥有 singleton 模式的 Page Ability 将永远使用唯一的实例。

这里涉及栈的概念，这是 HarmonyOS 管理 Page Ability 的模式。HarmonyOS App 同时只能显示一个 Page Ability，那么能显示哪一个 Page Ability 呢？HarmonyOS App 会使用一个栈来管理 App 中所有的 Page Ability，只有在栈顶的 Page Ability 才会显示。如果要让栈中第 2 个 Page Ability 显示，那么栈顶的 Page Ability 就必须先出栈，也就是销毁栈顶的 Page Ability，这是调用 terminateAbility 方法要完成的工作。

下面用图示来说明这一过程。图 2-8 中每一个矩形区域表示 App 中当前用于保存 Page Ability 的栈。1 中只有一个 Page Ability1，如果让 Page Ability2 显示，那么 Page Ability2 必须入栈，即形成冲栈的状态，让 Page Ability3 显示也需要完成同样的工作。在 3 中，如果让 Page Ability2 显示，那么 Page Ability3 必须先出栈，这就形成了 4 种栈的状态。

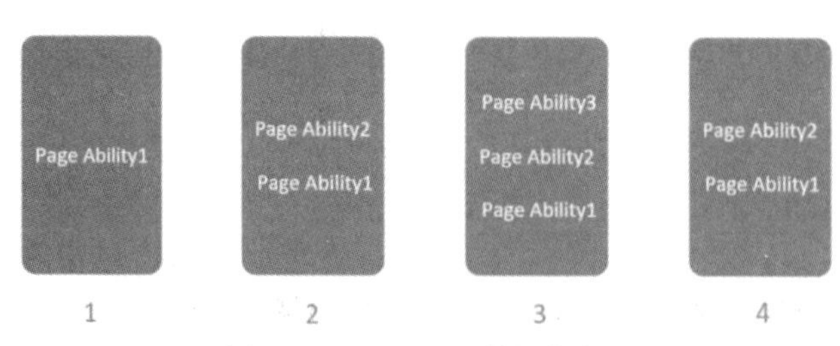

图 2-8　Page Ability 的创建过程

假设 Page Ability1 的 launchType 属性值是 standard，从 Page Ability1 中启动 Page Ability1，如果启动两次，就会再创建两个 Page Ability1 实例，这时栈的状态如图 2-9 所示。

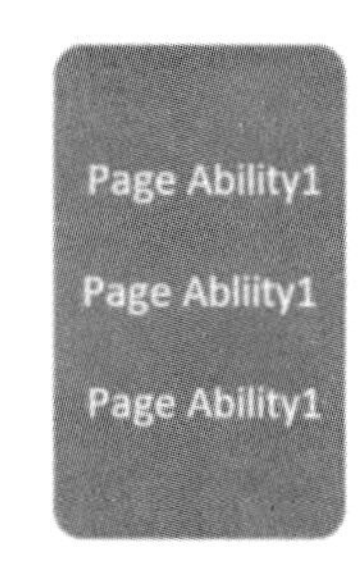

图 2-9 standard 模式下的栈状态

很明显，在栈中有 3 个 Page Ability1 实例。

假设 Page Ability1 的 launchType 属性值是 singleton，则不管显示多少次 Page Ability1，在栈中永远只有 1 个 Page Ability1 实例。所以如果想让某一个 Page Ability1 永远只有一个实例，可以将该 Page Ability 的 launchType 属性值设为 singleton。

【例 2.2】 本例中演示了 standard 和 singleton 的区别。

首先创建一个名为 LaunchTypeAbility 的 Page Ability，并编写下面的代码。

代码位置：src\main\java\com\unitymarvel\demo\ability\LaunchTypeAbility.java

```
package com.unitymarvel.demo.ability;

import com.unitymarvel.demo.ResourceTable;
import ohos.aafwk.ability.Ability;
import ohos.aafwk.content.Intent;
import ohos.agp.components.Button;
import ohos.agp.components.Component;
import ohos.agp.components.Text;


public class LaunchTypeAbility extends Ability {
   private static int count = 0;   // 计数器
   @Override
   public void onStart(Intent intent) {
      super.onStart(intent);
      super.setUIContent(ResourceTable.Layout_launch_type_layout);
      count++;
      Text text = (Text)findComponentById(ResourceTable.Id_text);
      if(text != null) {
         text.setText(String.valueOf(count));
      }
      Button buttonStartAbility=(Button)findComponentById(ResourceTable.Id_button_start_
      ability);
      if(buttonStartAbility != null) {
         buttonStartAbility.setClickedListener(new Component.ClickedListener() {
            @Override
            public void onClick(Component component) {
               Intent intent = new Intent();
               intent.setAction("action.harmonyos.demo.ability.testlaunchtype");
               // 显示另外一个 Page Ability
               startAbility(intent);
            }
         });
      }
   }
}
```

在 config.json 文件中配置 LaunchTypeAbility，将 LaunchTypeAbility 的 launchType 属性值设为 standard，代码如下：

```
{
   "skills": [
     {
       "actions": [
         "action.harmonyos.demo.ability.launchtype"
       ]
     }
   ],
   "orientation": "landscape",
   "formEnabled": false,
   "name": "com.unitymarvel.demo.ability.LaunchTypeAbility",
   "icon": "$media:icon",
   "label": "Page Ability的启动类型",
   "type": "page",
   "launchType": "standard"
}
```

然后再创建一个名为 TestLaunchTypeAbility 的 Page Ability。

代码位置：src\main\java\com\unitymarvel\demo\ability\TestLaunchTypeAbility.java

```java
package com.unitymarvel.demo.ability;

import com.unitymarvel.demo.ResourceTable;
import ohos.aafwk.ability.Ability;
import ohos.aafwk.content.Intent;
import ohos.agp.components.Button;
import ohos.agp.components.Component;
import ohos.agp.components.Text;


public class TestLaunchTypeAbility extends Ability {

   @Override
   public void onStart(Intent intent) {
      super.onStart(intent);
      super.setUIContent(ResourceTable.Layout_launch_type_layout);

      Button buttonStartAbility =
              (Button)findComponentById(ResourceTable.Id_button_start_ability);
      if(buttonStartAbility != null) {
         buttonStartAbility.setClickedListener(new Component.ClickedListener() {
            @Override
            public void onClick(Component component) {
               Intent intent = new Intent();
               intent.setAction("action.harmonyos.demo.ability.launchtype");
               // 显示 LaunchTypeAbility
               startAbility(intent);
            }
         });
      }
   }
}
```

本例包含两个 Page Ability——LaunchTypeAbility 和 TestLaunchTypeAbility。目前这两个 Page Ability 的 launchType 属性值都是 standard。两个 Page Ability 的关系是在 LaunchTypeAbility 中单击“显示 Page Ability”按钮跳转到 TestLaunchTypeAbility，然后在 TestLaunchTypeAbility 中单击“显示 Page Ability”按钮跳转到 LaunchTypeAbility，如图 2-10 所示。

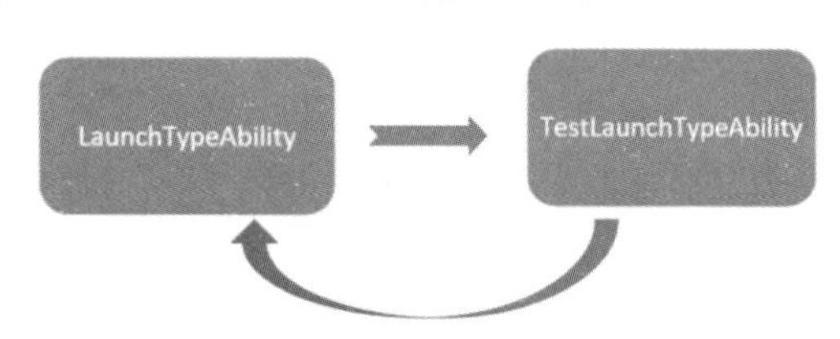

图 2-10 LaunchTypeAbility 与 TestLaunchTypeAbility 的关系

在 LaunchTypeAbility 类中定义了一个静态变量 count 作为计算器，如果每次显示 LaunchTypeAbility 时都创建一个新的实例，那么每创建一个新的实例 count 加 1，如果显示 3 次 LaunchTypeAbility，应该看到图 2-11 所示的窗口。

如果将 LaunchTypeAbility 的 launchType 属性值改成 singleton，那么不管显示多少次 LaunchTypeAbility，计数器 count 的值永远是 1，如图 2-12 所示。因为 LaunchTypeAbility 在创建一个实例后，不会再创建新的 LaunchTypeAbility 实例了，所以 onStart 方法自然就不会再次调用。

图 2-11 启动类型为 standard 的情况

图 2-12 启动类型为 singleton 的情况

2.5 Page Ability 的跨设备迁移

HarmonyOS 的核心特性（或称为卖点）之一就是软总线技术，而 Page Ability 的跨设备迁移是软总线技术的一个具体实现。所谓跨设备迁移 Page Ability，是指设备 A 中的特定 App 调用设备 B 中该 App 的 Page Ability。实现这个操作的前提是设备 A 和设备 B 都安装了同一个特定 App。如果设备 B 没有安装 App，设备 B 就会自动从华为应用商店下载这个 App，当然，这一过程对用户是完全静默的。下载完成后，设备 B 会自动启动 App 并显示相应的 Page Ability。这种技术不仅可以启动另一个设备上相应的 Page Ability，还可以向这个 Page Ability 传递数据。

这种技术的主要应用场景之一是，将在设备 A 上完成了一半的工作，迁移到设备 B 上继续完成。例如，在家中用平板电脑上写一封电子邮件，但临时有急事需要出门，这时可以将在平板电脑上写了一半的电子邮件迁移到手机上，在路上完成剩下的工作。

2.5.1 跨设备迁移前的准备工作

在进行跨设备迁移之前（后面几章讲的跨设备调用 Data Ability、Service Ability 也同样需要做这些准备），需要对 HarmonyOS 设备做一些准备工作。

（1）打开 HarmonyOS 设备的蓝牙。

（2）将 HarmonyOS 设备连入 Wi-Fi，而且所有参与跨设备迁移的 HarmonyOS 设备在同一个网段。

（3）用同一个华为账号登录所有参与跨设备迁移的 HarmonyOS 设备，如图 2-13 所示。

（4）选择“设置”→“更多连接”→“多设备协同”，进入“多设备协同”窗口，打开“多设备协同”开关，如图 2-14 所示。

（5）修改 HarmonyOS 设备名。选择“设置”→“蓝牙”→“设备名称”，进入“设备名称”窗口，输入新的名称，如图 2-15 所示。这一步不是必需的，但如果拥有多部 HarmonyOS 设备，可能很多 HarmonyOS 设备的名称是相同或相近的。为了更好区分不同的 HarmonyOS 设备，建议修改不同 HarmonyOS 设备名称。

图 2-13　用同一个华为账号登录

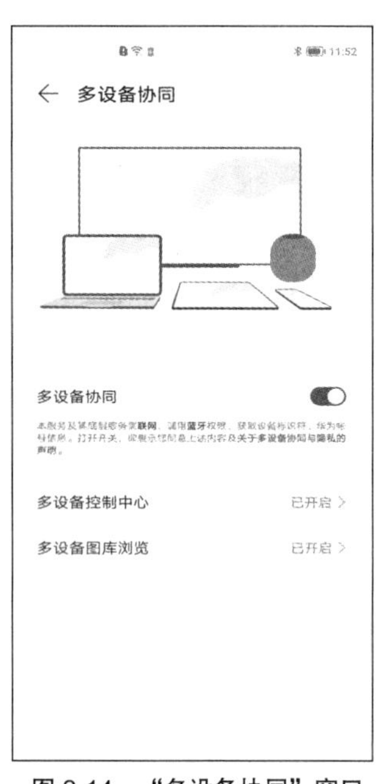

图 2-14　“多设备协同”窗口

图 2-15　修改 HarmonyOS 设备名称

2.5.2　获取设备列表

跨设备迁移是通过设备 ID 来区分不同设备的，所以首先要获取所有可用的设备的 ID。获取设备 ID 需要调用 DeviceManager.getDeviceList 方法，该方法返回一个 List 对象，类型是 DeviceInfo。DeviceInfo 类型描述了设备的相关信息，包括设备 ID、设备名称（即在上一节中设置的设备名称）等。实现代码如下：

```
List<DeviceInfo> deviceInfoList =
        DeviceManager.getDeviceList(DeviceInfo.FLAG_GET_ONLINE_DEVICE);
```

getDeviceList 方法有一个参数，是一个 int 类型的值，表示获取在某种状态下的设备的信息。可以设置的值有如下 3 个。

- DeviceInfo. FLAG_GET_ONLINE_DEVICE：获取所有在线设备的信息。
- DeviceInfo. FLAG_GET_OFFLINE_DEVICE：获取所有离线设备的信息。
- DeviceInfo. FLAG_GET_ALL_DEVICE：获取所有设备的信息。

通常会使用第 1 个值，获取所有在线设备的信息，因为只有设备在线，才能将 Page Ability 迁移到该设备上。

【例 2.3】 实现一个通用的显示可用设备列表的 Page Ability，单击某一个设备，会返回该设备的 ID，这个 Page Ability 在以后的章节中会经常使用。

下面先看一下这个 Page Ability 的布局文件的代码。

代码位置：src/main/resources/base/layout/device_ids.xml

```
<?xml version="1.0" encoding="utf-8"?>
<DirectionalLayout xmlns:ohos="http://schemas.huawei.com/res/ohos"
                ohos:width="match_parent"
                ohos:height="match_parent"
                ohos:background_element="#CCCCCC"
                ohos:orientation="vertical">
   <Text
      ohos:text_size="30vp"
      ohos:height="match_content"
      ohos:width="match_content"
      ohos:text="请选择设备 ID"/>
   <ListContainer ohos:id="$+id:listcontainer_deviceids"
      ohos:height="match_parent"
      ohos:top_margin="10vp"
      ohos:orientation="vertical"
      ohos:width="match_parent"
      ohos:weight="1"/>
</DirectionalLayout>
```

在 device_ids.xml 布局文件中放置了一个 ListContainer 组件，用于显示获取的所有可用设备的信息。下面是列表项的布局文件的代码。

代码位置：src/main/resources/base/layout/device_id_item.xml

```
<?xml version="1.0" encoding="utf-8"?>
<DirectionalLayout xmlns:ohos="http://schemas.huawei.com/res/ohos"
                ohos:top_margin="10vp"
                ohos:width="match_parent"
                ohos:height="match_content"
                ohos:orientation="vertical">
   <Text
      ohos:id="$+id:text_device_name"
      ohos:text_size="30vp"
      ohos:height="match_content"
      ohos:width="match_parent"/>
   <Text
      ohos:id="$+id:text_device_id"
      ohos:text_size="20fp"
      ohos:multiple_lines="true"
```

```
        ohos:text_color="#FF0000"
        ohos:top_margin="10vp"
        ohos:height="match_content"
        ohos:width="match_parent"/>
</DirectionalLayout>
```

在 device_id_item.xml 布局文件中放置了两个 Text 组件，分别用来显示设备名称和设备 ID。下面是 Java 代码的实现。

代码位置：src\main\java\com\unitymarvel\demo\ability\DeviceIdsAbility.java

```
package com.unitymarvel.demo.ability;
import com.unitymarvel.demo.ResourceTable;
import com.unitymarvel.demo.Tools;
import ohos.aafwk.ability.Ability;
import ohos.aafwk.content.Intent;
import ohos.agp.components.*;
import ohos.distributedschedule.interwork.DeviceInfo;
import ohos.distributedschedule.interwork.DeviceManager;
import java.util.ArrayList;
import java.util.List;

public class DeviceIdsAbility extends Ability {
   // 保存获取的所有设备的信息
   private List<DeviceInfo> deviceInfos;
   private ListContainer listContainerDeviceIds;
   // 获取所有可用设备的相关信息
   public static List<DeviceInfo> getAvailableDeviceIds() {
      List<DeviceInfo> deviceInfoList =
             DeviceManager.getDeviceList(DeviceInfo.FLAG_GET_ONLINE_DEVICE);
      if (deviceInfoList == null) {
          return new ArrayList<>();
      }
      if (deviceInfoList.size() == 0) {
          return new ArrayList<>();
      }
      return deviceInfoList;
   }
   @Override
   public void onStart(Intent intent) {
      super.onStart(intent);
      super.setUIContent(ResourceTable.Layout_device_ids);
      deviceInfos = getAvailableDeviceIds();
      listContainerDeviceIds =
      (ListContainer)findComponentById(ResourceTable.Id_listcontainer_deviceids);
      if(listContainerDeviceIds != null) {
         // 为 ListContainer 组件设置列表项监听器
         listContainerDeviceIds.setItemClickedListener(new ListContainer.
         ItemClickedListener() {
            @Override
            public void onItemClicked(ListContainer listContainer, Component component,
            int i, long l) {
               // 当单击某个列表项（设备）后，会获取该设备的 ID，并将这个 ID 作为 Page Ability 的结果返回
```

```
                String deviceId = deviceInfos.get(i).getDeviceId();
                Intent intent = new Intent();
                intent.setParam("deviceId", deviceId);
                setResult(100,intent);
                // 关闭当前的 Page Ability
                terminateAbility();

            }
        });
        // 为 ListContainer 组件设置 Provider
        listContainerDeviceIds.setItemProvider(new RecycleItemProvider() {
            @Override
            public int getCount() {
                return deviceInfos.size();
            }

            @Override
            public Object getItem(int i) {
                return deviceInfos.get(i);
            }

            @Override
            public long getItemId(int i) {
                return i;
            }

            @Override
            public Component getComponent(int i, Component component, ComponentContainer
            componentContainer) {
                if(component == null) {
                    /* 如果 component 值为 null，说明没有可以利用的列表项视图，需要从布局文件装载一
                    个新的视图对象*/
                    component= (DirectionalLayout)LayoutScatter.getInstance(DeviceIdsAbility.
                    this).parse(ResourceTable.Layout_device_id_item,null,false);
                }
                Text textDeviceName(Text)component.findComponentById(ResourceTable.
                Id_text_device_name);
                Text textDeviceId(Text)component.findComponentById(ResourceTable.
                Id_text_device_id);
                if(textDeviceName != null) {
                    // 显示设备名称
                    textDeviceName.setText(deviceInfos.get(i).getDeviceName());
                }
                if(textDeviceId != null) {
                    // 显示设备 ID
                    textDeviceId.setText(deviceInfos.get(i).getDeviceId());
                }
                return component;
            }
        });

    }

  }
}
```

在 DeviceIdsAbility 类中为 ListContainer 组件装载列表项时，利用了 getComponent 方法的第 2 个参数 component，通过该参数传入列表项的根视图。如果 component 值为 null，表明并没有可以利用的列表项视图，所以要创建一个新的列表项视图。如果值不为 null，表明可以利用其他还没有显示的列表项视图，使用时只需要替换该视图的 Text 组件中显示的信息即可。

最后在 config.json 文件中添加一些与分布式相关的权限。

```
"reqPermissions": [
      {
        "name": "ohos.permission.GET_DISTRIBUTED_DEVICE_INFO"
      },
      {
        "name": "com.huawei.permission.ACCESS_DISTRIBUTED_ABILITY_GROUP"
      },
      {
        "name": "ohos.permission.DISTRIBUTED_DATASYNC"
      }
]
```

运行程序，会看到图 2-16 所示的设备列表。

注意，通过 DeviceManager.getDeviceList 方法只能获取其他设备的信息，不能获取自身的信息，例如，有设备 A、设备 B 和设备 C。在设备 A 中只能获取设备 B 和设备 C 的信息，而不能获取设备 A 的信息。在设备 B 中和在设备 C 中获取设备信息情况相同。

图 2-16 设备列表

2.5.3 根据设备 ID 调用 Page Ability

要想实现跨设备访问一个 Page Ability，必须实现 IAbilityContinuation 接口，否则在访问过程中设备会抛出异常。该接口必须实现 4 个方法，它们的含义如下：

```
public interface IAbilityContinuation {
   // 开始迁移，如果返回 true，表示可以开始迁移
   boolean onStartContinuation();
   // 开始传递数据，如果返回 true，表示成功传递数据
   boolean onSaveData(IntentParams var1);
   // 开始恢复数据，如果返回 true，表示成功恢复数据
   boolean onRestoreData(IntentParams var1);
   // 已经完成 Page Ability 迁移
   void onCompleteContinuation(int var1);
}
```

假设将 Page Ability 从设备 A 迁移到设备 B，onStartContinuation 方法和 onSaveData 方法是在设备 A 上被调用的，而 onRestoreData 方法和 onCompleteContinuation 方法是在设备 B 上被调用的。为了迁移 Page Ability，需要在设备 A 上执行下面的代码：

```
continueAbility(deviceId);
```

其中，deviceId 是设备 ID。调用该方法后，在设备 A 上就会依次调用 onStartContinuation 方法

和 onSaveData 方法，在设备 B 上会依次调用 onRestoreData 方法和 onCompleteContinuation 方法。其中，onSaveData 方法和 onRestoreData 方法都有一个 IntentParams 类型的参数，使用该参数可以在设备 A 和设备 B 之间通过 Page Ability 传递数据（使用方式与 Intent 类似）。通常在 onRestoreData 方法中恢复 Page Ability 从设备 A 迁移到设备 B 时的数据。

【例 2.4】 在 Page Ability 上放置了一个 TextField 组件，并在该组件中输入文本，然后单击按钮，将该 Page Ability 迁移到另一部 HarmonyOS 手机上，并恢复迁移时的数据。

Page Ability 对应的布局代码如 cross_device_page_ability.xml 所示。

代码位置：src/main/resources/base/layout/cross_device_page_ability.xml

```
<?xml version="1.0" encoding="utf-8"?>
<DirectionalLayout xmlns:ohos="http://schemas.huawei.com/res/ohos"
                ohos:width="match_parent"
                ohos:height="match_parent"
                ohos:background_element="#CCCCCC"
                ohos:orientation="vertical">
   <Button
      ohos:id="$+id:button_cross_device_page_ability"
      ohos:background_element="$graphic:shape_text_bac"
      ohos:height="40vp"
      ohos:width="match_parent"
      ohos:text_color="#9b9b9b"
      ohos:text_size="17fp"
      ohos:text_alignment="center"
      ohos:text="跨设备迁移 Page Ability"
      ohos:right_margin="60vp"
      ohos:left_margin="60vp"
      ohos:top_margin="10vp"/>
   <TextField
      ohos:id="$+id:textfield_content"
      ohos:background_element="$graphic:border"
      ohos:height="match_parent"
      ohos:weight="1"
      ohos:width="match_parent"
      ohos:text_size="17fp"
      ohos:right_margin="5vp"
      ohos:left_margin="5vp"
      ohos:top_margin="10vp"
      ohos:multiple_lines="true"/>
</DirectionalLayout>
```

Java 实现代码如 CrossDevicePageAbility.java 所示。

代码位置：src\main\java\com\unitymarvel\demo\ability\CrossDevicePageAbility.java

```
package com.unitymarvel.demo.ability;

import com.unitymarvel.demo.ResourceTable;
import com.unitymarvel.demo.Tools;
import ohos.aafwk.ability.Ability;
import ohos.aafwk.ability.IAbilityContinuation;
```

```
import ohos.aafwk.content.Intent;
import ohos.aafwk.content.IntentParams;
import ohos.aafwk.content.Operation;
import ohos.agp.components.*;
import ohos.distributedschedule.interwork.DeviceInfo;
import java.util.ArrayList;
import java.util.List;

public class CrossDevicePageAbility extends Ability implements IAbilityContinuation {
   private List<DeviceInfo> deviceInfos;
   private ListContainer listContainerDeviceIds;
   private TextField textFieldContent;
   private String content;
   // 授权方法
   private void requestPermission() {
      // 实现 Page Ability 跨设备迁移，必须用 Java 代码申请下面的权限，否则不会有任何反应
      String[] permission = {"ohos.permission.DISTRIBUTED_DATASYNC"};
      List<String> applyPermissions = new ArrayList<>();
      for (String element : permission) {
         // 验证自身是否已经获得了该权限
         if (verifySelfPermission(element) != 0) {
             if (canRequestPermission(element)) {
                 // 如果未获得权限，将该权限添加到权限列表
                 applyPermissions.add(element);
             } else {
             }
         } else {
         }
      }
      // 申请相应权限
      requestPermissionsFromUser(applyPermissions.toArray(new String[0]), 0);
   }
   // 要想成功跨设备迁移 Page Ability，该方法必须返回 true
   @Override
   public boolean onStartContinuation() {
       return true;
   }
   @Override
   public boolean onSaveData(IntentParams intentParams) {
      // 保存要传递的数据
      intentParams.setParam("content",textFieldContent.getText());
      return true;
   }
   @Override
   public boolean onRestoreData(IntentParams intentParams) {
      // 获取传递过来的数据
      content = String.valueOf(intentParams.getParam("content"));
      return true;
   }
   @Override
   public void onCompleteContinuation(int i) {

   }
```

```
    @Override
    protected void onAbilityResult(int requestCode, int resultCode, Intent resultData)
    {
        // 当选择设备后，利用返回的设备 ID 迁移 Page Ability
        if(resultCode == 100 && requestCode == 99) {
            // 获取设备 ID
            String deviceId = resultData.getStringParam("deviceId");
            Tools.showTip(this, deviceId);
            // 跨设备迁移 Page Ability
            continueAbility(deviceId);
        }
    }

    @Override
    public void onStart(Intent intent) {
        super.onStart(intent);
        super.setUIContent(ResourceTable.Layout_cross_device_page_ability);
        // 申请权限
        requestPermission();
        Button button =
        (Button)findComponentById(ResourceTable.Id_button_cross_device_page_ability);
        if(button != null) {
            button.setClickedListener(new Component.ClickedListener() {
                @Override
                public void onClick(Component component) {
                    // 显示列表窗口
                    Intent intentPageAbility = new Intent();
                    Operation operation = new Intent.OperationBuilder()
                            .withBundleName("com.unitymarvel.demo")
                            .withAbilityName("com.unitymarvel.demo.ability.DeviceIdsAbility")
                            .build();
                    intentPageAbility.setOperation(operation);
                    startAbilityForResult(intentPageAbility,99);
                    }
            });
         }

        textFieldContent =
        (TextField)findComponentById(ResourceTable.Id_textfield_content);
        if(textFieldContent != null) {
            // 恢复 TextField 组件中的数据
            textFieldContent.setText(content);
        }
    }
}
```

阅读这段代码，需要了解下面几点。

- 若要实现 Page Ability 迁移，并成功传递数据，onStartContinuation 方法、onSaveData 方法和 onRestoreData 方法都必须返回 true，如果读者使用 IDE 的自动生成代码功能，这几个方法都会默认返回 false，需要将它们的返回值改成 true。
- HarmonyOS 中有一些权限，并不是在 config.json 文件中声明就可以了，还需要使用 Java

代码申请，例如，Page Ability 跨设备迁移就需要使用 Java 代码申请 ohos.permission. DISTRIBUTED_DATASYNC 权限。如果是第一次申请，会弹出图 2-17 所示的授权对话框，单击“始终允许”按钮关闭该对话框，在第 2 次申请权限时，将不会弹出该对话框。

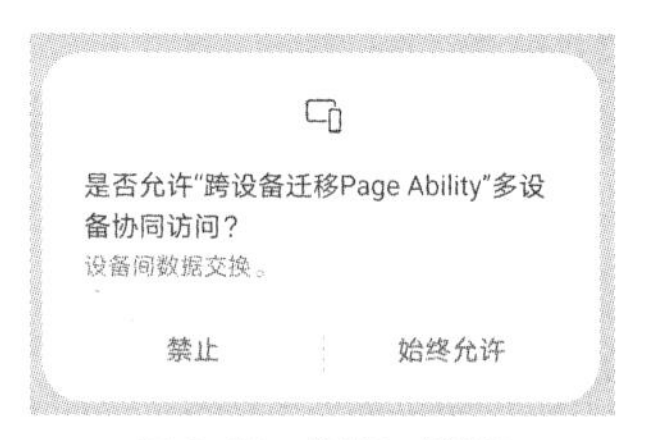

图 2-17　授权对话框

- 因为 onRestoreData 方法在 onStart 方法之前调用，所以不能直接在 onRestoreData 方法中使用组件对象，因为组件对象通常都是在 onStart 方法中创建的。因此当 onRestoreData 方法被调用时，这些组件对象还都为空。正确的做法是，在 onRestoreData 方法中将要恢复的数据用成员变量保存，然后在 onStart 方法中创建组件对象，最后通过这些成员变量恢复组件中的数据。
- 本例考虑了 Page Ability 可能跨多部 HarmonyOS 设备迁移的情况，所以使用了上一节实现的设备列表窗口。在开始跨设备迁移 Page Ability 之前，会先弹出一个设备列表窗口，当用户选择一个设备后，会返回该设备的 ID，然后在 onAbilityResult 方法中获取返回的设备 ID，最后使用 continueAbility 方法迁移 Page Ability。

图 2-18　跨设备迁移 Page Ability 的效果

运行程序，关闭授权对话框，并在 TextField 组件中输入一些内容，最后单击“跨设备迁移 Page Ability”按钮，将会弹出一个设备列表窗口。选择相应的设备后，会在选中的设备中弹出同样的 Page Ability，并且 TextField 组件中的数据与原设备上的完全相同，如图 2-18 所示。注意，只要被调用方安装了 App，不管设备是否已经启动了 App，都会自动弹出这个被迁移的 Page Ability。

2.6 AbilitySlice

AbilitySlice 是另外一种实现 UI 的技术。使用 AbilitySlice 显示 UI，类似使用同一个浏览器页面显示不同的网页。本节将详细讲解 Ability Slice 的主要用法。

AbilitySlice 类只是普通的 Java 类，并不需要在 config.json 文件中配置。但 AbilitySlice 类必须从 ohos.aafwk.ability.AbilitySlice 类继承，下面就是一个典型的 AbilitySlice 类。

```
public class MainAbilitySlice extends AbilitySlice {
 @Override
   public void onStart(Intent intent) {
      super.onStart(intent);
   }
}
```

AbilitySlice 与 Page Ability 拥有相同的生命周期方法，使用方法也类似。Page Ability 可以直接装载布局文件，也可以导航到 AbilitySlice，然后由 AbilitySlice 负责装载布局文件。Page Ability 导航到 AbilitySlice 的代码如下：

```
public class MainAbility extends Ability {
   @Override
   public void onStart(Intent intent) {
      super.onStart(intent);
      // 导航到Ability Slice
      super.setMainRoute(MainAbilitySlice.class.getName());
   }
}
```

从这段代码可以看出，从 Page Ability 导航到 AbilitySlice，需要使用 setMainRoute 方法。该方法需要获得 AbilitySlice 类的全名，如 com.harmonyos.MainAbilitySlice。

如果 App 中拥有多个 AbilitySlice，就需要在这些 AbilitySlice 之间导航，假设有一个名为 MySlice 的 AbilitySlice，要从当前 AbilitySlice 导航到 MySlice，并传递两个属性 age 和 name 的值，可以使用下面的代码：

```
Intent intent = new Intent();
// 向 MySlice 传递数据
intent.setParam("age",20);
intent.setParam("name","bill");
// 导航到 MySlice
present(new MySlice(),intent);
```

从这段代码可以看出，导航到 AbilitySlice 需要调用 present 方法。该方法的第 1 个参数是 AbilitySlice 对象，如本例中的 new MySlice()。第 2 个参数是 Intent 对象，该对象的作用之一是向 MySlice 传递参数值。AbilitySlice 之间使用 Intent 对象传递数据，这一点与 Page Ability 相同。

如果想实现从 AbilitySlice 返回数据，需要使用 presentForResult 方法，该方法的原型如下：

```
public final void presentForResult(AbilitySlice targetSlice, Intent intent, int requestCode)
```

presentForResult 方法的前两个参数与 present 方法的参数完全相同，而第 3 个参数 requestCode 是请求码，用于区分调用者。因为参与导航的 AbilitySlice 可能不止一个，所以需要使用 requestCode 区分到底是哪一个 AbilitySlice 导航到另外一个 AbilitySlice 的。

如果想实现从 MainAbilitySlice 导航到 MyAlice，并从 MyAlice 返回信息给 MainAbilitySlice，应该在 MainAbilitySlice 中重写 onResult 方法，代码如下：

```
public class MainAbilitySlice extends AbilitySlice {
   ...
   @Override
   protected void onResult(int requestCode, Intent resultIntent) {
      if (requestCode == 100) {
         // 接收 Ability Slice 返回的信息
      } else if(requestCode == 200) {
         // 接收 Ability Slice 返回的信息
      }
   }
}
```

在这段代码中，requestCode 的值有 2 个——100 和 200。这说明至少调用了 2 次 presentForResult 方法导航到另外一个 AbilitySlice。如果这两个 requestCode 对应的业务逻辑不同，那么需要使用 requestCode 进行判断，并根据结果分别处理。

在目标 AbilitySlice（本例是 MySlice）中，需要使用 terminate 方法关闭当前的 AbilitySlice。如果要返回数据，需要在 terminate 方法前面使用 setResult 方法设置要返回的 Intent 对象，代码如下：

```
Intent intent = new Intent();
intent.setParam("value",20);
setResult(intent);
terminate();
```

【例 2.5】 本例完整地演示了从 Page Ability 导航到 AbilitySlice，然后从 AbilitySlice 导航到另外一个 AbilitySlice，最后接收返回值的全过程。

本例及以后的所有案例，除非有特殊需要，都将使用 HarmonyOS 手机进行演示。

代码位置：src/main/java/com/unitymarvel/demo/ability/BasicSliceAbility.java

```
package com.unitymarvel.demo.ability;

import com.unitymarvel.demo.slice.ability.BasicSlice;
import ohos.aafwk.ability.Ability;
import ohos.aafwk.content.Intent;

public class BasicSliceAbility extends Ability {

   @Override
   public void onStart(Intent intent) {
      super.onStart(intent);
      // 导航到 BasicSlice
      super.setMainRoute(BasicSlice.class.getName());
   }
}
```

BasicSlice 类是一个 AbilitySlice 类，负责导航到 MySlice，并接收 MySlice 的返回值，最后将返回值显示在 Text 组件中。

代码位置：src/main/java/com/unitymarvel/demo/slice/ability/BasicSlice.java

```
package com.unitymarvel.demo.slice.ability;

import com.unitymarvel.demo.ResourceTable;
import ohos.aafwk.ability.AbilitySlice;
import ohos.aafwk.content.Intent;
import ohos.agp.components.Button;
import ohos.agp.components.Component;
import ohos.agp.components.Text;

public class BasicSlice extends AbilitySlice {
   private Text textResult;
   @Override
   public void onStart(Intent intent) {
```

```
        super.onStart(intent);
        super.setUIContent(ResourceTable.Layout_basic_slice_layout);
        textResult = (Text)findComponentById(ResourceTable.Id_text_result);
        Button buttonPresentSlice =
        (Button)findComponentById(ResourceTable.Id_button_present_slice);
        if(buttonPresentSlice != null) {
            buttonPresentSlice.setClickedListener(new Component.ClickedListener() {
                @Override
                public void onClick(Component component) {
                    Intent intent = new Intent();
                    // 设置要传递给 MySlice 的数据
                    intent.setParam("data","hello world");
                    // 导航到 MySlice，并设置要传递的数据以及 requestCode（本例是 100）
                    presentForResult(new MySlice(),intent,100);
                }
            });
        }
    }
    // 接收 MySlice 返回的数据
    @Override
    protected void onResult(int requestCode, Intent resultIntent) {
        switch (requestCode){
            case 100:
                // 将 MySlice 返回的值显示在 Text 组件中
                textResult.setText(resultIntent.getStringParam("data"));
                break;
        }
    }
}
```

MySlice 类是一个 AbilitySlice 类，负责接收 BasicSlice 传递过来的数据，并将数据返回给 BasicSlice。

代码位置：src/main/java/com/unitymarvel/demo/slice/ability/MySlice.java

```
package com.unitymarvel.demo.slice.ability;

import com.unitymarvel.demo.ResourceTable;
import ohos.aafwk.ability.AbilitySlice;
import ohos.aafwk.content.Intent;
import ohos.agp.components.Button;
import ohos.agp.components.Component;
import ohos.agp.components.Text;

public class MySlice extends AbilitySlice {

    @Override
    public void onStart(Intent intent) {
        super.onStart(intent);
        super.setUIContent(ResourceTable.Layout_my_slice_layout);
        Text text = (Text)findComponentById(ResourceTable.Id_text);
        if(text != null) {
            // 接收从 BasicSlice 传递过来的数据
            String data = intent.getStringParam("data");
```

```
            // 将数据显示在 Text 组件中
            text.setText(data);
        }
        Button buttonCloseSlice =
            (Button)findComponentById(ResourceTable.Id_button_close_slice);
        if(buttonCloseSlice != null) {
            buttonCloseSlice.setClickedListener(new Component.ClickedListener() {
                @Override
                public void onClick(Component component) {
                    Intent intent = new Intent();
                    // 设置要返回给 BasicSlice 的数据
                    intent.setParam("data", "I love you.");
                    setResult(intent);
                    // 关闭当前的 Ability Slice
                    terminate();
                }
            });
        }
    }
}
```

现在运行程序，单击“导航到另一个 Ability Slice”按钮，就会显示 MySlice，效果如图 2-19 所示。然后单击“关闭 Slice”按钮，就会返回 BasicSlice，效果如图 2-20 所示。

图 2-19 MySlice 效果

图 2-20 BasicSlice 效果

2.7 生命周期

生命周期是 Page Ability 的重要特征之一。所谓生命周期，就是在 Page Ability 从创建到在前台显示，然后切换到后台，最后到销毁的过程所经历的各个状态的总称。每一个状态通过一个方法通知用户，以便用户在该状态完成必要的工作，这些方法称为生命周期方法。例如，通过 HarmonyOS

模板创建的工程会为 Page Ability 自动生成一个 onStart 方法，这就是 Page Ability 的生命周期方法之一，表示 Page Ability 在初始化后，激活前的状态。

除了 onStart 方法，还有很多生命周期方法，图 2-21 显示了 Page Ability 的生命周期中所有状态、生命周期方法以及它们之间的关系。

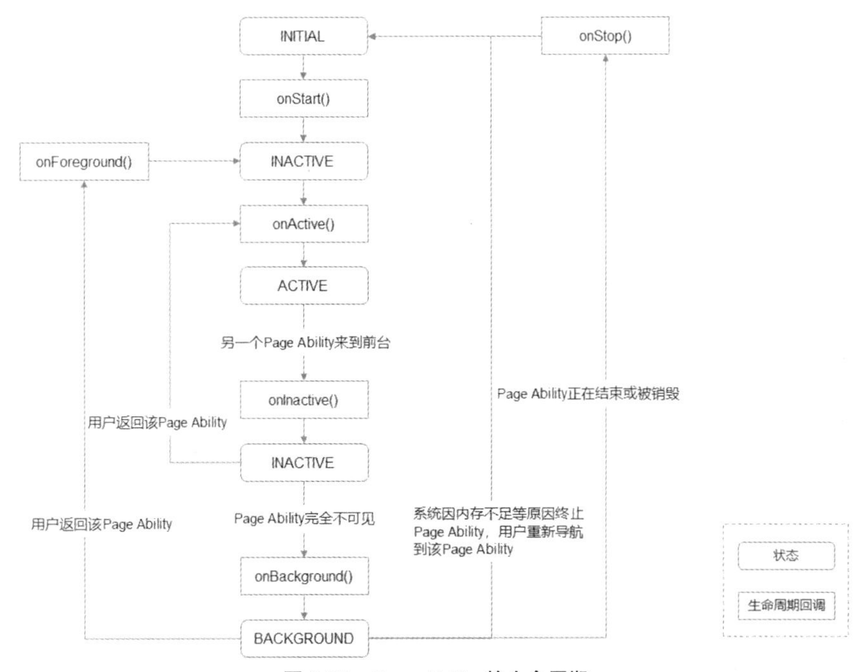

图 2-21 Page Ability 的生命周期

Page Ability 的生命周期方法共有 6 个，下面就具体描述一下这些方法。

- onStart 方法。当系统首次创建 Page Ability 实例时，触发 onStart 方法。对于一个 Page Ability 实例，onStart 方法在其生命周期中仅触发一次，Page Ability 在调用 onStart 方法后进入 INACTIVE 状态。开发人员需要重写该方法，并在该方法中完成必要的初始化工作，如导航到 AbilitySlice。

```
@Override
public void onStart(Intent intent) {
   super.onStart(intent);
   super.setMainRoute(MySlice.class.getName());
}
```

- onActive 方法。Page Ability 会在进入 INACTIVE 状态后切换到前台，然后系统调用 onActive 方法。在调用完 onActive 方法后，Page Ability 就会进入 ACTIVE 状态，该状态是 App 与用户交互的状态，也就是用户最终看到的 UI 显示的状态。Page Ability 将保持在此状态，除非某些操作让 Page Ability 失去焦点，例如，用户单击返回键或导航到其他 Page Ability。当发生这种情况时，Page Ability 会回到 INACTIVE 状态，系统将调用 onInactive 方法。如果希望 Page Ability 回到 ACTIVE 状态，需要系统再次调用 onActive 方法。因此，开发人员通常需要成对实现 onActive 方法和 onInactive 方法，并在 onActive 方法中获取在 onInactive 方法中被释放的资源。

- onInactive 方法。当 Page Ability 失去焦点时，系统将调用 onInactive 方法，然后 Page Ability 会进入 INACTIVE 状态。开发人员可以在 onInactive 方法中完成 Page Ability 失去焦点后需要完成的业务逻辑。
- onBackground 方法。如果 Page Ability 不再对用户可见，系统将调用 onBackground 方法通知开发人员，通常会在该方法中释放 Page Ability 打开的一些资源，如文件、数据库等，或执行较为耗时的状态保存操作。在调用完 onBackground 方法后，Page Ability 就会进入 BACKGROUND 状态。
- onForeground 方法。处于 BACKGROUND 状态的 Page Ability 仍然驻留在内存中，当重新回到前台时（比如用户重新导航到此 Page Ability），系统将先调用 onForeground 方法通知开发人员，然后 Page Ability 回到 INACTIVE 状态。开发人员应该在 onForeground 方法中重新申请在 onBackground 方法中释放的资源，最后 Page Ability 回到 ACTIVE 状态，系统将调用 onActive 方法通知用户。
- onStop 方法。系统将要销毁 Page Ability 时，会触发 onStop 方法，通知用户释放系统资源。销毁 Page Ability 的可能原因包括以下几个方面。
  - 用户通过系统管理能力关闭指定 Page Ability，例如使用任务管理器关闭 Page Ability。
  - 调用 Page Ability 的 terminateAbility 方法来关闭当前 Page Ability。
  - 配置变更导致系统暂时销毁 Page Ability 并重建。
  - 系统出于资源管理目的，自动触发对处于 BACKGROUND 状态 Page Ability 的销毁。

不仅 Page Ability 有生命周期，AbilitySlice 同样有生命周期，而且 AbilitySlice 的生命周期依赖于 Page Ability 的生命周期。Page Ability 与 AbilitySlice 具有相同的生命周期状态和同名的生命周期方法。当 Page Ability 的生命周期发生变化时，它的 AbilitySlice 也会发生相同的生命周期变化。此外，AbilitySlice 还具有独立于 Page Ability 的生命周期变化，这发生在同一 Page Ability 中存在 AbilitySlice 间导航的情况下，此时 Page Ability 的生命周期状态不会改变。

AbilitySlice 生命周期方法与 Page Ability 的同名生命周期方法在使用上和功能上都相同，这里不再赘述。如果 Page Ability 与 AbilitySlice 同时使用，那么通常 Page Ability 会导航到 AbilitySlice，然后 AbilitySlice 通过 setUIContent 方法装载 XML 布局文件，并显示在窗口中，代码如下：

```
@Override
protected void onStart(Intent intent) {
   super.onStart(intent);
   setUIContent(ResourceTable.Layout_main_layout);
}
```

AbilitySlice 实例的创建和管理通常由 App 负责，系统仅在特定情况下会创建 AbilitySlice 实例，例如，通过导航启动某个 AbilitySlice 时，是由系统负责实例化的。但是在同一个 Page 中不同 AbilitySlice 间导航时由 App 负责实例化。

那么 Page Ability 与 AbilitySlice 的生命周期有什么关联呢？当 AbilitySlice 处于前台且具有焦

点时，其生命周期状态随着所属 Page 的生命周期状态的变化而变化。一个 Page Ability 可能拥有多个 AbilitySlice，例如 MyAbility 拥有 FooAbilitySlice 和 BarAbilitySlice。假设当前 FooAbilitySlice 处于前台并获得了焦点，且即将导航到 BarAbilitySlice，则在此期间生命周期的状态按以下顺序变化。

（1）FooAbilitySlice 从 ACTIVE 状态变为 INACTIVE 状态。

（2）BarAbilitySlice 首先从 INITIAL 状态变为 INACTIVE 状态，然后变为 ACTIVE 状态（假定此前 BarAbilitySlice 未启动）。

（3）FooAbilitySlice 从 INACTIVE 状态变为 BACKGROUND 状态。

对应两个 AbilitySlice 的生命周期方法回调顺序为 FooAbilitySlice.onInactive()→BarAbilitySlice.onStart()→BarAbilitySlice.onActive()→FooAbilitySlice.onBackground()。

在整个流程中，MyAbility 始终处于 ACTIVE 状态。但是，当 Page Ability 被系统销毁时，其所有已实例化的 AbilitySlice 将被联动销毁，而不只是销毁处于前台的 AbilitySlice。

2.8 总结与回顾

本章讲解了 Page Ability 的相关内容，通过 Page Ability 可以创建 HarmonyOS 的窗口，它在开发 App 的 UI 上起了关键的作用。但 Page Ability 不只是用来开发 App 的界面，它可以使 HarmonyOS 具有跨设备迁移的独特功能。

当然，Page Ability 拥有非常多的 API，以及复杂的生命周期。在本章中我们已经对这些内容进行了深入的讲解，这为读者后续学习打下了很好的基础。

第3章　布局

布局是用于将众多组件按用户的要求在屏幕上摆放的技术。HarmonyOS 为了让 App 适应尽可能多的屏幕，提供了强大的布局支持。本章主要介绍如何在 XML 布局文件中使用 HarmonyOS 的布局。

通过阅读本章，读者可以掌握：

- 方向布局（DirectionalLayout）；
- 依赖布局（DependentLayout）；
- 栈布局（StackLayout）；
- 表格布局（TableLayout）；
- 位置布局（PositionLayout）；
- 如何动态装载布局文件。

3.1 方向布局

顾名思义，方向布局是用于控制组件方向的布局。方向包括水平方向和垂直方向，默认是垂直布局。方向布局使用<DirectionalLayout>标签表示，通过 ohos:orientation 属性设置方向，vertical 表示垂直布局，horizontal 表示水平布局。

其他属性与大多数组件一样，例如，ohos:width 表示组件的宽度，ohos:height 表示组件的高度。组件的尺寸可以设置为具体的值，也可以设置为 match_parent 和 match_content 中的一个。match_parent 表示组件在水平或垂直方向尽可能占用剩余的空间，match_content 则表示组件根据内容自动调整自身的尺寸。

【例 3.1】 通过方向布局在窗口中放置 6 个按钮，其中前 3 个使用水平方向布局，后 3 个使用垂直方向布局，并且这两个方向的布局是垂直方向。

代码位置： src/main/resources/base/layout/directional_layout_demo.xml

```
<?xml version="1.0" encoding="utf-8"?>
<DirectionalLayout
   xmlns:ohos="http://schemas.huawei.com/res/ohos"
   ohos:height="match_parent"
   ohos:width="match_parent"
```

```
    ohos:background_element="#CCCCCC"
    ohos:orientation="vertical"
    ohos:padding="32"
    >
    <DirectionalLayout
        ohos:height="100vp"
        ohos:width="match_parent"
        ohos:orientation="horizontal">
        <Button
            ohos:id="$+id:button1"
            ohos:height="match_parent"
            ohos:width="match_content"
            ohos:background_element="#00FFFF"
            ohos:text="按钮 1"
            ohos:text_alignment="center"
            ohos:text_size="30vp"
            />

        <Button
            ohos:id="$+id:button2"
            ohos:height="match_parent"
            ohos:width="match_parent"
            ohos:background_element="#00FFFF"
            ohos:left_margin="10vp"
            ohos:text="按钮 2"
            ohos:text_alignment="center"
            ohos:text_size="20vp"
            ohos:weight="1"
            />

        <Button
            ohos:id="$+id:button3"
            ohos:height="match_parent"
            ohos:width="match_parent"
            ohos:background_element="#00FFFF"
            ohos:left_margin="10vp"
            ohos:text="按钮 3"
            ohos:text_alignment="center"
            ohos:text_size="30vp"
            ohos:weight="3"
            />
    </DirectionalLayout>

    <DirectionalLayout
        ohos:height="match_parent"
        ohos:width="match_parent"
        ohos:orientation="vertical">

        <Button
            ohos:id="$+id:button4"
            ohos:height="match_content"
            ohos:width="match_parent"
            ohos:background_element="#00FFFF"
            ohos:text="按钮 4"
```

```
            ohos:text_alignment="center"
            ohos:text_size="30vp"
            ohos:top_margin="20vp"
            />

        <Button
            ohos:id="$+id:button5"
            ohos:height="match_parent"
            ohos:width="match_parent"
            ohos:background_element="#00FFFF"
            ohos:text="按钮 5"
            ohos:text_alignment="center"
            ohos:text_size="30vp"
            ohos:top_margin="20vp"
            ohos:weight="3"
            />

        <Button
            ohos:id="$+id:button6"
            ohos:height="match_parent"
            ohos:width="match_parent"
            ohos:background_element="#00FFFF"
            ohos:text="按钮 6"
            ohos:text_alignment="center"
            ohos:text_size="30vp"
            ohos:top_margin="20vp"
            ohos:weight="1"
            />
    </DirectionalLayout>
</DirectionalLayout>
```

运行程序，会看到图 3-1 所示方向布局的效果。

directional_layout_demo.xml 文件中使用了一个重要的属性——ohos:weight，该属性的值是一个数值类型，通常是正整数，当 ohos:width 属性或 ohos:height 属性的值为 match_parent 时使用该属性，其目的是确认水平方向或垂直方向相邻组件的尺寸比例。例如，有相邻两个按钮 button1 和 button2，它们的 ohos:width 属性的值都是 match_parent。如果 button1 的 ohos:weight 属性值设为 1，button2 的 ohos:weight 属性值设为 3，而且 button1 和 button2 在水平方向布局中，那么 button1 占了水平方向 1/4 的空间，而 button2 占了水平方向 3/4 的空间。如果水平方向还有其他非 match_parent 尺寸的组件，那么系统会先安排这些组件，然后 button1 和 button2 会在剩余的空间继续按比例进行划分。

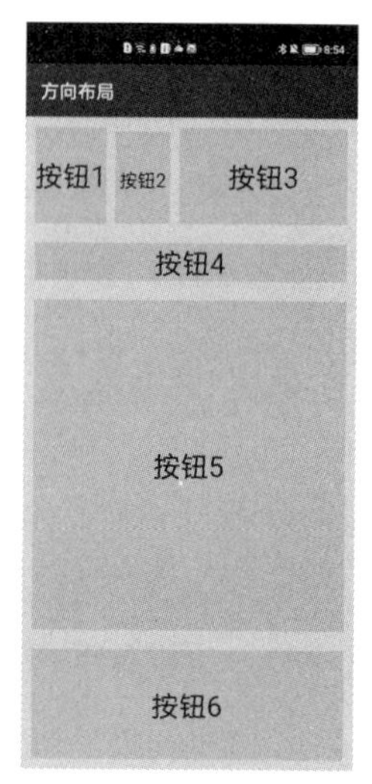

图 3-1　方向布局的效果

3.2 依赖布局

依赖布局可以利用组件之间的相对关系进行布局，使用<DependentLayout>标签描述。组件之间的关系分为如下两种：

- 同级别组件的关系；
- 父容器与子组件的关系。

这两类关系分别由两组属性控制。控制同级别组件关系的属性如表 3-1 所示。

表 3-1 控制同级别组件关系的属性

布局属性	描述
above	处于指定组件的上方
below	处于指定组件的下方
start_of	处于指定组件的起始侧
end_of	处于指定组件的结束侧
left_of	处于指定组件的左侧
right_of	处于指定组件的右侧

表 3-1 中的 6 个属性需要指定组件的 ID。

控制父容器与子组件关系的属性如表 3-2 所示。

表 3-2 控制父容器与子组件关系的属性

布局属性	描述
align_parent_left	处于父容器的左侧
align_parent_right	处于父容器的右侧
align_parent_start	处于父容器的起始侧
align_parent_end	处于父容器的结束侧
align_parent_top	处于父容器的上方
align_parent_bottom	处于父容器的下方
center_in_parent	处于父容器的中心

表 3-2 中属性的值都是布尔类型，只能设置为 true 或 false。

【例 3.2】 通过依赖布局在窗口中放置 5 个按钮，其中前 3 个使用控制同级别组件关系的属性，后 2 个使用控制父容器与子组件关系的属性。

代码位置：src/main/resources/base/layout/dependent_layout_demo.xml

```
<?xml version="1.0" encoding="utf-8"?>
<DependentLayout
   xmlns:ohos="http://schemas.huawei.com/res/ohos"
   ohos:height="match_parent"
   ohos:width="match_parent"
   ohos:background_element="#CCCCCC"
   ohos:padding="32">
   <Button
      ohos:id="$+id:button1"
      ohos:height="match_content"
      ohos:width="match_content"
      ohos:background_element="#00FFFF"
```

```
            ohos:center_in_parent="true"
            ohos:text="按钮 1"
            ohos:text_alignment="center"
            ohos:text_size="30vp"
            />
        <Button
            ohos:height="match_content"
            ohos:width="match_content"
            ohos:background_element="#00FFFF"
            ohos:center_in_parent="true"
            ohos:left_margin="20vp"
            ohos:right_of="$id:button1"
            ohos:text="按钮 2"
            ohos:text_size="30vp"
            />
        <Button
            ohos:height="match_content"
            ohos:width="match_content"
            ohos:background_element="#00FFFF"
            ohos:below="$id:button1"
            ohos:center_in_parent="true"
            ohos:left_margin="20vp"
            ohos:text="按钮 3"
            ohos:text_size="30vp"
            ohos:top_margin="20vp"
            />
        <Button
            ohos:height="match_content"
            ohos:width="match_content"
            ohos:align_parent_right="true"
            ohos:background_element="#00FFFF"
            ohos:left_margin="20vp"
            ohos:text="按钮 4"
            ohos:text_size="30vp"
            ohos:top_margin="20vp"
            />
        <Button
            ohos:height="match_content"
            ohos:width="match_content"
            ohos:align_parent_bottom="true"
            ohos:align_parent_left="true"
            ohos:background_element="#00FFFF"
            ohos:bottom_margin="20vp"
            ohos:left_margin="20vp"
            ohos:text="按钮 5"
            ohos:text_size="30vp"
            />
    </DependentLayout>
```

运行程序，会看到图 3-2 所示依赖布局的效果。

图 3-2 依赖布局的效果

“按钮 2”和“按钮 3”分别使用 right_of 属性和 below 属性控制与“按钮 1”的相对位置。这两个属性的值都是"$id:button1"。注意，这里$和 id 之间不要加“+”号。“+”号是在组件的 ID 不存在时自动创建的，通常用于设置组件的 id 属性值。而其他属性，如果需要引用组件的 ID，那么这个 ID 必须是存在的，因此就不需要使用“+”号了。

3.3 栈布局

栈布局用来实现组件之间的层叠布局，使用<StackLayout>标签描述。这种布局有点儿像 Photoshop 中的图层，第一个放置的组件在最底层，后面放置的组件叠在前一个组件的上面。

【例 3.3】 通过栈布局在窗口中放置 3 个不同尺寸的按钮，这 3 个按钮根据放置的顺序层叠在一起。

代码位置：src/main/resources/base/layout/stack_layout_demo.xml

```
<?xml version="1.0" encoding="utf-8"?>
<StackLayout
   xmlns:ohos="http://schemas.huawei.com/res/ohos"
   ohos:height="match_parent"
   ohos:width="match_parent"
   ohos:background_element="#CCCCCC"
   ohos:padding="32">
   <Button
      ohos:height="300vp"
      ohos:width="300vp"
      ohos:background_element="#FF0000"
      ohos:text="按钮 1"
      ohos:text_size="30vp"
      />
   <Button
      ohos:height="200vp"
      ohos:width="200vp"
      ohos:background_element="#00FF00"
      ohos:text="按钮 2"
      ohos:text_size="30vp"
      />
   <Button
      ohos:height="100vp"
      ohos:width="100vp"
      ohos:background_element="#0000FF"
      ohos:text="按钮 2"
      ohos:text_size="30vp"
      />
</StackLayout>
```

运行程序，会看到图 3-3 所示栈布局的效果。

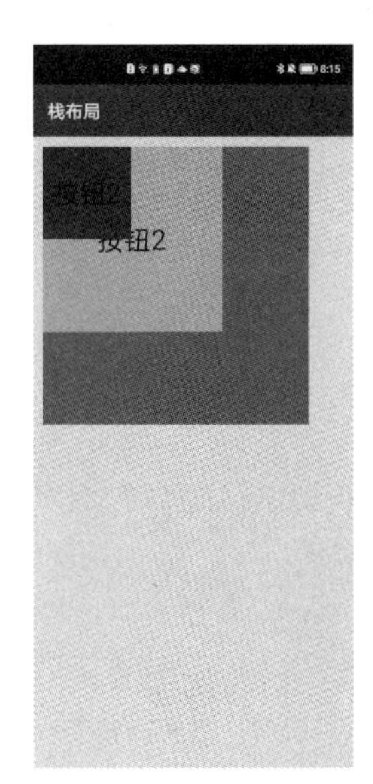

图 3-3 栈布局的效果

3.4 表格布局

表格布局可以将组件按行列形式摆放，即可以将多个组件分别放置一个表格的不同单元格中。表格布局使用<TableLayout>标签描述，它最重要的属性是 ohos:row 和 ohos:col，这两个属性分别指定了行和列的个数。

【例 3.4】 通过表格布局在窗口中放置 6 个图像组件，将这些图像组件以 3 行 2 列形式分别摆放。

代码位置：src/main/resources/base/layout/table_layout_demo.xml

```
<?xml version="1.0" encoding="utf-8"?>
<TableLayout
    xmlns:ohos="http://schemas.huawei.com/res/ohos"
    ohos:height="match_parent"
    ohos:width="match_parent"
    ohos:background_element="#CCCCCC"
    ohos:column_count="2"
    ohos:row_count="3">
    <Image
        ohos:height="120vp"
        ohos:width="120vp"
        ohos:bottom_margin="5vp"
        ohos:image_src="$media:device"
        ohos:left_margin="25vp"
        ohos:right_margin="25vp"
        ohos:scale_mode="zoom_center"/>
    <Image
        ohos:height="120vp"
        ohos:width="120vp"
        ohos:bottom_margin="5vp"
        ohos:image_src="$media:multimedia"
        ohos:left_margin="25vp"
        ohos:right_margin="25vp"
        ohos:scale_mode="zoom_center"/>
    <Image
        ohos:height="120vp"
        ohos:width="120vp"
        ohos:bottom_margin="5vp"
        ohos:image_src="$media:network"
        ohos:left_margin="25vp"
        ohos:right_margin="25vp"
        ohos:scale_mode="zoom_center"/>
    <Image
        ohos:height="120vp"
        ohos:width="120vp"
        ohos:bottom_margin="5vp"
        ohos:image_src="$media:storage"
        ohos:left_margin="25vp"
        ohos:right_margin="25vp"
        ohos:scale_mode="zoom_center"/>
    <Image
        ohos:height="120vp"
        ohos:width="120vp"
```

```
        ohos:bottom_margin="5vp"
        ohos:image_src="$media:others"
        ohos:left_margin="25vp"
        ohos:right_margin="25vp"
        ohos:scale_mode="zoom_center"/>
    <Image
        ohos:height="120vp"
        ohos:width="120vp"
        ohos:bottom_margin="5vp"
        ohos:image_src="$media:components"
        ohos:left_margin="25vp"
        ohos:right_margin="25vp"
        ohos:scale_mode="zoom_center"/>
</TableLayout>
```

运行程序，会看到图 3-4 所示表格布局的效果。

图 3-4 表格布局的效果

3.5 位置布局

位置布局用于为组件指定特定的位置。尽管在移动 App 中位置布局不经常使用，但在一些特定的场景下，如需要用代码控制组件的移动、将组件摆放在没有特定规律的位置，就需要使用位置布局。位置布局使用<PositionLayout>标签描述，但要通过 setContentPosition 方法设置组件的 X 值和 Y 值，也可以使用 setContentPositionX 方法和 setContentPositionY 方法分别设置组件的 X 值和 Y 值。

【例 3.5】 通过位置布局在窗口中摆放 3 个按钮，并为这 3 个按钮指定不同的位置。

代码位置：src/main/resources/base/layout/position_layout_demo.xml

```
<?xml version="1.0" encoding="utf-8"?>
<PositionLayout
    xmlns:ohos="http://schemas.huawei.com/res/ohos"
    ohos:height="match_parent"
    ohos:width="match_parent"
    ohos:background_element="#CCCCCC"
    ohos:padding="32">
    <Button
        ohos:id="$+id:button1"
        ohos:height="match_content"
        ohos:width="match_content"
        ohos:background_element="#00FFFF"
        ohos:text="按钮 1"
        ohos:text_size="30vp"/>
    <Button
        ohos:id="$+id:button2"
        ohos:height="match_content"
        ohos:width="match_content"
        ohos:background_element="#00FFFF"
        ohos:text="按钮 2"
        ohos:text_size="30vp"/>
```

```
    <Button
        ohos:id="$+id:button3"
        ohos:height="match_content"
        ohos:width="match_content"
        ohos:background_element="#00FFFF"
        ohos:text="按钮 3"
        ohos:text_size="30vp"/>
</PositionLayout>
```

下面的代码用来设置这 3 个按钮的位置。

代码位置：src/main/java/com/unitymarvel/demo/layout/PositionLayoutDemo.java

```
package com.unitymarvel.demo.layout;
import ohos.aafwk.ability.Ability;
import ohos.aafwk.content.Intent;
import ohos.agp.components.Button;
public class PositionLayoutDemo extends Ability {
    @Override
    public void onStart(Intent intent) {
        super.onStart(intent);
        super.setUIContent(com.unitymarvel.demo.ResourceTable.Layout_position_layout_demo);
        Button button1 = (Button)findComponentById(com.unitymarvel.demo.ResourceTable.
        Id_button1);
        if(button1 != null) {
            // 设置 button1 的位置
            button1.setContentPosition(100,200);
        }
        Button button2 = (Button)findComponentById(com.unitymarvel.demo.ResourceTable.
        Id_button2);
        if(button2 != null) {
            // 设置 button2 的 X 值
            button2.setContentPositionX(500);
            // 设置 button2 的 Y 值
            button2.setContentPositionY(800);
            // 设置 button2 的宽度
            button2.setWidth(300);
            // 设置 button2 的高度
            button2.setHeight(200);
        }
        Button button3 = (Button)findComponentById(com.unitymarvel.
        demo.ResourceTable.Id_button3);
        if(button3 != null) {
            // 设置 button3 的位置
            button3.setContentPosition(600,1200);
        }
    }
}
```

运行程序，会看到图 3-5 所示位置布局的效果。

图 3-5 位置布局的效果

3.6 动态装载布局

只是用 setUIContent 方法通过 XML 布局文件装载布局，尽管可以获取布局中组件的对象，但无法获取布局根节点的对象。还有就是，如果在 UI 中某些部分的布局相同或类似，只是显示的信息不同的情况下要复用这些布局，就必须将一个 XML 布局文件当作一个可复用的部分，因此就需要获取整个 XML 布局文件的对象。要实现这些场景，就需要使用本节介绍的 LayoutScatter 类，该类的 parse 方法可以动态装载整个 XML 布局文件，并返回一个 Component 类型的对象。要获得具体类型的对象，如 PositionLayout，还需要进行类型转换。

parse 方法的原型如下：

```
public Component parse(int xmlId, ComponentContainer root, boolean attachToRoot)
```

parse 方法的参数解释如下。

- xmlId：XML 布局文件的 ID，如 ResourceTable.Layout_scatter_demo。
- root：将装载的布局文件作为 root 指定的容器组件的子组件，如果要独立装载一个 XML 布局文件，该参数值应该为 null。
- attachToRoot：如果不想将装载的 XML 布局文件与 root 指定的容器组件关联，该参数值应该为 false。

如果想装载一个独立的 XML 布局文件，应该使用如下的代码：

```
LayoutScatter.getInstance(this).parse(ResourceTable.Layout_scatter_demo,null,false;
```

【例 3.6】 在窗口中显示 3 组图像和按钮，因为这 3 组布局的样式相同，所以都使用同一个 XML 布局文件 image_item.xml，并且使用 parse 方法动态装载该 XML 布局文件。为了将这 3 组布局动态添加到根布局中，本例同样使用 parse 方法动态装载 scatter_demo.xml 布局文件。

首先创建 scatter_demo.xml 文件，并输入如下代码。

代码位置：src/main/resources/base/layout/scatter_demo.xml

```
<?xml version="1.0" encoding="utf-8"?>
<PositionLayout
   xmlns:ohos="http://schemas.huawei.com/res/ohos"
   ohos:height="match_parent"
   ohos:width="match_parent"
   ohos:background_element="#CCCCCC"
   ohos:padding="32">
</PositionLayout>
```

很明显，scatter_demo.xml 文件中除了根节点什么都没有，文件中的内容会动态添加。

然后，创建 image_item.xml 文件，并输入如下内容。

代码位置：src/main/resources/base/layout/image_item.xml

```
<?xml version="1.0" encoding="utf-8"?>
```

```
<DirectionalLayout
    xmlns:ohos="http://schemas.huawei.com/res/ohos"
    ohos:height="120vp"
    ohos:width="120vp"
    ohos:alignment="center"
    ohos:orientation="vertical">
    <Image
        ohos:id="$+id:image"
        ohos:height="90vp"
        ohos:width="90vp"/>
    <Button
        ohos:id="$+id:button"
        ohos:height="30vp"
        ohos:width="120vp"
        ohos:text_size="20vp"
        ohos:background_element="#00FFFF"/>
</DirectionalLayout>
```

image_item.xml 文件是每一个图像和按钮的组合用到的布局文件，通过动态装载该布局文件可以达到布局复用的目的。

最后，通过 parse 方法动态装载上面两个 XML 布局文件。

代码位置：src/main/java/com/unitymarvel/demo/layout/LayoutScatterDemo.java

```
package com.unitymarvel.demo.layout;

import com.unitymarvel.demo.ResourceTable;
import ohos.aafwk.ability.Ability;
import ohos.aafwk.content.Intent;
import ohos.agp.components.*;

public class LayoutScatterDemo extends Ability {
    @Override
    public void onStart(Intent intent) {
        super.onStart(intent);
        // 动态装载 scatter_demo.xml 文件
        PositionLayout positionLayout = (PositionLayout) LayoutScatter.getInstance(this).
                  parse(ResourceTable.Layout_scatter_demo,null,false);
        super.setUIContent(positionLayout);
        // 第 1 次动态装载 image_item.xml 文件
        DirectionalLayout directionalLayoutItem1 = (DirectionalLayout) LayoutScatter
            .getInstance(this).parse(ResourceTable.Layout_image_item,null,false);
        // 在 image_item.xml 中搜索图像组件
        Image image = (Image)directionalLayoutItem1.findComponentById(ResourceTable.
        Id_image);
        if(image != null) {
            image.setScaleMode(Image.ScaleMode.ZOOM_CENTER);
            image.setPixelMap(ResourceTable.Media_components);
        }
        // 在 image_item.xml 中搜索按钮组件
```

```
        Button button = (Button)directionalLayoutItem1.findComponentById(ResourceTable.
        Id_button);
        if(button != null) {
            button.setText("Components");
        }
        // 第 2 次动态装载 image_item.xml 文件
        DirectionalLayout directionalLayoutItem2 = (DirectionalLayout) LayoutScatter.
              getInstance(this).parse(ResourceTable.Layout_image_item,null,false);
        image = (Image)directionalLayoutItem2.findComponentById(ResourceTable.Id_image);
        if(image != null) {
            image.setScaleMode(Image.ScaleMode.ZOOM_CENTER);
            image.setPixelMap(ResourceTable.Media_ability);
        }
        button = (Button)directionalLayoutItem2.
                    findComponentById(ResourceTable.Id_button);
        if(button != null) {
            button.setText("Ability");
        }
        // 第 3 次动态装载 image_item.xml 文件
        DirectionalLayout directionalLayoutItem3 = (DirectionalLayout) LayoutScatter.
               getInstance(this).parse(ResourceTable.Layout_image_item,null,false);
        image = (Image)directionalLayoutItem3.findComponentById(ResourceTable.Id_image);
        if(image != null) {
            image.setScaleMode(Image.ScaleMode.ZOOM_CENTER);
            image.setPixelMap(ResourceTable.Media_device);
        }
        button = (Button)directionalLayoutItem3.
                        findComponentById(ResourceTable.Id_button);
        if(button != null) {
            button.setText("Device");
        }
        // 将 3 个动态装载的 image_item.xml 布局文件对应的组件添加到容器组件
        positionLayout.addComponent(directionalLayoutItem1);
        positionLayout.addComponent(directionalLayoutItem2);
        positionLayout.addComponent(directionalLayoutItem3);
        // 设置组件的绝对位置
        directionalLayoutItem1.setContentPosition(300,200);
        directionalLayoutItem2.setContentPosition(600,700);
        directionalLayoutItem3.setContentPosition(300,1200);
    }
}
```

运行程序，会看到图 3-6 所示动态装载布局的效果。

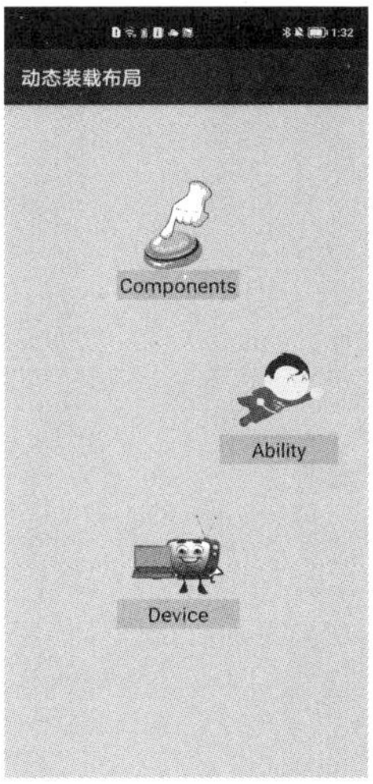

图 3-6 动态装载布局的效果

3.7 总结与回顾

本章深入介绍了 HarmonyOS 中的核心布局，其中常用布局为方向布局和依赖布局。目前，绝大多数窗口都会使用这两种布局。通过方向布局可以让组件按照水平方向或垂直方向摆放，而通过依赖布局可以确定组件的相对位置。其他布局尽管不太常用，但在某些特殊情况下也会用到。例如，栈布局类似于 Photoshop 的图层，第一个放置的组件会在最底层，最后一个放置的组件会在最顶层，每一层组件会压在前一层组件上，可以利用这个特性实现嵌套的效果；还有表格布局，如果要实现像计算器按钮那样的布局，表格布局是最好的选择；尽管位置布局不太常用，但当需要将组件摆放在固定位置时，也会用到这个布局。

本章的最后介绍了布局的动态装载，因为有时需要根据特定条件装载不同的布局文件，或更换布局文件，所以需要用 Java 代码动态装载布局，这样可以让 UI 更灵活。

读者可以结合本章的示例学习这些布局。不过到目前为止，还只是介绍了 HarmonyOS 的一些基础知识，后面将逐步介绍 HarmonyOS 中的组件和 HarmonyOS 提供的 API 的用法。

第 4 章　UI 组件

只有布局是没用的，组件才是 HarmonyOS 的核心，布局是为组件服务的。为了让读者可以开发出更强大的 HarmonyOS App，本章将介绍 HarmonyOS 中的常用组件，并提供大量代码供读者练习。

通过阅读本章，读者可以掌握：

- 展示组件；
- 交互组件；
- 高级组件。

4.1 展示组件

展示组件主要用于向用户展示一些信息，如文本、图像等。本节将介绍 HarmonyOS 中的主要展示组件。

4.1.1 Text 组件

Text 组件（即文本组件）用于展示文本信息，在 Text 组件中，我们可以对文本尺寸、文本颜色、组件背景颜色、文本对齐方式等基本属性进行设置。对于同样的属性，我们可以在 XML 布局文件中静态设置，也可以使用 Java 代码动态设置。

【例 4.1】 通过 XML 布局和 Java 代码两种方式设置 Text 组件的一些常用属性。

代码位置：src/main/resources/base/layout/text_demo.xml

```
<?xml version="1.0" encoding="utf-8"?>
<DirectionalLayout
   xmlns:ohos="http://schemas.huawei.com/res/ohos"
   ohos:height="match_parent"
   ohos:width="match_parent"
   ohos:background_element="#CCCCCC"
   ohos:orientation="vertical">
   <Text
      ohos:id="$+id:text1"
      ohos:height="match_content"
      ohos:width="300vp"
      ohos:text_size="20vp"
```

```
        ohos:text="沧海横流，方显英雄本色。"/>
    <Text
        ohos:id="$+id:text2"
        ohos:height="match_content"
        ohos:width="match_content"
        ohos:text_size="25vp"
        ohos:top_margin="20vp"
        ohos:background_element="#FF0000"
        ohos:text_color="#FFFF00"
        ohos:text="挽狂澜于既倒 扶大厦之将倾"/>
    <Text
        ohos:id="$+id:text3"
        ohos:height="match_content"
        ohos:width="300vp"
        ohos:text_size="15vp"
        ohos:top_margin="20vp"
        />
</DirectionalLayout>
```

通过这段布局代码我们在窗口中设置了 3 个 Text 组件，其中前两个 Text 组件完全使用 XML 方式完成布局和属性设置，最后一个 Text 组件除了通过 XML 方式设置一部分属性，还通过下面的 Java 代码设置和修改一部分属性的值。

代码位置：src/main/java/com/unitymarvel/demo/components/TextDemo.java

```
package com.unitymarvel.demo.components;

import com.unitymarvel.demo.ResourceTable;
import ohos.aafwk.ability.Ability;
import ohos.aafwk.content.Intent;
import ohos.agp.colors.RgbPalette;
import ohos.agp.components.*;
import ohos.agp.components.element.ShapeElement;
import ohos.agp.utils.Color;
import ohos.agp.utils.TextAlignment;

public class TextDemo extends Ability {
    @Override
    public void onStart(Intent intent) {
        super.onStart(intent);
        super.setUIContent(com.unitymarvel.demo.ResourceTable.Layout_text_demo);
        Text text3 = (Text)findComponentById(ResourceTable.Id_text3);
        if(text3 != null) {
            // 设置 Text 组件的文本
            text3.setText("喑呜则山岳崩颓，叱咤则风云变色。");
            ShapeElement element = new ShapeElement();
            element.setRgbColor(RgbPalette.BLUE);
            // 设置 Text 组件的背景颜色
            text3.setBackground(element);
            // 设置 Text 组件的宽度
            text3.setWidth(StackLayout.LayoutConfig.MATCH_PARENT);
            // 设置 Text 组件的文本颜色
            text3.setTextColor(Color.YELLOW);
```

```
            // 设置 Text 组件的文本对齐方式
            text3.setTextAlignment(TextAlignment.CENTER);
        }
    }
}
```

运行程序，会看到图 4-1 所示的显示效果。

图 4-1 Text 组件的显示效果

4.1.2 Image 组件

Image 组件（即图像组件）用于显示图像，图像来源可以是 HAP 内部的资源目录，也可以是外部目录，如 App 的私有目录、sdcard 目录等。

【例 4.2】 在窗口中放置 3 个 Image 组件，分别用于显示 3 个图像。其中前两个图像直接通过 XML 方式布局，最后一个图像使用 Java 代码设置相应的属性，并且在单击该图像时图像会顺时针旋转 60° 。

代码位置：src/main/resources/base/layout/image_demo.xml

```
<?xml version="1.0" encoding="utf-8"?>
<DirectionalLayout
    xmlns:ohos="http://schemas.huawei.com/res/ohos"
    ohos:height="match_parent"
    ohos:width="match_parent"
    ohos:background_element="#CCCCCC"
    ohos:alignment="center"
    ohos:orientation="vertical">
    <Image
        ohos:id="$+id:image1"
        ohos:height="150vp"
        ohos:width="150vp"
        ohos:image_src="$media:device"/>
    <Image
        ohos:id="$+id:image2"
        ohos:top_margin="10vp"
        ohos:height="150vp"
        ohos:width="150vp"
        ohos:image_src="$media:device"
        ohos:scale_mode="zoom_center"/>
    <Image
        ohos:id="$+id:image3"
        ohos:top_margin="10vp"
        ohos:height="150vp"
        ohos:width="150vp"
        ohos:scale_mode="zoom_center"/>
</DirectionalLayout>
```

代码位置：src/main/java/com/unitymarvel/demo/components/ImageDemo.java

```
package com.unitymarvel.demo.components;

import com.unitymarvel.demo.ResourceTable;
```

```
import ohos.aafwk.ability.Ability;
import ohos.aafwk.content.Intent;
import ohos.agp.components.Component;
import ohos.agp.components.Image;

public class ImageDemo extends Ability {
   @Override
   public void onStart(Intent intent) {
      super.onStart(intent);
      super.setUIContent(ResourceTable.Layout_image_demo);
      Image image3 = (Image)findComponentById(ResourceTable.Id_image3);
      if(image3 != null) {
         // 装载资源中的图像
         image3.setPixelMap(ResourceTable.Media_device);
         image3.setClickedListener(new Component.ClickedListener() {
            @Override
            public void onClick(Component component) {
                // 让图像顺时针旋转 60°
                image3.setRotation(60);
            }
         });
      }
   }
}
```

运行程序，并单击第 3 个图像，会发现图像顺时针旋转了 60° ，显示效果如图 4-2 所示。

图 4-2 Image 组件的显示效果

从 Image 组件的显示效果可以看出，第 1 个 Image 组件使用默认显示模式，也就是图像按实际尺寸显示，不进行缩放和拉伸。因为 Image 组件的尺寸小于图像的实际尺寸，所以图像没有显示完全。而第 2 个和第 3 个 Image 组件由于设置的显示模式为 zoom_center，该模式会让图像在不超出 Image 组件的前提下，保持原始比例并以最大尺寸显示。

4.1.3 ProgressBar 组件

ProgressBar 组件（即进度条组件）用于展示任务的进度，支持水平进度条和垂直进度条。ProgressBar 组件默认设置为水平进度条，如果将 orientation 属性设为 vertical，则变为垂直进度条。ProgressBar 组件最常用的属性是 progress、max 和 min，分别用于设置进度条的当前进度、最大值和最小值。

【例 4.3】 在窗口中放置 3 个 ProgressBar 组件，其中前两个 ProgressBar 组件是水平进度条，最后一个 ProgressBar 组件是垂直进度条，而且垂直进度条的大多数属性是通过 Java 代码设置的。

代码位置：src/main/resources/base/layout/progressbar_demo.xml

```
<?xml version="1.0" encoding="utf-8"?>
<DirectionalLayout
   xmlns:ohos="http://schemas.huawei.com/res/ohos"
   ohos:height="match_parent"
```

```
    ohos:width="match_parent"
    ohos:background_element="#CCCCCC"
    ohos:alignment="center"
    ohos:orientation="vertical">
    <ProgressBar
        ohos:id="$+id:progressbar1"
        ohos:top_margin="20vp"
        ohos:height="match_content"
        ohos:width="match_parent"
        ohos:progress="30"
        ohos:max="100"
        ohos:min="0"/>
    <ProgressBar
        ohos:id="$+id:progressbar2"
        ohos:top_margin="20vp"
        ohos:height="match_content"
        ohos:width="match_parent"
        ohos:progress_color="#FF0000"
        ohos:progress_width="10vp"
        ohos:progress="50"
        ohos:max="100"
        ohos:min="0"/>
    <ProgressBar
        ohos:id="$+id:progressbar3"
        ohos:top_margin="20vp"
        ohos:height="match_parent"
        ohos:width="match_content"
        ohos:orientation="vertical"/>
</DirectionalLayout>
```

下面的 Java 代码设置了 progressbar3 的部分属性。

代码位置：src/main/java/com/unitymarvel/demo/components/ProgressBarDemo.java

```
package com.unitymarvel.demo.components;

import com.unitymarvel.demo.ResourceTable;
import ohos.aafwk.ability.Ability;
import ohos.aafwk.content.Intent;
import ohos.agp.colors.RgbPalette;
import ohos.agp.components.ProgressBar;
import ohos.agp.components.element.ShapeElement;
import ohos.agp.utils.Color;

public class ProgressBarDemo extends Ability {
    @Override
    public void onStart(Intent intent) {
        super.onStart(intent);
        super.setUIContent(ResourceTable.Layout_progressbar_demo);
        ProgressBar progressBar3 =
            (ProgressBar)findComponentById(ResourceTable.Id_progressbar3);
        if(progressBar3 != null) {
            // 设置进度条颜色
```

```
            progressBar3.setProgressColor(Color.BLUE);
            // 设置进度条宽度
            progressBar3.setProgressWidth(200);
            // 设置进度条当前进度
            progressBar3.setProgressValue(45);
            // 设置进度条最大值
            progressBar3.setMaxValue(100);
            // 设置进度条最小值
            progressBar3.setMinValue(0);
        }
    }
}
```

运行程序，会看到图 4-3 所示的显示效果。

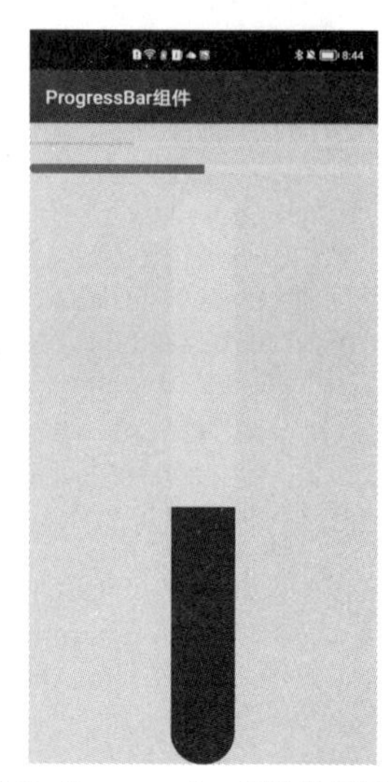

图 4-3 ProgressBar 组件的显示效果

4.1.4 RoundProgressBar 组件

RoundProgressBar 组件（即圆形进度条组件）用于显示圆形风格的进度条，RoundProgressBar 类是 ProgressBar 的子类，它调用的 API 与 ProgressBar 基本一致。

【例 4.4】 在窗口中放置两个 RoundProgressBar 组件，第 1 个 RoundProgressBar 组件通过 XML 方式设置属性，第 2 个 RoundProgressBar 组件主要通过 Java 代码设置属性。

代码位置：src/main/resources/base/layout/roundprogressbar_demo.xml

```
<?xml version="1.0" encoding="utf-8"?>
<DirectionalLayout xmlns:ohos="http://schemas.huawei.com/res/ohos"
      ohos:height="match_parent"
      ohos:width="match_parent"
      ohos:background_element="#CCCCCC"
      ohos:alignment="center"
      ohos:orientation="vertical">
   <RoundProgressBar
      ohos:id="$+id:roundprogressbar1"
      ohos:top_margin="20vp"
      ohos:height="100vp"
      ohos:width="100vp"
      ohos:progress="40"
      ohos:progress_width="10vp"
      ohos:progress_color="#FF0000"
      ohos:background_element="#0000FF"
      ohos:max="100"
      ohos:min="0"/>
   <RoundProgressBar
      ohos:id="$+id:roundprogressbar2"
      ohos:top_margin="20vp"
      ohos:height="200vp"
      ohos:width="200vp"/>
</DirectionalLayout>
```

下面的 Java 代码设置了第 2 个 RoundProgressBar 组件的一些属性。

代码位置：src/main/java/com/unitymarvel/demo/components/RoundProgressBarDemo.java

```
package com.unitymarvel.demo.components;

import com.unitymarvel.demo.ResourceTable;
import ohos.aafwk.ability.Ability;
import ohos.aafwk.content.Intent;
import ohos.agp.components.RoundProgressBar;
import ohos.agp.utils.Color;

public class RoundProgressBarDemo extends Ability {
    @Override
    public void onStart(Intent intent) {
        super.onStart(intent);
        super.setUIContent(ResourceTable.Layout_roundprogressbar_demo);
        RoundProgressBar roundProgressBar2 =
            (RoundProgressBar)findComponentById(ResourceTable.Id_roundprogressbar2);
        if(roundProgressBar2 != null) {
            // 设置进度条当前进度
            roundProgressBar2.setProgressValue(67);
            // 设置进度条宽度
            roundProgressBar2.setProgressWidth(80);
            // 设置进度条颜色
            roundProgressBar2.setProgressColor(Color.BLUE);
            // 设置进度条最大值
            roundProgressBar2.setMaxValue(100);
            // 设置进度条最小值
            roundProgressBar2.setMinValue(0);
        }
    }
}
```

运行程序，会看到图 4-4 所示的显示效果。

图 4-4 RoundProgressBar 组件的显示效果

4.1.5 Clock 组件

Clock 组件（即时钟组件）用于以数字形式显示当前时间，支持 12 小时和 24 小时这两种格式，默认设置为 24 小时格式。

【例 4.5】 在窗口中放置两个 Clock 组件，第 1 个 Clock 组件以 24 小时格式显示当前时间，第 2 个 Clock 组件以 12 小时格式显示当前时间。

代码位置：src/main/resources/base/layout/clock_demo.xml

```
<?xml version="1.0" encoding="utf-8"?>
<DirectionalLayout
    xmlns:ohos="http://schemas.huawei.com/res/ohos"
    ohos:height="match_parent"
    ohos:width="match_parent"
    ohos:alignment="center"
    ohos:background_element="#CCCCCC"
    ohos:orientation="vertical">
    <Clock
```

```
        ohos:text_color="#FF0000"
        ohos:text_size="50vp"
        ohos:height="match_content"
        ohos:width="match_content"/>
    <Clock
        ohos:id="$+id:clock"
        ohos:text_color="#0000FF"
        ohos:top_margin="20vp"
        ohos:text_size="60vp"
        ohos:height="match_content"
        ohos:width="match_content"/>
</DirectionalLayout>
```

下面的 Java 代码将第 2 个 Clock 组件设置为 12 小时格式，并获取了当前的时间。

```
package com.unitymarvel.demo.components;

import com.unitymarvel.demo.ResourceTable;
import com.unitymarvel.demo.Tools;
import ohos.aafwk.ability.Ability;
import ohos.aafwk.content.Intent;
import ohos.agp.components.Clock;
import java.text.SimpleDateFormat;

public class ClockDemo extends Ability {
    @Override
    public void onStart(Intent intent) {
        super.onStart(intent);
        super.setUIContent(com.unitymarvel.demo.ResourceTable.
        Layout_clock_demo);
        Clock clock = (Clock)findComponentById(ResourceTable.
        Id_clock);
        if(clock != null) {
            // 将 Clock 组件设置为 12 小时格式
            clock.set24HourModeEnabled(false);
            SimpleDateFormat format = new SimpleDateFormat
            ("HH 时 mm 分 ss 秒");
            // 获取、格式化和输出当前时间
            Tools.print("now time:" + format.format(clock.
            getTime()));
        }
    }
}
```

运行程序，会看到图 4-5 所示的显示效果。

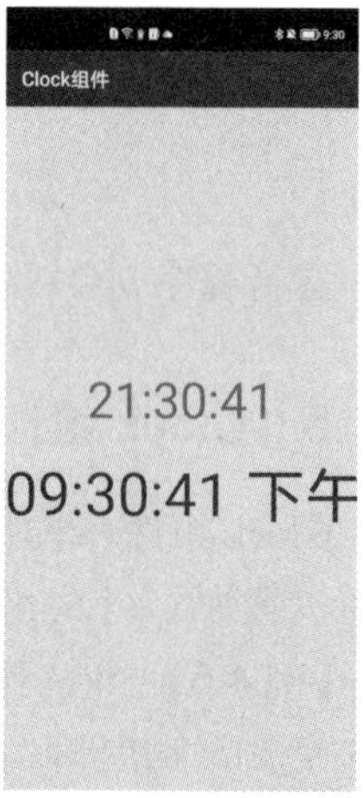

图 4-5 Clock 组件的显示效果

4.2 交互组件

交互组件是 App 中重要的组成部分，只有使用交互组件，用户才能与 App 进行互动，从而让 App 拥有更多、更强大的功能。交互组件有多个类别，如按钮组件、输入组件、滑动组件等。本节将介绍 HarmonyOS 中常用的交互组件。

4.2.1 Button 组件

Button 组件（即按钮组件）在 App 中十分常见，绝大多数 App 都会使用 Button 组件。Button 组件允许用户设置按钮文本、背景颜色、图标等，也可以接收用户的单击（触摸）事件。

【例 4.6】 在窗口中放置 5 个 Button 组件，并为这几个按钮设置图标、显示形状等属性，为第 4 个按钮设置单击事件响应，当单击这个按钮时，会将按钮文本替换成“+”。

首先完成 XML 布局文件。

代码位置：src/main/resources/base/layout/button_demo.xml

```
<?xml version="1.0" encoding="utf-8"?>
<DirectionalLayout xmlns:ohos="http://schemas.huawei.com/res/ohos"
      ohos:height="match_parent"
      ohos:width="match_parent"
      ohos:alignment="center"
      ohos:background_element="#CCCCCC"
      ohos:orientation="vertical">
   <Button
      ohos:id="$+id:button1"
      ohos:height="match_content"
      ohos:width="match_content"
      ohos:text_size="20vp"
      ohos:background_element="#00FFFF"
      ohos:text="按钮 1"/>
   <Button
      ohos:id="$+id:button2"
      ohos:height="match_content"
      ohos:width="match_content"
      ohos:top_margin="20vp"
      ohos:text_size="30vp"
      ohos:background_element="$graphic:button_element"
      ohos:element_left="$media:icon"
      ohos:text="按钮 2"/>
   <Button
      ohos:id="$+id:button3"
      ohos:height="match_content"
      ohos:width="300vp"
      ohos:top_margin="20vp"
      ohos:text_size="30vp"
      ohos:background_element="$graphic:oval_button_element"
      ohos:element_left="$media:icon"
      ohos:text="按钮 3"/>
   <Button
      ohos:id="$+id:button4"
      ohos:height="100vp"
      ohos:width="100vp"
      ohos:top_margin="20vp"
      ohos:text_size="30vp"
      ohos:background_element="$graphic:circle_button_element"
      ohos:text="按钮 4"/>
   <Button
      ohos:id="$+id:button5"
```

```
        ohos:height="match_content"
        ohos:width="300vp"
        ohos:top_margin="20vp"
        ohos:text_size="30vp"
        ohos:background_element="$graphic:capsule_button_element"
        ohos:element_right="$media:icon"
        ohos:text="按钮 5"/>
</DirectionalLayout>
```

在设置 button2 的 ohos:background_element 属性时使用了一个 button_element，这是一个 graphic 资源，通常是 XML 形式的，对应资源文件要放置 graphic 目录下。

代码位置：src/main/resources/base/graphic/button_element.xml

```
<?xml version="1.0" encoding="utf-8"?>
<shape xmlns:ohos="http://schemas.huawei.com/res/ohos"
      ohos:shape="rectangle">
   <corners
      ohos:radius="10"/>
   <solid
      ohos:color="#FF0000"/>
</shape>
```

在 button_element.xml 文件中设置了 shape，将矩形区域设置为圆角，同时还设置了颜色。因为该文件被用于设置 ohos:background_element 属性，所以这个颜色就是按钮的背景颜色。

在设置 button3、button4 和 button5 的 ohos:background_element 属性时也使用了类似的 graphic 资源，下面是这些资源的源代码。

代码位置：src/main/resources/base/graphic/oval_button_element.xml

```
<?xml version="1.0" encoding="utf-8"?>
<shape xmlns:ohos="http://schemas.huawei.com/res/ohos"
      ohos:shape="oval">
   <solid
      ohos:color="#FFCC7D"/>
</shape>
```

代码位置：src/main/resources/base/graphic/circle_button_element.xml

```
<?xml version="1.0" encoding="utf-8"?>
<shape xmlns:ohos="http://schemas.huawei.com/res/ohos"
      ohos:shape="oval">
   <solid
      ohos:color="#DD337D"/>
</shape>
```

代码位置：src/main/resources/base/graphic/capsule_button_element.xml

```
<?xml version="1.0" encoding="utf-8"?>
<shape xmlns:ohos="http://schemas.huawei.com/res/ohos"
      ohos:shape="rectangle">
   <corners
```

```
        ohos:radius="100"/>
    <solid
        ohos:color="#FF007D"/>
</shape>
```

其中，oval_button_element.xml 文件 和 circle_button_element.xml 文件的 shape 设置除了颜色，其他属性值的都一样，这说明圆形按钮只是椭圆按钮的一个特例。只要将 Button 组件的 width 和 height 设置成相同的值，椭圆按钮就变成圆形按钮了。

下面的 Java 代码设置了 button4 的单击事件，在该事件中重新设置了 button4 的文本和文本尺寸。

代码位置：src/main/java/com/unitymarvel/demo/components/ButtonDemo.java

```
package com.unitymarvel.demo.components;

import com.unitymarvel.demo.ResourceTable;
import ohos.aafwk.ability.Ability;
import ohos.aafwk.content.Intent;
import ohos.agp.components.Button;
import ohos.agp.components.Component;

public class ButtonDemo extends Ability {
    @Override
    public void onStart(Intent intent) {
        super.onStart(intent);
        super.setUIContent(ResourceTable.Layout_button_demo);
        Button button4 = (Button)findComponentById(ResourceTable.Id_button4);
        if(button4 != null) {
            // 设置 button4 的单击事件
            button4.setClickedListener(new Component.ClickedListener() {
                @Override
                public void onClick(Component component) {
                    // 设置 button4 的文本
                    button4.setText("+");
                    // 设置 button4 的文本尺寸（单位：px）
                    button4.setTextSize(300);
                }
            });
        }
    }
}
```

运行程序，单击第 4 个按钮，会看到图 4-6 所示的显示效果。

图 4-6 Button 组件的显示效果

4.2.2 ToggleButton 组件

ToggleButton 组件（即切换按钮组件）也是按钮组件，ToggleButton 类是 Button 的子类。只是 ToggleButton 组件主要用于切换状态，当处于“开”状态时，默认会显示 ON；当处于“关”状态时，默认会显示 OFF。当然，默认显示的文本和颜色可以修改。

【例 4.7】 在窗口中放置 3 个 ToggleButton 组件，并修改第 2 个 ToggleButton 组件的默认显示文本，为第 3 个 ToggleButton 组件设置状态变化的事件。

首先完成 XML 布局文件。

代码位置： src/main/resources/base/layout/togglebutton_demo.xml

```
<?xml version="1.0" encoding="utf-8"?>
<DirectionalLayout
   xmlns:ohos="http://schemas.huawei.com/res/ohos"
   ohos:height="match_parent"
   ohos:width="match_parent"
   ohos:alignment="center"
   ohos:background_element="#CCCCCC"
   ohos:orientation="vertical">
   <ToggleButton
      ohos:id="$+id:togglebutton1"
      ohos:height="match_content"
      ohos:width="match_content"
      ohos:text_size="30vp"
      ohos:background_element="#00FFFF"/>
   <ToggleButton
      ohos:id="$+id:togglebutton2"
      ohos:height="match_content"
      ohos:width="match_content"
      ohos:text_size="60vp"
      ohos:top_margin="30vp"
      ohos:text_state_on="开"
      ohos:text_state_off="关"
      ohos:background_element="#00FFFF"/>
   <ToggleButton
      ohos:id="$+id:togglebutton3"
      ohos:height="match_content"
      ohos:width="match_content"
      ohos:text_size="40vp"
      ohos:top_margin="30vp"
      ohos:text_color="#FFFF00"
      ohos:background_element="#FF00FF"/>
</DirectionalLayout>
```

下面的 Java 代码设置了 togglebutton3 的状态变化事件，并在状态变化时重新设置 togglebutton3 的文本。

代码位置： src/main/java/com/unitymarvel/demo/components/ToggleButtonDemo.java

```
package com.unitymarvel.demo.components;

import com.unitymarvel.demo.ResourceTable;
import ohos.aafwk.ability.Ability;
import ohos.aafwk.content.Intent;
import ohos.agp.components.AbsButton;
import ohos.agp.components.ToggleButton;
import ohos.agp.utils.Color;

public class ToggleButtonDemo extends Ability {
```

```
    @Override
    public void onStart(Intent intent) {
        super.onStart(intent);
        super.setUIContent(ResourceTable.Layout_togglebutton_demo);
        ToggleButton toggleButton3 =
                (ToggleButton)findComponentById(ResourceTable.Id_togglebutton3);
        if(toggleButton3 != null) {
            // 设置 ToggleButton 组件关闭状态下显示的文本
            toggleButton3.setStateOffText("已经关闭");
            // 设置 ToggleButton 组件打开状态下显示的文本
            toggleButton3.setStateOnText("已经打开");
            // 设置 ToggleButton 组件关闭状态下显示的文本的颜色
            toggleButton3.setTextColorOff(Color.BLUE);
            // 设置 ToggleButton 组件打开状态下显示的文本的颜色
            toggleButton3.setTextColorOn(Color.BLACK);
            // 设置用于监听 ToggleButton 组件状态变化的事件
            toggleButton3.setCheckedStateChangedListener(new AbsButton.
            CheckedStateChangedListener() {
                @Override
                public void onCheckedChanged(AbsButton absButton, boolean b) {
                    // 处于打开状态，b 为 true，否则为 false
                    if(b) {
                        toggleButton3.setStateOnText("已经打开（Open）");
                    } else {
                        toggleButton3.setStateOffText("已经关闭（Close）");
                    }
                }
            });
        }
    }
}
```

运行程序，然后单击最后一个 ToggleButton 组件，会看到图 4-7 所示的显示效果。

图 4-7　ToggleButton 组件的显示效果

4.2.3　TextField 组件

TextField 组件（即文本编辑组件）用于输入文本信息。当 TextField 组件处于焦点时，会自动弹出软键盘，以便让用户输入文本信息。TextField 组件允许开发人员设置必要的属性，如文字颜色、文字尺寸、背景颜色、文本提示等。在 TextField 组件中默认只能输入单行文本，而将 ohos:multiple_lines 属性值设为 true，可以让 TextField 组件允许用户输入多行文本。TextField 组件还允许开发人员通过监听器捕捉文本的变化，实现这个功能的监听器是 Component.ComponentStateChangedListener。

【例 4.8】 在窗口中放置 3 个 TextField 组件和 1 个 Text 组件，并为第 2 个 TextField 组件和第 3 个 TextField 组件设置边框，而且使用 Java 代码监听第 3 个 TextField 组件文本的变化，并将变化后的结果显示在 Text 组件中。

首先完成 XML 布局文件。

代码位置：src/main/resources/base/layout/textfield_demo.xml

```
<?xml version="1.0" encoding="utf-8"?>
<DirectionalLayout
   xmlns:ohos="http://schemas.huawei.com/res/ohos"
   ohos:height="match_parent"
   ohos:width="match_parent"
   ohos:alignment="center"
   ohos:background_element="#CCCCCC"
   ohos:orientation="vertical" ohos:padding="20vp">
   <TextField
       ohos:id="$+id:textfield1"
       ohos:height="match_content"
       ohos:width="match_parent"
       ohos:text="柳暗花明又一村"
       ohos:text_size="30vp"/>
   <TextField
       ohos:id="$+id:textfield2"
       ohos:height="match_content"
       ohos:width="match_parent"
       ohos:top_margin="20vp"
       ohos:background_element="$graphic:border"
       ohos:text_color="#FF0000"
       ohos:padding="2vp"
       ohos:text="hello world"
       ohos:text_size="30vp"/>
   <TextField
       ohos:id="$+id:textfield3"
       ohos:height="200vp"
       ohos:width="match_parent"
       ohos:top_margin="20vp"
       ohos:padding="2vp"
       ohos:background_element="$graphic:border"
       ohos:text_color="#0000FF"
       ohos:multiple_lines="true"
       ohos:text_size="30vp"/>
   <Text
       ohos:id="$+id:text"
       ohos:top_margin="20vp"
       ohos:text_color="#0000FF"
       ohos:text_size="30vp"
       ohos:height="match_parent"
       ohos:width="match_content"/>
</DirectionalLayout>
```

textfield2 和 textfield3 的 ohos:background_element 属性都使用了一个 graphic 资源$graphic:border，该资源对应 border.xml 文件。

代码位置：src/main/resources/base/graphic/border.xml

```
<?xml version="1.0" encoding="UTF-8" ?>
<shape xmlns:ohos="http://schemas.huawei.com/res/ohos"
```

```
        ohos:shape="rectangle">
    <solid ohos:color="#FFFFFF"/>
    <stroke ohos:width="1vp" ohos:color="#4fa5d5"/>
</shape>
```

border.xml 文件用于设置一个宽度为 1vp[①]的边框。

下面的 Java 代码监听了第 3 个 TextField 组件文本的变化，并将文本变化的结果显示在 Text 组件中。

代码位置：src/main/java/com/unitymarvel/demo/components/TextFieldDemo.java

```
package com.unitymarvel.demo.components;

import com.unitymarvel.demo.ResourceTable;
import ohos.aafwk.ability.Ability;
import ohos.aafwk.content.Intent;
import ohos.agp.components.Component;
import ohos.agp.components.Text;
import ohos.agp.components.TextField;

public class TextFieldDemo extends Ability {
    @Override
    public void onStart(Intent intent) {
        super.onStart(intent);
        super.setUIContent(ResourceTable.Layout_textfield_demo);
        Text text = (Text)findComponentById(ResourceTable.Id_text);
        TextField textField3 =
               (TextField)findComponentById(ResourceTable.Id_textfield3);
      if(textField3 != null) {
          // 设置文本提示信息
          textField3.setHint("请输入详细信息");
          // 设置文本变化状态监听器
          textField3.setComponentStateChangedListener(new Component.
          ComponentStateChangedListener() {
              @Override
              public void onComponentStateChanged(Component component, int i) {
                  if(text != null) {
                      // 将文本变化的结果显示在 Text 组件中
                      text.setText(textField3.getText());
                  }
              }
          });
      }


    }
}
```

运行程序，会显示图 4-8 所示的效果。将焦点放到 textfield3 中，会弹出软键盘，这时可以输入文本，如图 4-9 所示。关闭软键盘后，会看到 Text 组件与 textfield3 的文本同步显示，如图 4-10 所示。

① vp 是虚拟单位，在实际设备上，会将其转换为对应的 px。vp 类似于 Android 中的 dp。

图4-8 TextField组件的显示效果

图4-9 弹出的软键盘

图4-10 Text与textfield3文本同步显示

4.2.4 RadioButton组件和Checkbox组件

RadioButton组件（即单选组件）与Checkbox组件（即多选组件）的风格和功能都类似，尽管这两个组件在名字上一点都不同。单选组件用于在同一时刻只能有一个组件被选中的情况（当然，也可以都没被选中），多选组件用于在同一时刻可能有0到n个组件被选中的情况。

Checkbox组件是互相独立的，但RadioButton组件则不同，多个RadioButton组件是互斥的，也就是说，选中了你，就不能选中我。那么现在的问题是，如果窗口中有多组RadioButton组件，系统怎么知道哪些RadioButton组件是一组呢？为了解决这个问题，HarmonyOS提供了RadioContainer组件，该组件是RadioButton组件的容器。包含在RadioContainer组件中的所有RadioButton组件被看作一组，这些组件是互斥的。

【例4.9】 在窗口中放置3个RadioButton组件和3个Checkbox组件。在选中这些组件后，在窗口下方的Text组件中显示当前选择的结果。

首先完成XML布局文件。

代码位置：src/main/resources/base/layout/check_demo.xml

```
<?xml version="1.0" encoding="utf-8"?>
<DirectionalLayout
   xmlns:ohos="http://schemas.huawei.com/res/ohos"
   ohos:height="match_parent"
   ohos:width="match_parent"
   ohos:background_element="#CCCCCC"
   ohos:orientation="vertical" ohos:padding="20vp">

   <RadioContainer
      ohos:id="$+id:radiocontainer"
      ohos:height="match_content"
```

```
    ohos:width="match_parent"
    ohos:orientation="vertical">
    <RadioButton
        ohos:id="$+id:radiobutton1"
        ohos:height="match_content"
        ohos:width="match_content"
        ohos:text="单选 1"
        ohos:text_color_off="#0000FF"
        ohos:text_size="30vp"/>
    <RadioButton
        ohos:id="$+id:radiobutton2"
        ohos:height="match_content"
        ohos:width="match_content"
        ohos:text="单选 2"
        ohos:text_color_off="#FF0000"
        ohos:text_size="30vp"
        />
    <RadioButton
        ohos:id="$+id:radiobutton3"
        ohos:height="match_content"
        ohos:width="match_content"
        ohos:text="单选 3"
        ohos:text_color_off="#FF00FF"
        ohos:text_size="30vp"
        />
</RadioContainer>
<Checkbox
    ohos:id="$+id:checkbox1"
    ohos:top_margin="20vp"
    ohos:height="match_content"
    ohos:width="match_content"
    ohos:text_color_on="#0000FF"
    ohos:text_size="30vp"
    ohos:text="多选 1"/>
<Checkbox
    ohos:id="$+id:checkbox2"
    ohos:top_margin="20vp"
    ohos:height="match_content"
    ohos:width="match_content"
    ohos:text_size="30vp"
    ohos:text_color_on="#FF0000"
    ohos:text="多选 2"/>
<Checkbox
    ohos:id="$+id:checkbox3"
    ohos:top_margin="20vp"
    ohos:height="match_content"
    ohos:width="match_content"
    ohos:text_color_on="#FF00FF"
    ohos:text_size="30vp"
    ohos:text="多选 3"/>
<Text
    ohos:id="$+id:text"
    ohos:top_margin="20vp"
    ohos:text_size="30vp"
```

```
        ohos:height="match_parent"
        ohos:width="match_content"/>
</DirectionalLayout>
```

RadioButton 组件与 Checkbox 组件在设置文本颜色时，应该使用 ohos:text_color_on 和 ohos:text_color_off，分别用于设置选中和未选中这两种状态下的文本颜色。

下面的 Java 代码监听了 3 个 RadioButton 组件和 3 个 Checkbox 组件的状态，并将选择结果显示在 Text 组件中。

```
package com.unitymarvel.demo.components;

import com.unitymarvel.demo.ResourceTable;
import ohos.aafwk.ability.Ability;
import ohos.aafwk.content.Intent;
import ohos.agp.components.*;

public class CheckDemo extends Ability {
   private Text text;
   private RadioButton radioButton1,radioButton2,radioButton3;
   private Checkbox checkbox1, checkbox2, checkbox3;
   @Override
   public void onStart(Intent intent) {
      super.onStart(intent);
      super.setUIContent(ResourceTable.Layout_check_demo);
      // 通用的监听器，监听了 3 个 RadioButton 组件和 3 个 Checkbox 组件的状态变化
      AbsButton.CheckedStateChangedListener listener = new
                  AbsButton.CheckedStateChangedListener() {
         @Override
         public void onCheckedChanged(AbsButton absButton, boolean b) {
            String value = "";
            // 当任何一个组件的状态发生变化后，扫描所有的选择组件，并重新生成选择结果
            if(radioButton1.isChecked()) {
               value += radioButton1.getText() + " ";
            }
            if(radioButton2.isChecked()) {
               value += radioButton2.getText() + " ";
            }
            if(radioButton3.isChecked()) {
               value += radioButton3.getText() + " ";
            }
            if(checkbox1.isChecked()) {
               value += checkbox1.getText() + " ";
            }
            if(checkbox2.isChecked()) {
               value += checkbox2.getText() + " ";
            }
            if(checkbox3.isChecked()) {
               value += checkbox3.getText();
            }
            // 将选择结果显示在 Text 组件中
            text.setText(value);
         }
      };
      text = (Text)findComponentById(ResourceTable.Id_text);
```

```
        radioButton1 = (RadioButton)findComponentById(ResourceTable.Id_radiobutton1);

        if(radioButton1 != null) {
            // 将第 1 个 RadioButton 组件设置为选中状态
            radioButton1.setChecked(true);
            // 绑定状态监听器
            radioButton1.setCheckedStateChangedListener(listener);
        }
        radioButton2 = (RadioButton)findComponentById(ResourceTable.Id_radiobutton2);
        if(radioButton2 != null) {
            radioButton2.setCheckedStateChangedListener(listener);
        }
        radioButton3 = (RadioButton)findComponentById(ResourceTable.Id_radiobutton3);
        if(radioButton3 != null) {
            radioButton3.setCheckedStateChangedListener(listener);
        }
        checkbox1 = (Checkbox)findComponentById(ResourceTable.Id_checkbox1);
        if(checkbox1 != null){
            checkbox1.setCheckedStateChangedListener(listener);
        }
        checkbox2 = (Checkbox)findComponentById(ResourceTable.Id_checkbox2);
        if(checkbox2 != null){
            checkbox2.setCheckedStateChangedListener(listener);
        }
        checkbox3 = (Checkbox)findComponentById(ResourceTable.Id_checkbox3);
        if(checkbox3 != null){
            checkbox3.setCheckedStateChangedListener(listener);
        }
    }
}
```

运行程序，并且改变相应选择组件的状态，会看到图 4-11 所示的显示效果。

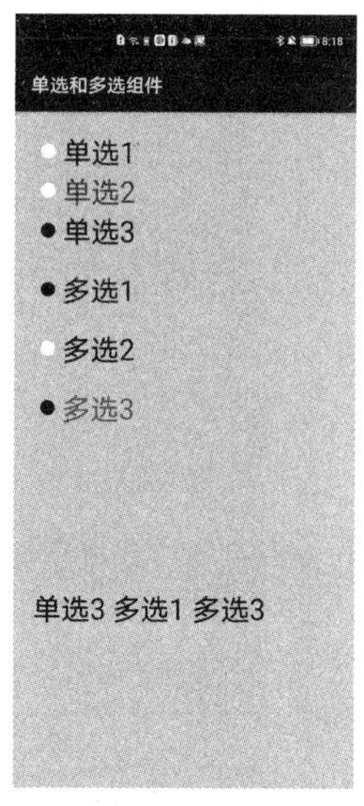

图 4-11 选择组件的显示效果

4.2.5 Switch 组件

Switch 组件（即开关组件）用于切换状态，与 ToggleButton 组件的功能类似，只是样式不同。Switch 组件也有两个状态——on 和 off。并且 Switch 组件拥有一个滑块，如果组件处于 off 状态，滑块在左侧；如果组件处于 on 状态，滑块在右侧。

【例 4.10】 在窗口中放置两个 Switch 组件，当第 1 个 Switch 组件改变状态时，设置第 2 个 Switch 组件的状态。除了用第 1 个 Switch 组件控制第 2 个 Switch 组件的开关状态，还用 Java 代码改变第 2 个 Switch 组件的样式，使其更美观。

首先完成 XML 布局文件。

代码位置： src/main/resources/base/layout/switch_demo.xml

```
<?xml version="1.0" encoding="utf-8"?>
<DirectionalLayout
```

```
    xmlns:ohos="http://schemas.huawei.com/res/ohos"
    ohos:height="match_parent"
    ohos:width="match_parent"
    ohos:background_element="#CCCCCC"
    ohos:orientation="vertical" ohos:padding="20vp">
    <Switch
        ohos:id="$+id:switch1"
        ohos:height="30vp"
        ohos:width="60vp"/>
    <Switch
        ohos:id="$+id:switch2"
        ohos:top_margin="20vp"
        ohos:height="30vp"
        ohos:width="60vp"/>

</DirectionalLayout>
```

在这段布局代码中并没有为Switch组件设置太多属性，大多数属性都是用Java代码设置的。下面来实现Java代码控制部分。

```
package com.unitymarvel.demo.components;

import com.unitymarvel.demo.ResourceTable;
import ohos.aafwk.ability.Ability;
import ohos.aafwk.content.Intent;
import ohos.agp.colors.RgbColor;
import ohos.agp.components.*;
import ohos.agp.components.element.ShapeElement;
import ohos.agp.components.element.StateElement;

public class SwitchDemo extends Ability {
    private Switch switch1,switch2;
    @Override
    public void onStart(Intent intent) {
        super.onStart(intent);
        super.setUIContent(ResourceTable.Layout_switch_demo);

        switch1 = (Switch)findComponentById(ResourceTable.Id_switch1);
        if(switch1 != null) {
            switch1.setCheckedStateChangedListener(new AbsButton.CheckedStateChangedListener() {
                @Override
                public void onCheckedChanged(AbsButton absButton, boolean b) {
                    // 根据 switch1 的状态设置 switch2 的状态
                    switch2.setChecked(b);
                }
            });
        }
        switch2 = (Switch)findComponentById(ResourceTable.Id_switch2);
        if(switch2 != null) {
            ShapeElement elementThumbOn = new ShapeElement();
            elementThumbOn.setShape(ShapeElement.OVAL);
            elementThumbOn.setRgbColor(RgbColor.fromArgbInt(0xFF1E90FF));
            elementThumbOn.setCornerRadius(50);
            // 关闭状态下滑块的样式
```

```
            ShapeElement elementThumbOff = new ShapeElement();
            elementThumbOff.setShape(ShapeElement.OVAL);
            elementThumbOff.setRgbColor(RgbColor.fromArgbInt(0xFFFFFFFF));
            elementThumbOff.setCornerRadius(50);
            // 开启状态下轨迹的样式
            ShapeElement elementTrackOn = new ShapeElement();
            elementTrackOn.setShape(ShapeElement.RECTANGLE);
            elementTrackOn.setRgbColor(RgbColor.fromArgbInt(0xFF87CEFA));
            elementTrackOn.setCornerRadius(50);
            // 关闭状态下轨迹的样式
            ShapeElement elementTrackOff = new ShapeElement();
            elementTrackOff.setShape(ShapeElement.RECTANGLE);
            elementTrackOff.setRgbColor(RgbColor.fromArgbInt(0xFF808080));
            elementTrackOff.setCornerRadius(50);
            // 设置轨迹样式
            switch2.setTrackElement(trackElementInit(elementTrackOn,elementTrackOff));
            // 设置滑块样式
            switch2.setThumbElement(thumbElementInit(elementThumbOn,elementThumbOff));
        }
    }
    // 关联开关时轨迹的状态
    private StateElement trackElementInit(ShapeElement on, ShapeElement off){
        StateElement trackElement = new StateElement();
        trackElement.addState(new int[]{ComponentState.
        COMPONENT_STATE_CHECKED}, on);
        trackElement.addState(new int[]{ComponentState.
        COMPONENT_STATE_EMPTY}, off);
        return trackElement;
    }
    // 关联开关时滑块的状态
    private StateElement thumbElementInit(ShapeElement on,
    ShapeElement off) {
        StateElement thumbElement = new StateElement();
        thumbElement.addState(new int[]{ComponentState.
        COMPONENT_STATE_CHECKED}, on);
        thumbElement.addState(new int[]{ComponentState.
        COMPONENT_STATE_EMPTY}, off);
        return thumbElement;
    }
}
```

运行程序，切换第 1 个 Switch 组件的状态，会发现第 2 个 Switch 组件的状态也改变了，显示效果如图 4-12 所示。

图 4-12 Switch 组件的显示效果

4.3 高级组件

本节将介绍一些 HarmonyOS 中比较复杂的组件，如列表组件、标签列表组件等。

4.3.1 ListContainer 组件

ListContainer 组件（即列表组件）是 HarmonyOS 中最复杂的组件之一，用于以垂直方向或水平方向显示列表。每一个列表项可以任意定义，其实每一个列表项就是一个独立的组件，可以直接通过 XML 布局文件装载。

要想让 ListContainer 组件显示数据，必须为 ListContainer 组件提供下面两样东西：

- 数据源；
- ItemProvider。

要显示一个列表，肯定需要数据源。数据源可以来自数组、列表、文件、数据库、网络等。

ItemProvider 其实是一类对象，该类对象可以将数据源与 ListContainer 组件连接起来。ItemProvider 会根据数据源中的每一条数据调用回调方法（getComponent 方法），并通过回调方法返回当前数据对应的 Component 对象，该对象就是要显示当前数据的列表项，通常直接在 XML 布局文件中装载，可以使用任何 UI 组件。开发人员只需要考虑返回什么 UI 组件即可，ListContainer 组件会将这些返回的 UI 组件放到合适的位置，并负责管理这些 UI 组件。

【例 4.11】 在窗口中放置两个 ListContainer 组件，第 1 个 ListContainer 组件显示的列表项以水平方向排列，第 2 个 ListContainer 组件显示的列表项以垂直方向排列。每一个列表项都会显示一个图像和对应的文本。

首先完成 XML 布局文件。

代码位置：src/main/resources/base/layout/listcontainer_demo.xml

```xml
<?xml version="1.0" encoding="utf-8"?>
<DirectionalLayout xmlns:ohos="http://schemas.huawei.com/res/ohos"
                ohos:width="match_parent"
                ohos:height="match_parent"
                ohos:background_element="#CCCCCC"
                ohos:orientation="vertical">
   <Text
      ohos:id="$+id:text1"
      ohos:text_size="30vp"
      ohos:height="match_content"
      ohos:width="match_content"/>
   <ListContainer
      ohos:id="$+id:listcontainer_horizontal"
      ohos:height="200vp"
      ohos:top_margin="20vp"
      ohos:orientation="horizontal"
      ohos:width="match_parent"/>
   <Text
      ohos:id="$+id:text2"
      ohos:text_size="30vp"
      ohos:top_margin="20vp"
      ohos:height="match_content"
      ohos:width="match_content"/>
   <ListContainer
      ohos:id="$+id:list_container_vertical"
      ohos:height="match_parent"
      ohos:width="match_parent"
      ohos:top_margin="20vp"/>
</DirectionalLayout>
```

在 listcontainer_demo.xml 文件中除了有两个 ListContainer 组件，还有两个 Text 组件，当单击这两个 ListContainer 组件的列表项时，会显示相应的信息。

ListContainer 组件需要通过 ohos:orientation 属性设置列表项显示是水平方向还是垂直方向，默认是垂直方向。

接下来的 Java 代码是本例的核心代码，在这段代码中，会为两个 ListContainer 组件设置数据源以及列表项单击事件的监听器。

代码位置：src/main/java/com/unitymarvel/demo/components/ListContainerDemo.java

```
package com.unitymarvel.demo.components;

import com.unitymarvel.demo.ResourceTable;
import ohos.aafwk.ability.Ability;
import ohos.aafwk.content.Intent;
import ohos.agp.components.*;
import java.util.ArrayList;
import java.util.List;

public class ListContainerDemo extends Ability {
   private Text text;
   private RadioButton radioButton1,radioButton2,radioButton3;
   private Checkbox checkbox1, checkbox2, checkbox3;
   private Text text1,text2;
   @Override
   public void onStart(Intent intent) {
      super.onStart(intent);
      super.setUIContent(ResourceTable.Layout_listcontainer_demo);
      ListContainer listcontainerHorizontal =
      (ListContainer) findComponentById(ResourceTable.Id_listcontainer_horizontal);
      ListContainer listcontainerVertical =
      (ListContainer) findComponentById(ResourceTable.Id_list_container_vertical);
      text1 = (Text)findComponentById(ResourceTable.Id_text1);
      text2 = (Text)findComponentById(ResourceTable.Id_text2);
      // 产生两个 ListContainer 组件使用的数据源的文本部分
      List<String> mList = new ArrayList<>();
      for (int i = 0; i < 20; i++) {
         mList.add("Item" + (i + 1));
      }
      // 为水平方向 ListContainer 组件设置列表项单击事件监听器
      listcontainerHorizontal.setItemClickedListener(new ListContainer.ItemClickedListener() {
         @Override
         public void onItemClicked(ListContainer listContainer, Component component,
         int i, long l) {
             // 单击水平列表项时将相应的信息显示在第 1 个 Text 组件中
             text1.setText("单击了第" + (i + 1) + "项");
         }
      });
      // 为水平方向 ListContainer 组件指定 ItemProvider
      listcontainerHorizontal.setItemProvider(new RecycleItemProvider() {
```

```
        // 获取数据源中数据的个数
        @Override
        public int getCount() {
            return mList.size();
        }
        // 获取数据源中每一个数据对象
        @Override
        public Object getItem(int i) {
            return mList.get(i);
        }
        // 获取数据源中每一个数据的 ID
        @Override
        public long getItemId(int i) {
            return i;
        }
        // 这个方法非常重要，每一个列表项对应的 UI 组件都是由该方法返回的
        // i 表示当前列表项的索引，从 0 开始，component 表示当前的列表项是否有复用的 UI 组件
        // 如果没有，则值为 null。componentContainer 表示列表项组件的父组件
        @Override
        public Component getComponent(int i, Component component, ComponentContainer
        componentContainer) {
            DirectionalLayout view = (DirectionalLayout)
            LayoutScatter.getInstance(getContext()).parse(ResourceTable.Layout_list_item_
            horizontal, null, false);
            // 动态装载列表项布局文件
            Image s = (Image) view.getComponentAt(0);
            // 为每一个列表项的图像部分指定一个 Media 资源
            s.setPixelMap(ResourceTable.Media_face);
            Text t1 = (Text) view.findComponentById(ResourceTable.Id_item_title1);

            // 为每一个列表项指定文本
            t1.setText(" 标题" + (i + 1));
            // 返回列表项 UI 组件
            return view;
        }
    });
    // 为垂直方向 ListContainer 组件设置 ItemProvider
    listcontainerVertical.setItemProvider(new ListComponentAdapter<String>(this,
    mList, ResourceTable.Layout_list_item) {
        // 当需要为列表项组件设置值时调用该方法
        @Override
        public void onBindViewHolder(CommentViewHolder commonViewHolder, String item,
        int position) {
            Image image = commonViewHolder.getImageView(ResourceTable.Id_item_image);
            // 设置列表项的图像部分
            image.setPixelMap(ResourceTable.Media_face);
            Text t1 = commonViewHolder.getTextView(ResourceTable.Id_item_title);
            // 设置列表项的文本部分
            t1.setText((position + 1) + " 标题");
        }

        @Override
        public void onItemClick(Component component, String item, int position) {
```

```
                text2.setText("单击了第" + (position + 1) + "项");

            }
        });
    }
}
```

有的读者可能会发现，第 2 个 ListContainer 组件指定的 ItemProvider 与第 1 个 ListContainer 组件指定的 ItemProvider 不同。第 1 个 ListContainer 组件直接绑定了 RecycleItemProvider 对象，并实现了 getComponent 方法。在该方法中动态装载了名为 list_item_horizontal.xml 的水平列表项布局文件，并设置了布局中的图像和文本。这种实现方式从表面上看没什么问题，但如果列表项非常多，而且每一个列表项的布局都相同，这就会造成资源浪费。

造成资源浪费的原因是，不管是水平列表，还是垂直列表，手机一屏能显示的列表项都是有限的，这会导致有的列表项暂时无法看到，只有滚动列表，才能看到隐藏的列表项。如果按第 1 个 ListContainer 组件的方式，那么每显示一个列表项，就会通过 getComponent 方法创建一个新的 Component 对象。假设这些数据非常多，有 100 万个或更多，那么只要用户不断往下滚动，系统就会不断创建 Component 对象，关键是每一个 Component 对象只是数据不同，布局风格完全相同，而且那些暂时看不到的列表项也会占用内存空间。

为了解决这个问题，就要求暂时未显示的列表项可以复用已经创建好的 Component 对象，也就是说，将已经创建好的 Component 对象中的组件重新设置数据后，直接使用。getComponent 方法的第 2 个参数 component 就是用于完成这个功能的。component 参数值就是可以利用的 Component 对象。如果没有可以利用的 Component 对象（通常是数据源中数据的个数小于或等于一屏可以显示的最大列表项个数），component 参数值就会为 null。

第 2 个 ListContainer 组件与一个 ListComponentAdapter 对象绑定，就是为了解决这个问题。ListComponentAdapter 类提供了 onBindViewHolder 方法，在该方法中并不会直接操作 Component 对象，而是通过 CommentViewHolder 对象直接操作 Component 对象中的具体 UI 组件，本例中具体 UI 组件是 Image 组件和 Text 组件。包含这些组件的 Component 对象可能是新创建的，也可能是复用以前创建的，但这一切在 onBindViewHolder 方法中是完全透明的，开发人员并不需要关心这些 Component 对象的来源，只需要使用就可以了。

下面先看一下实现 ListComponentAdapter 类的代码。

代码位置：src/main/java/com/unitymarvel/demo/components/ListComponentAdapter.java

```java
package com.unitymarvel.demo.components;

import ohos.agp.components.Component;
import ohos.agp.components.ComponentContainer;
import ohos.agp.components.LayoutScatter;
import ohos.agp.components.RecycleItemProvider;
import ohos.app.Context;
import java.util.List;
// ListcomponentAdapter 类使用了泛型，这个泛型就是列表项文字部分的类型，本例是 String
public abstract class ListComponentAdapter<T> extends RecycleItemProvider {
```

```
    private Context context;
    private List<T> listBean;
    private int xmlId;
    private LayoutScatter layoutScatter;
    // 用于绑定列表项数据的抽象方法
    public abstract void onBindViewHolder(CommentViewHolder commonViewHolder, T item,
    int position);
    // 构造方法
    public ListComponentAdapter(Context context, List<T> list, int xmlId) {
        this.context = context;
        this.listBean = list;
        this.xmlId = xmlId;
        layoutScatter = LayoutScatter.getInstance(context);
    }

    @Override
    public int getCount() {
        return listBean.size();
    }
    @Override
    public T getItem(int i) {
        return listBean.get(i);
    }
    @Override
    public long getItemId(int i) {
        return i;
    }
    // 获取 i 指定的列表项的 Component 对象
    @Override
    public Component getComponent(int i, Component component, ComponentContainer
    componentContainer) {
        // 获取 CommentViewHolder 对象，该对象封装了 Image 组件和 Text 组件
        // 该方法根据是否有可复用的 Component 对象，返回相应的 Component 对象
        CommentViewHolder commentViewHolder =
        CommentViewHolder.getCommentViewHolder(context, component, xmlId);
        T t = listBean.get(i);
        // 调用 OnBindViewHolder 方法为 Image 组件和 Text 组件绑定数据
        onBindViewHolder(commentViewHolder, t, i);
        commentViewHolder.convertView.setClickedListener(component1 -> onItemClick
       (component,t,i));
        // 返回当前列表项的 Component 对象
        return commentViewHolder.convertView;
    }
    public void  onItemClick(Component component, T item, int position){

    }
}
```

最后看一下 CommentViewHolder 类的实现，该类考虑了列表项 Component 对象的复用问题。

代码位置：src/main/java/com/unitymarvel/demo/components/CommentViewHolder.java

```
package com.unitymarvel.demo.components;

import ohos.agp.components.Component;
```

```
import ohos.agp.components.Image;
import ohos.agp.components.LayoutScatter;
import ohos.agp.components.Text;
import ohos.app.Context;
import java.util.HashMap;
import java.util.Map;

public class CommentViewHolder {
   // 使用单例模式
   public static CommentViewHolder getCommentViewHolder(Context context, Component
   convertView, int resource) {
      // 没有可复用的 Component 对象，创建一个新的 Component 对象
      if (convertView == null) {
          convertView = LayoutScatter.getInstance(context).parse(resource, null, false);
          return new CommentViewHolder(convertView);
      } else {  // 有可复用的 Component 对象
      // 从 Component 对象本身获得要返回的 CommentViewHolder 对象
          return (CommentViewHolder) convertView.getTag();
      }
   }

   public Component convertView;

   // 对成员变量赋值，并把当前 CommentViewHolder 用 tag 存起来以便复用
   public CommentViewHolder(Component convertView) {
      this.convertView = convertView;
      convertView.setTag(this);
   }
   // 获取列表项中的 Text 组件
   public Text getTextView(int resId) {
      return  (Text)convertView.findComponentById(resId);
   }
   // 获取列表项中的 Image 组件
   public Image getImageView(int resId) {
      return (Image)convertView.findComponentById(resId);
   }

}
```

CommentViewHolder 类的关键点如下两个。

- getCommentViewHolder 方法返回了 CommentViewHolder 对象，该对象封装了 Component 对象,这里的 Component 对象会根据有无复用的列表项而选择使用以前创建的 Component 对象，或新创建 Component 对象。
- 每一个返回的 Component 对象,都会通过该对象的 tag 保存与其对应的 CommentViewHolder 对象，这样就可以直接通过 Component 对象返回封装该对象的 CommentViewHolder 对象了。

运行程序，并且单击水平列表和垂直列表中的某个列表项，会看到类似图4-13的显示效果。

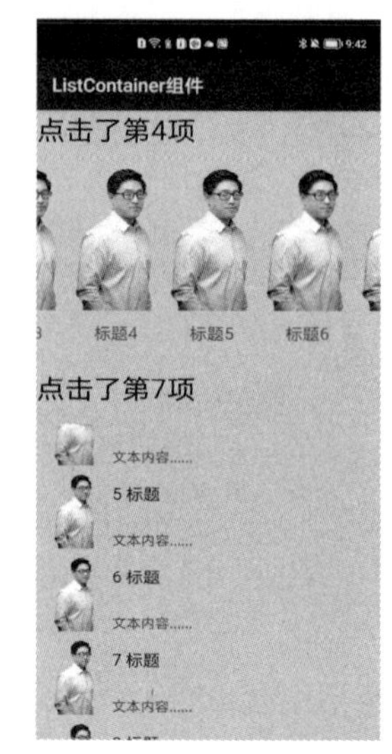

图 4-13 ListContainer 组件的显示效果

4.3.2 TabList 组件

TabList 组件（即标签列表组件）允许用户左右滑动视图来切换当前显示的视图。也就是说，可以将多个视图添加到 TabList 组件中，一次只显示一个视图，可以通过手势左右滑动切换不同的视图。如果一屏无法放下所有的 UI 时，通常会采用分屏的风格，而 TabList 组件就是实现分屏风格的最佳组件。

【例4.12】 在窗口中放置1个TabList组件和3个RadioButton组件，当选中某一个RadioButton组件时，TabList组件会显示与其对应的视图。

先看一下XML布局文件的实现。

代码位置：src/main/resources/base/layout/tablist_demo.xml

```
<?xml version="1.0" encoding="UTF-8"?>
<DirectionalLayout xmlns:ohos="http://schemas.huawei.com/res/ohos">
                  ohos:orientation="vertical"
                  ohos:background_element="#dd85cfcb"
                  ohos:height="match_parent"
                  ohos:width="match_parent"
   <PageSlider
       ohos:id="$+id:pagerslider"
       ohos:width="match_parent"
       ohos:height="0vp"
       ohos:weight="1"
       ohos:background_element="#CCCCCC"/>
   <RadioContainer
       ohos:id="$+id:radiocontainer"
       ohos:height="60vp"
       ohos:width="match_parent"
       ohos:orientation="horizontal">
   <RadioButton
       ohos:width="0vp"
       ohos:weight="1"
       ohos:height="match_parent"
       ohos:text="首页"
       ohos:text_alignment="center"
       ohos:text_size="19fp" />
   <RadioButton
       ohos:width="0vp"
       ohos:weight="1"
       ohos:height="match_parent"
       ohos:text="商城"
       ohos:text_alignment="center"
       ohos:text_size="19fp"/>
```

```
    <RadioButton
        ohos:width="0vp"
        ohos:weight="1"
        ohos:height="match_parent"
        ohos:text="设置"
        ohos:text_alignment="center"
        ohos:text_size="19fp"/>
    </RadioContainer>
</DirectionalLayout>
```

下面的 Java 代码用来设置 TabList 组件和 RadioContainer 组件的相关事件。

代码位置：src/main/java/com/unitymarvel/demo/components/TabListDemo.java

```
package com.unitymarvel.demo.components;

import com.unitymarvel.demo.ResourceTable;
import ohos.aafwk.ability.Ability;
import ohos.aafwk.content.Intent;
import ohos.agp.components.*;
import java.util.ArrayList;

public class TabListDemo extends Ability {

    private TabList tabList;
    private ArrayList<Component> pageview;
    private RadioContainer radioContainer;

    @Override
    protected void onStart(Intent intent) {
        super.onStart(intent);
        Component rootView = LayoutScatter.getInstance(getContext()).parse(ResourceTable.
        Layout_tablist_demo, null, false);
        super.setUIContent((ComponentContainer) rootView);
        radioContainer = (RadioContainer)
        findComponentById(ResourceTable.Id_radiocontainer);

        PageSlider view_pager = (PageSlider)
        findComponentById(ResourceTable.Id_pagerslider);
        LayoutScatter dc = LayoutScatter.getInstance(getContext());
        // 下面的 3 行代码用于装载 TabList 组件中使用的 3 个页面的视图对象（DependentLayout 对象）
        DependentLayout view0 = (DependentLayout) dc.parse(ResourceTable.Layout_tab_page0,
        null, false);
        DependentLayout view1 = (DependentLayout) dc.parse(ResourceTable.Layout_tab_page1,
        null, false);
        DependentLayout view2 = (DependentLayout) dc.parse(ResourceTable.Layout_tab_page2,
        null, false);
        // 将 view 装入数组中
        pageview = new ArrayList<Component>();
        pageview.add(view0);
        pageview.add(view1);
        pageview.add(view2);
```

```
        // 创建TabList使用的PageSliderProvider对象，用于提供数据源和显示的页面
        PageSliderProvider mPagerAdapter = new PageSliderProvider() {
            @Override
            // 获取当前窗口中的页面数
            public int getCount() {
                return pageview.size();
            }
            // 返回一个对象，这个对象表明了PagerAdapter适配器选择哪个对象放在当前的ViewPager中
            @Override
            public Object createPageInContainer(ComponentContainer componentContainer,
            int i) {
                componentContainer.addComponent(pageview.get(i));
                return pageview.get(i);
            }

            // 是从ViewGroup中移出当前View
            @Override
            public void destroyPageFromContainer(ComponentContainer componentContainer,
            int i, Object o) {
                ((PageSlider) componentContainer).removeComponent(pageview.get(i));
            }

            // 判断是否由对象生成页面
            @Override
            public boolean isPageMatchToObject(Component component, Object o) {
                return false;
            }

        };
        // 为TabList组件绑定适配器
        view_pager.setProvider(mPagerAdapter);
        view_pager.addPageChangedListener(new PageSlider.PageChangedListener() {
            @Override
            public void onPageSliding(int i, float v, int i1) {

            }

            @Override
            public void onPageSlideStateChanged(int i) {

            }
            // 当左右滑动TabList组件中的页面时，让对应的RadioButton组件也被选中
            @Override
            public void onPageChosen(int i) {
                ((RadioButton) radioContainer.getComponentAt(i)).setChecked(true);
            }
        });
        // 为RadioContainer组件指定选项变化监听器
        radioContainer.setMarkChangedListener((radioContainer, i) -> view_pager.
        setCurrentPage(i));

    }
}
```

运行程序，会看到图 4-14 所示的显示效果。选中窗口下方的 3 个 RadioButton 组件，TabList 组件中的页面会不断切换，同样也可以用手指左右滑动 TabList 组件中的页面，这时与该页面对应的 RadioButton 组件也会被选中。

图 4-14 TabList 组件的显示效果

4.3.3 Picker 组件

Picker 组件（即滑动选择器组件）用于在窗口中显示一个列表，用户可以用手指上下滑动列表，并选择其中的某个列表项。

Picker 组件的绝大多数属性既可以在 XML 布局中设置，也可以使用 Java 代码设置。例如，要使用 Picker 组件显示一个列表，首先要为 Picker 组件指定最大值和最小值。在 XML 布局文件中需要使用 max_value 属性和 min_value 属性来指定最大值和最小值。在 Java 代码中需要使用 setMaxValue 方法和 setMinValue 方法来指定最大值和最小值。Picker 组件默认会根据最大值和最小值显示一个数字列表，如图 4-15 所示。

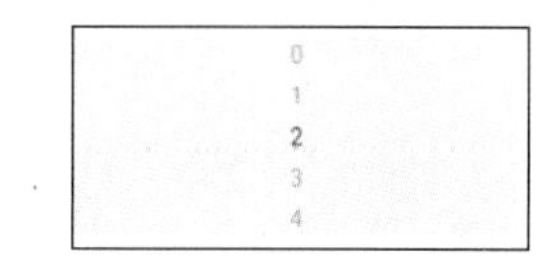

图 4-15 Picker 组件的默认显示效果

如果要监听 Picker 组件当前选择值的变化，需要使用 setValueChangedListener 方法设置监听器。监听器类型是 Picker.ValueChangedListener，这是一个接口，该接口有一个 onValueChanged 方法，原型如下：

```
public void onValueChanged(Picker picker, int oldVal, int newVal);
```

onValueChanged 方法有 3 个参数，这些参数的含义如下。

- picker：Picker 组件对象。
- oldVal：上一个被选择的列表项的值。
- newVal：当前被选择的列表项的值。

从 oldVal 和 newVal 的数据类型（即 int）可以看出，Picker 组件显示的列表项的值默认只能是整数。不过可以在显示时将列表项的值变成字符串类型，这就要依赖 setFormatter 方法了。该方法可以为 Picker 组件设置一个 Picker.Formatter 类型的对象。Picker.Formatter 是一个接口，该接口中有一个 format 方法，原型如下：

```
public String format(int i);
```

其中，i 就是 Picker 组件当前列表项的值，format 方法的返回值就是与当前列表项的值对应的要显示的字符串。所以在 format 方法中通常会根据 i 的值返回相应的字符串。也就是说，Picker 组件在显示列表中每一个值时，会调用 format 方法，并将当前值传递给 format 方法，而显示的值就是 format 方法返回的值。

【例 4.13】 在窗口中放置两个 Picker 组件，第 1 个 Picker 组件直接显示当前列表项的值，第 2 个 Picker 组件根据当前列表项的值显示特定的字符串。当滑动这两个 Picker 组件中的列表时，会在窗口上方的 Text 组件中显示上一个选中的值（即旧值）和当前选中的值（即新值）。

先看一下 XML 布局文件的实现。

代码位置：src/main/resources/base/layout/picker_demo.xml

```
<?xml version="1.0" encoding="utf-8"?>
<DirectionalLayout
   xmlns:ohos="http://schemas.huawei.com/res/ohos"
   ohos:height="match_parent"
   ohos:width="match_parent"
   ohos:background_element="#CCCCCC"
   ohos:orientation="vertical" ohos:padding="20vp">
   <Text
      ohos:id="$+id:text"
      ohos:text_size="20fp"
      ohos:height="30vp"
      ohos:width="match_parent"/>
   <Picker
      ohos:id="$+id:picker1"
      ohos:height="match_content"
      ohos:width="300vp"
      ohos:background_element="#E1FFFF"
      ohos:layout_alignment="horizontal_center"
      ohos:top_margin="20vp"
      ohos:min_value="2"
      ohos:max_value="10"
      ohos:normal_text_size="30fp"
      ohos:selected_text_size="30fp"/>
   <Picker
      ohos:id="$+id:picker2"
      ohos:height="match_content"
      ohos:width="300vp"
      ohos:background_element="#E1FFFF"
      ohos:layout_alignment="horizontal_center"
      ohos:top_margin="20vp"
      ohos:min_value="1"
      ohos:max_value="7"
      ohos:normal_text_size="30fp"
      ohos:selected_text_size="30fp"/>
</DirectionalLayout>
```

在这个布局文件中，两个 Picker 组件都通过 min_value 属性和 max_value 属性设置了最小值和最大值，不过第 1 个 Picker 组件会通过 setMinValue 方法和 setMaxValue 方法修改最小值和最大值。下面是完整的 Java 实现代码。

代码位置：src/main/java/com/unitymarvel/demo/components/PickerDemo.java

```
package com.unitymarvel.demo.components;

import com.unitymarvel.demo.ResourceTable;
import ohos.aafwk.ability.Ability;
import ohos.aafwk.content.Intent;
import ohos.agp.colors.RgbColor;
import ohos.agp.components.element.ShapeElement;
```

```
import ohos.agp.utils.Color;
import ohos.agp.components.Picker;
import ohos.agp.components.Text;
import ohos.agp.text.Font;

public class PickerDemo extends Ability {
    private Text text;
    private Picker picker1,picker2;
    @Override
    public void onStart(Intent intent) {
        super.onStart(intent);
        super.setUIContent(com.unitymarvel.demo.ResourceTable.Layout_picker_demo);
        text = (Text)findComponentById(ResourceTable.Id_text);

        picker1 = (Picker)findComponentById(ResourceTable.Id_picker1);
        if(picker1 != null) {
            // 设置picker1的最小值
            picker1.setMinValue(10);
            // 设置picker1的最大值
            picker1.setMaxValue(30);
            // 为picker1指定当前值变化监听器
            picker1.setValueChangedListener(new Picker.ValueChangedListener() {
                @Override
                public void onValueChanged(Picker picker, int oldVal, int newVal) {
                    // 当值变化时，将旧值和新值显示在Text组件中
                    text.setText("旧值: " + oldVal + "       新值: " + newVal);
                }
            });
        }
        picker2 = (Picker)findComponentById(ResourceTable.Id_picker2);
        if(picker2 != null) {
            // 为picker2设置显示的文本
            picker2.setFormatter(new Picker.Formatter() {
                @Override
                public String format(int i) {
                    String value = "unknown";
                    // 根据当前值i显示对应的文本
                    switch (i) {
                        case 1:
                            value = "星期一";
                            break;
                        case 2:
                            value = "星期二";
                            break;
                        case 3:
                            value = "星期三";
                            break;
                        case 4:
                            value = "星期四";
                            break;
                        case 5:
                            value = "星期五";
                            break;
                        case 6:
```

```
                value = "星期六";
                break;
            case 7:
                value = "星期日";
                break;
        }
        return value;
    }
});
// 为picker2设置当前值变化监听器
picker2.setValueChangedListener(new Picker.ValueChangedListener() {
    @Override
    public void onValueChanged(Picker picker, int oldVal, int newVal) {
        text.setText("旧值: " + oldVal + "      新值: " + newVal);
    }
});
/* 通过setDisplayedData方法也可以设置要显示的值，只是通过该方法的参数指定的字符串数组
的元素个数必须与列表中元素的数量相同*/
//picker2.setDisplayedData(new String[]{"Mon", "Tue", "Wed", "Thu", "Fri",
"Sat", "Sun"});
// 设置picker2的文本样式
picker2.setNormalTextFont(Font.DEFAULT_BOLD);
picker2.setNormalTextColor(new Color(Color.getIntColor("#FFA500")));
picker2.setSelectedTextFont(Font.DEFAULT_BOLD);
picker2.setSelectedTextColor(new Color(Color.getIntColor("#00FFFF")));
// 设置picker2的着色器颜色
picker2.setShaderColor(new Color(Color.getIntColor("#1E90FF")));

// 设置picker2中所选文本边距与普通文本边距的比例
picker2.setSelectedNormalTextMarginRatio(10.0f);
// 设置所选文本的上下边框
ShapeElement shape = new ShapeElement();
shape.setShape(ShapeElement.RECTANGLE);
shape.setRgbColor(RgbColor.fromArgbInt(0xFFFF00D0));
// 设置picker2是否是循环显示数据
picker2.setDisplayedLinesElements(shape,shape);
picker2.setWheelModeEnabled(true);
        }
    }
}
```

运行程序，会显示图4-16所示的效果。滑动这两个Picker组件会发现窗口上方的Text组件显示的旧值和新值的文本变化了。

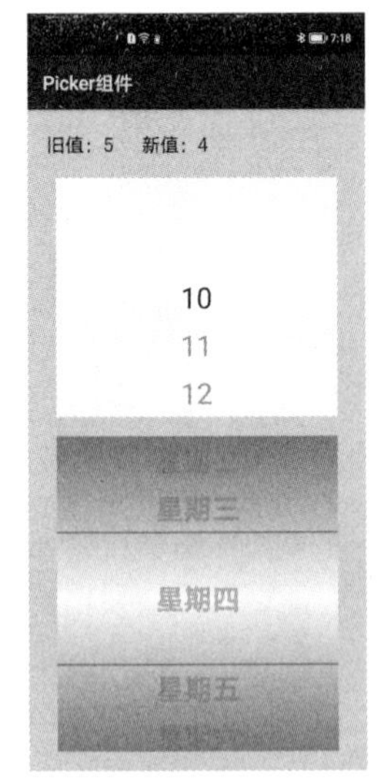

图4-16 Picker组件的显示效果

4.3.4 DatePicker组件

DatePicker组件（即日期选择器组件）用于选择日期，分成3个列显示，其中第1列显示年、第2列显示月、第3列显示日，显示效果如图4-17所示。

DatePicker组件与Picker组件类似，可以监听选项的变化，监听

动作由 DatePicker.ValueChangedListener 接口的 onValueChanged 方法完成。该方法的原型如下：

```
public void onValueChanged(DatePicker datePicker, int year, int monthOfYear, int dayOfMonth);
```

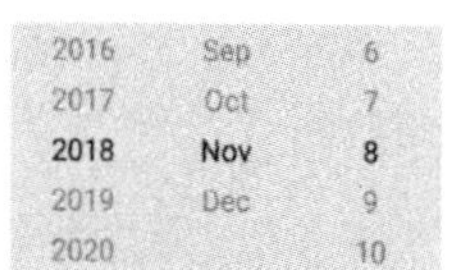

图 4-17　DatePicker 组件的显示效果

下面是 onValueChanged 方法参数的含义。

- datePicker：DatePicker 组件对象。
- year：当前选中的年。
- monthOfYear：当前选中的月。
- dayOfMonth：当前选中的日。

除了可以通过 onValueChanged 方法的参数获取当前选择的日期，还可以通过下面的几个方法获取当前选中的年、月、日。

```
int year = datePicker.getYear();              // 获取选择的年
int month = datePicker.getMonth();            // 获取选择的月
int day =  datePicker.getDayOfMonth();        // 获取选择的日
```

要让 DatePicker 组件显示特定的日期，还可以通过 DatePicker.updateDate 方法更新当前选择的日期，该方法的原型如下：

```
public void updateDate(int year, int month, int dayOfMonth);
```

其中，year、month 和 dayOfMonth 分别表示年、月、日。

【例 4.14】　在窗口中放置两个 DatePicker 组件，并且监听两个 DatePicker 组件的值的变化。当日期变化后，会将变化后的日期显示在窗口上方的 Text 组件中。然后设置第 2 个 DatePicker 组件的默认显示日期。最后定制第 2 个 DatePicker 组件的样式。

先看一下 XML 布局文件的实现。

代码位置：src/main/resources/base/layout/datepicker_demo.xml

```
<?xml version="1.0" encoding="utf-8"?>
<DirectionalLayout
   xmlns:ohos="http://schemas.huawei.com/res/ohos"
   ohos:height="match_parent"
   ohos:width="match_parent"
   ohos:background_element="#CCCCCC"
   ohos:orientation="vertical" ohos:padding="20vp">
   <Text
      ohos:id="$+id:text"
      ohos:text_size="20fp"
      ohos:height="30vp"
      ohos:width="match_parent"/>
   <DatePicker
      ohos:id="$+id:datepicker1"
      ohos:height="match_content"
      ohos:width="300vp"
      ohos:background_element="#E1FFFF"
      ohos:top_margin="20vp"
      ohos:normal_text_size="30fp"
      ohos:selected_text_size="30fp"/>
```

```
    <DatePicker
        ohos:id="$+id:datepicker2"
        ohos:height="match_content"
        ohos:width="300vp"
        ohos:background_element="#E1FFFF"
        ohos:top_margin="20vp"
        ohos:year_fixed="true"
        ohos:operated_text_color="#FF0000"
        ohos:normal_text_color="#00FFFF"
        ohos:selected_text_color="#FF00FF"
        ohos:normal_text_size="30fp"
        ohos:selected_text_size="30fp"/>
</DirectionalLayout>
```

布局文件中涉及一些定制样式和动作的属性，下面是这些属性的含义。

- ohos:operated_text_color：正在操作的列的当前值的文本颜色。如果正在滑动显示月的列，那么当前月的文本就使用这个颜色。
- ohos:normal_text_color：默认文本显示的颜色。
- ohos:selected_text_color：被选中文本显示的颜色。
- ohos:normal_text_size：默认文本的字号。
- ohos:selected_text_size：被选中文本的字号。
- ohos:year_fixed：让第 1 列（显示年的列）固定。也就是说，被固定的列无法滑动了。

下面看一下 Java 实现代码。

代码位置：src/main/java/com/unitymarvel/demo/components/DatePickerDemo.java

```
package com.unitymarvel.demo.components;
import com.unitymarvel.demo.ResourceTable;
import ohos.aafwk.ability.Ability;
import ohos.aafwk.content.Intent;
import ohos.agp.components.DatePicker;
import ohos.agp.components.Text;
import ohos.agp.utils.Color;
import java.text.DateFormat;
import java.text.SimpleDateFormat;
import java.util.Calendar;

public class DatePickerDemo extends Ability {
    private DatePicker datePicker1, datePicker2;
    private Text text;
    @Override
    public void onStart(Intent intent) {
        super.onStart(intent);
        super.setUIContent(com.unitymarvel.demo.ResourceTable.Layout_datepicker_demo);
        Text text = (Text)findComponentById(ResourceTable.Id_text);
        datePicker1 = (DatePicker)findComponentById(ResourceTable.Id_datepicker1);
        if(datePicker1 != null) {
            // 监听 datePicker1 中值的变化
            datePicker1.setValueChangedListener(new DatePicker.ValueChangedListener() {
                @Override
                public void onValueChanged(DatePicker datePicker, int year,
                int monthOf Year, int dayOfMonth) {
```

```
                // 将 datePicker1 变化后的日期显示在 Text 组件中
                text.setText(year + "年 " + monthOfYear + " 月 " + dayOfMonth + " 日");

            }
        });
    }
    datePicker2 = (DatePicker)findComponentById(ResourceTable.Id_datepicker2);
    if(datePicker2 != null) {
        // 设置 datePicker2 的当前日期
        datePicker2.updateDate(2019,10,1);
        datePicker2.setValueChangedListener(new DatePicker.ValueChangedListener() {
            @Override
            public void onValueChanged(DatePicker datePicker, int year, int monthOfYear,
            int dayOfMonth) {
                // 将 datePicker2 变化后的日期显示在 Text 组件中
                text.setText(datePicker.getYear() + "year " + datePicker.getMonth() +
                " month " + datePicker.getDayOfMonth + " day");

            }
        });
    }
    // 设置着色器颜色
    datePicker2.setShaderColor(new Color(Color.getIntColor("#22CED1")));
    DateFormat df = new SimpleDateFormat("yyyy-MM-dd");
    try {
        java.util.Date date = df.parse("2010-10-20");
        Calendar calendar = Calendar.getInstance();
        calendar.setTime(date);
        long timestamp = calendar.getTimeInMillis();
        // 设置显示的最小日期（以秒为单位的时间戳）
        datePicker2.setMinDate(timestamp / 1000);

        date = df.parse("2030-10-20");
        calendar = Calendar.getInstance();
        calendar.setTime(date);
        timestamp = calendar.getTimeInMillis();
        // 设置显示的最大日期（以秒为单位的时间戳）
        datePicker2.setMaxDate(timestamp / 1000);

    } catch (Exception e) {

    }
  }
}
```

这段代码中使用 setMinDate 方法和 setMaxDate 方法分别设置了 DatePicker 组件的最小日期和最大日期，这两个日期必须用以秒为单位的时间戳设置。这样在 DatePicker 组件中就只能选择这两个日期之间的日期，其他日期都不会显示在 DatePicker 组件中。

运行程序，滑动这两个 DatePicker 组件，会看到 Text 组件显示了当前选择的日期，效果如图 4-18 所示。

图 4-18 DatePicker 组件的显示效果

4.3.5 TimePicker 组件

TimePicker 组件（即时间选择器组件）与 DatePicker 组件类似，只是 TimePicker 组件用于显示时间（时、分、秒），而 DatePicker 组件用于显示日期（年、月、日）。TimePicker 组件的显示效果如图 4-19 所示。

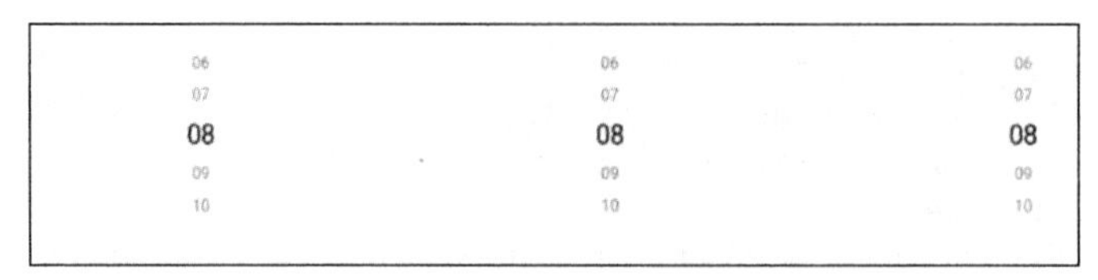

图 4-19 TimePicker 组件的显示效果

如果想监听 TimePicker 中时间的变化，需要使用 setTimeChangedListener 方法设置监听器，该监听器是 TimePicker.TimeChangedListener，这是一个接口，其中有一个 onTimeChanged 方法，当选择的时间变化时，该方法会被调用。onTimeChanged 方法的原型如下：

```
public void onTimeChanged(TimePicker timePicker, int hour, int minute, int second);
```

下面是 onTimeChanged 方法参数的含义。

- timePicker：TimePicker 组件本身。
- hour：当前选中的时。
- minute：当前选中的分。
- second：当前选中的秒。

除了通过 onTimeChanged 方法的参数获取当前选择的时间，还可以通过下面的几个方法获取当前选中的时、分、秒：

```
int hour = timePicker.getHour();              // 获取选择的时
int minute = timePicker.getMinute();          // 获取选择的分
int second = timePicker.getSecond();          // 获取选择的秒
```

如果要让 TimePicker 组件显示特定的时间，可以用下面 3 个方法分别设置时、分、秒：

```
timePicker.setHour(hour);                     // 设置当前的时
timePicker.setMinute(minute);                 // 设置当前的分
timePicker.setSecond(second);                 // 设置当前的秒
```

【例 4.15】 在窗口中放置两个 TimePicker 组件，并且监听这两个 TimePicker 组件的值的变化。当时间变化后，会将变化后的时间显示在窗口上方的 Text 组件中。然后设置第 2 个 TimePicker 组件的默认显示时间。最后定制第 2 个 TimePicker 组件的样式。

先看一下 XML 布局文件的实现。

代码位置：src/main/resources/base/layout/timepicker_demo.xml

```
<?xml version="1.0" encoding="utf-8"?>
<DirectionalLayout
    xmlns:ohos="http://schemas.huawei.com/res/ohos"
    ohos:height="match_parent"
```

```
    ohos:width="match_parent"
    ohos:background_element="#CCCCCC"
    ohos:orientation="vertical" ohos:padding="20vp">
    <Text
        ohos:id="$+id:text"
        ohos:text_size="20fp"
        ohos:height="30vp"
        ohos:width="match_parent"/>
    <TimePicker
        ohos:id="$+id:timepicker1"
        ohos:height="match_content"
        ohos:width="300vp"
        ohos:background_element="#E1FFFF"
        ohos:top_margin="20vp"
        ohos:normal_text_size="30fp"
        ohos:selected_text_size="30fp"/>
    <TimePicker
        ohos:id="$+id:timepicker2"
        ohos:height="match_content"
        ohos:width="300vp"
        ohos:background_element="#E1FFFF"
        ohos:top_margin="20vp"
        ohos:operated_text_color="#FF0000"
        ohos:normal_text_color="#00FFFF"
        ohos:selected_text_color="#FF00FF"
        ohos:selected_normal_text_margin_ratio="10"
        ohos:normal_text_size="30fp"
        ohos:selected_text_size="30fp"
        ohos:am_pm_order="1"/>
</DirectionalLayout>
```

布局文件中 TimePicker 组件的属性与 DatePicker 组件的同名属性功能相同，这里不再阐述。但 TimePicker 组件还使用了一个 ohos:am_pm_order 属性，该属性只有在 TimePicker 组件被设置为 12 小时制时才起作用，用于调整 AM/PM 的显示位置。如果该属性值指定为 1，AM/PM 在后面显示；如果不指定该属性的值，则 AM/PM 在前面显示。

下面看一下 Java 实现代码。

代码位置：src/main/java/com/unitymarvel/demo/components/TimePickerDemo.java

```
package com.unitymarvel.demo.components;

import com.unitymarvel.demo.ResourceTable;
import ohos.aafwk.ability.Ability;
import ohos.aafwk.content.Intent;
import ohos.agp.components.Text;
import ohos.agp.components.TimePicker;

public class TimePickerDemo extends Ability {
    private TimePicker timePicker1, timePicker2;
    private Text text;
    @Override
    public void onStart(Intent intent) {
        super.onStart(intent);
```

```
        super.setUIContent(ResourceTable.Layout_timepicker_demo);
        Text text = (Text)findComponentById(ResourceTable.Id_text);
        timePicker1 = (TimePicker)findComponentById(ResourceTable.Id_timepicker1);
        if(timePicker1 != null) {
            // 指定 TimePicker 组件的监听器
            timePicker1.setTimeChangedListener(new TimePicker.TimeChangedListener() {
                @Override
                public void onTimeChanged(TimePicker timePicker, int hour, int minute, int
                second) {
                    // 将 timePicker1 当前选择的时间显示在 Text 组件中
                    text.setText("时: " + hour + "     分: " + minute + "     秒: " + second);
                }
            });
        }
        timePicker2 = (TimePicker)findComponentById(ResourceTable.Id_timepicker2);
        if(timePicker2 != null) {
            timePicker2.setTimeChangedListener(new TimePicker.TimeChangedListener() {
                @Override
                public void onTimeChanged(TimePicker timePicker, int i, int i1, int i2) {
                    int hour = timePicker.getHour();
                    int minute = timePicker.getMinute();
                    int second = timePicker.getSecond();
                    // 将 timePicker2 当前选择的时间显示在 Text 组件中
                    text.setText("Hour: " + hour + "  Minute: " + minute + "  Second: " +
                    second);
                }
            });
        }
        // 下面 3 行代码调整 timePicker2 当前显示的时、分、秒
        timePicker2.setHour(8);
        timePicker2.setMinute(30);
        timePicker2.setSecond(15);
        // 将 timePicker2 设置为 12 小时制
        timePicker2.set24Hour(false);

    }
}
```

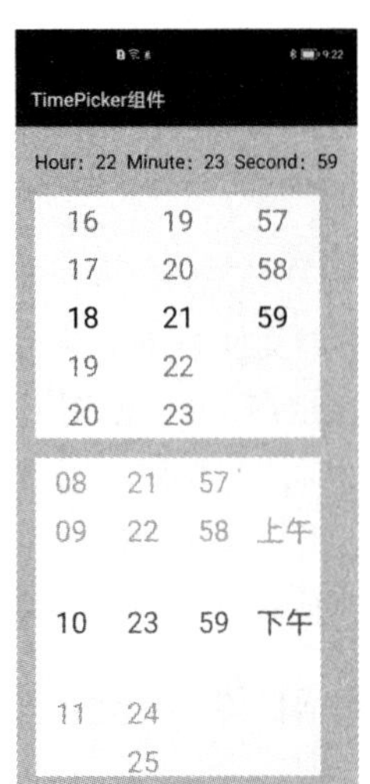

图 4-20 TimePicker 组件的显示效果

运行程序，滑动两个 TimePicker 组件中的时间，会看到类似图 4-20 的显示效果。

TimePicker 组件还允许隐藏部分列，例如，下面的代码可以隐藏用于显示时的列（第 1 列）：

```
timePicker2.showHour(false);
```

隐藏用于显示时的列的效果如图 4-21 所示。

图 4-21 隐藏用于显示时的列的效果

4.3.6 ScrollView 组件

ScrollView 组件（即滚动视图组件）通过水平滚动或垂直滚动的方式，允许开发人员在有限的区域显示更多的内容。通常会在 ScrollView 组件中放置一个容器（如各种布局组件），当容器的尺

寸超过 ScrollView 组件的尺寸时，ScrollView 组件就会通过水平滚动或垂直滚动的方式显示该容器的剩余部分。

ScrollView 组件垂直滚动的显示效果如图 4-22 所示。

图 4-22 ScrollView 组件垂直滚动的显示效果

除了可以通过手指（或鼠标）控制 ScrollView 组件的滚动，还可以通过下面几个方法用代码控制 ScrollView 组件的水平滚动和垂直滚动。

```
// 用相对位置在水平方向上和垂直方向上分别移动 dx 像素和 dy 像素
public void fluentScrollBy(int dx, int dy);
// 用相对位置在水平方向上移动 dx 像素
public void fluentScrollByX(int dx);
// 用相对位置在垂直方向上移动 dy 像素
public void fluentScrollByY(int dy) ;
// 用绝对位置在水平方向上和垂直方向上分别移动到 x 和 y
public void fluentScrollTo(int x, int y);
// 用绝对位置沿水平方向移动到 x
public void fluentScrollXTo(int x);
// 用绝对位置沿垂直方向移动到 y
public void fluentScrollYTo(int y);
```

这 6 个方法可以分成两组，包含 By 的方法用相对位置移动，也就是说，只要还有没显示完整的部分，调用这 3 个方法就会从当前位置继续向右或下移动（移动的尺度可能等于 dx 或 dy，也可能比 dx 或 dy 小）；包含 To 的方法用绝对位置移动，只将 ScrollView 组件沿水平方向或垂直方向移动到特定的位置。要注意，这 6 个方法中参数值的单位都是像素。

【例 4.16】 在窗口中放置两个 ScrollView 组件，分别以垂直滚动或水平滚动的方式显示多个图像，并且通过下方的两个按钮分别让这两个 ScrollView 组件平滑向右或向下滚动（单击一下按钮，就会滚动一次），直到无法滚动为止。

先看一下 XML 布局文件的实现。

代码位置： src/main/resources/base/layout/scrollview_demo.xml

```
<?xml version="1.0" encoding="utf-8"?>
<DirectionalLayout xmlns:ohos="http://schemas.huawei.com/res/ohos"
                ohos:width="match_parent"
                ohos:height="match_parent"
                ohos:background_element="#CCCCCC"
                ohos:orientation="vertical">
   <ScrollView
      ohos:id="$+id:scrollview1"
      ohos:height="300vp"
      ohos:width="match_parent"
      ohos:background_element="#FFDEAD"
      ohos:top_margin="20vp"
      ohos:bottom_padding="20vp"
      ohos:layout_alignment="horizontal_center">
      <DirectionalLayout
         ohos:id="$+id:directionlayout1"
         ohos:height="match_content"
         ohos:width="match_content">
```

```
        </DirectionalLayout>
    </ScrollView>
    <ScrollView
        ohos:id="$+id:scrollview2"
        ohos:height="match_content"
        ohos:width="300vp"
        ohos:background_element="#FFDEAD"
        ohos:top_margin="20vp"
        ohos:bottom_padding="20vp"
        ohos:layout_alignment="horizontal_center">
        <DirectionalLayout
            ohos:id="$+id:directionlayout2"
            ohos:height="match_content"
            ohos:width="match_content"
            ohos:orientation="horizontal">
        </DirectionalLayout>
    </ScrollView>
    <Button
        ohos:id="$+id:button1"
        ohos:background_element="$graphic:shape_text_bac"
        ohos:height="40vp"
        ohos:width="match_parent"
        ohos:text_color="#9b9b9b"
        ohos:text_size="17fp"
        ohos:text_alignment="center"
        ohos:text="根据像素数水平平滑滚动"
        ohos:right_margin="60vp"
        ohos:left_margin="60vp"
        ohos:top_margin="10vp"
        />
    <Button
        ohos:id="$+id:button2"
        ohos:background_element="$graphic:shape_text_bac"
        ohos:height="40vp"
        ohos:width="match_parent"
        ohos:text_color="#9b9b9b"
        ohos:text_size="17fp"
        ohos:text_alignment="center"
        ohos:text="根据像素数垂直平滑滚动"
        ohos:right_margin="60vp"
        ohos:left_margin="60vp"
        ohos:top_margin="10vp"
        />
</DirectionalLayout>
```

在 scrollview_demo.xml 布局文件中放置了两个 ScrollView 组件，这两个 ScrollView 组件中各有一个 DirectionalLayout 组件，但该组件中并没有任何子组件。后面会使用 Java 代码分别动态向这两个 DirectionalLayout 组件中添加 5 个 Image 组件。因为第 1 个 DirectionalLayout 组件是垂直方向的，第 2 个 DirectionalLayout 组件是水平方向的，所以就会导致第 1 个 ScrollView 组件会沿垂直方向滚动，第 2 个 ScrollView 组件沿水平方向滚动。

下面看一下 Java 实现代码。

代码位置：src/main/java/com/unitymarvel/demo/components/ScrollViewDemo.java

```
package com.unitymarvel.demo.components;

import com.unitymarvel.demo.ResourceTable;
import ohos.aafwk.ability.Ability;
import ohos.aafwk.content.Intent;
import ohos.agp.components.*;

public class ScrollViewDemo extends Ability {
   private DirectionalLayout directionalLayout1,directionalLayout2;
   // 指定 5 个图像文件的 ID
   private int[] imageIds = new int[]{ResourceTable.Media_p1,
                                ResourceTable.Media_p2,
                                ResourceTable.Media_p3,
                                ResourceTable.Media_p4,
                                ResourceTable.Media_p5};
   @Override
   public void onStart(Intent intent) {
      super.onStart(intent);
      super.setUIContent(com.unitymarvel.demo.ResourceTable.Layout_scrollview_demo);
      directionalLayout1 =
      (DirectionalLayout)findComponentById(ResourceTable.Id_directionlayout1);
      if(directionalLayout1 != null) {
         // 动态向第 1 个 DirectionalLayout 组件中添加 5 个 Image 组件
         for(int i = 0;i < imageIds.length;i++) {
             // 动态装载 scrollview_item.xml 布局文件
             DirectionalLayout item = (DirectionalLayout)LayoutScatter.getInstance
             (this).parse(ResourceTable.Layout_scrollview_item,null,false);
             if(item != null) {
                Image image = (Image)item.findComponentById(ResourceTable.Id_image);
                if(image != null) {
                   // 动态设置当前 Image 组件显示的图像
                   image.setPixelMap(imageIds[i]);
                }
                // 将当前装载的 DirectionalLayout 组件添加到 directionalLayout1 中
                directionalLayout1.addComponent(item);
             }
         }
      }
      directionalLayout2 =
      (DirectionalLayout)findComponentById(ResourceTable.Id_directionlayout2);
      if(directionalLayout2 != null) {
         // 动态向第 2 个 DirectionalLayout 组件中添加 5 个 Image 组件
         for(int i = 0;i < imageIds.length;i++) {
             // 动态装载 scrollview_item.xml 布局文件
             DirectionalLayout item = (DirectionalLayout)LayoutScatter.getInstance
             (this).parse(ResourceTable.Layout_scrollview_item,null,false);
             if(item != null) {
                Image image = (Image)item.findComponentById(ResourceTable.Id_image);
                if(image != null) {
                   // 动态设置当前 Image 组件显示的图像
```

```
                        image.setPixelMap(imageIds[i]);
                    }
                    // 将当前装载的 DirectionalLayout 组件添加到 directionalLayout2 中
                    directionalLayout2.addComponent(item);
                }
            }
        }
        ScrollView scrollView1 =
                (ScrollView)findComponentById(ResourceTable.Id_scrollview1);
        Button button1 = (Button)findComponentById(ResourceTable.Id_button1);
        if(button1 != null) {
            button1.setClickedListener(new Component.ClickedListener() {
                @Override
                public void onClick(Component component) {
                    // 让 scrollView1 每次在垂直方向上向下滚动 300 像素
                    scrollView1.fluentScrollByY(300);
                }
            });
        }
        scrollView1.setReboundEffect(true);

        ScrollView scrollView2 =
                (ScrollView)findComponentById(ResourceTable.Id_scrollview2);
        Button button2 = (Button)findComponentById(ResourceTable.Id_button2);
        if(button2 != null) {
            button2.setClickedListener(new Component.ClickedListener() {
                @Override
                public void onClick(Component component) {
                    // 让 scrollView1 每次在水平方向上向右滚动 500 像素
                    scrollView2.fluentScrollByX(500);
                }
            });
        }
    }
}
```

在运行程序之前，需要在 media 目录下放入 5 个图像文件，文件名分别为 p1.png、p2.png、p3.png、p4.png、p5.png。

运行程序后，可以用手指滑动两个 ScrollView 组件，也可以单击窗口下方的两个按钮来滑动这两个 ScrollView 组件，显示效果如图 4-23 所示。

图 4-23　ScrollView 组件水平滚动和垂直滚动的显示效果

4.4 总结与回顾

前面几章只讲解了一些基本组件的使用方式，如按钮组件、文本框组件等。本章逐步介绍了

HarmonyOS 的核心组件，使用这些组件，读者可以开发出任何复杂的 App 界面。如果说布局使用的基本组件“组织了一个 App 界面的五官的位置”，那么这些核心组件则相当于为了美化 App 界面使用的“化妆品和各种首饰”，让 App 界面变得丰富多彩。

本章首先介绍了 HarmonyOS 中最重要、最常用的两类组件——展示组件和交互组件。展示组件主要用于展示信息，如 Text 组件用于展示文本，Image 组件用于展示图像，ProgressBar 组件用于展示进度等。交互组件除了展示信息外，还有其他功能——与用户交互。例如，Button 组件用于响应手指的触摸事件，在事件中可以响应动作，并给用户反馈；TextField 组件用于输入文本信息，用户输入一个字符，在窗口中就会立刻显示这个字符，这也是给用户的反馈。

本章最后将一些复杂组件归类为高级组件，这些组件主要包括一些容器组件以及一些有完整功能的组件。容器组件主要包括 ListContainer 组件、TabList 组件和 ScrollView 组件。如果要设置日期，可以使用 DatePicker 组件；如果要设置时间，可以使用 TimePicker 组件。

尽管这些组件通常不会出现在同一个窗口或 App 界面中，但为了开发出更多样的 App，需要掌握这些组件的基本用法。下一章将深入讲解如何使用 API 显示和控制对话框。

第 5 章　对话框

本章将介绍如何实现在 HarmonyOS 中弹出对话框。对话框本质上就是窗口。HarmonyOS 为我们提供了现成的对话框类，只需要简单调用，就可以实现各种样式的对话框。

本章的源代码位置是 src/main/java/com/unitymarvel/demo/dialog/MyDialog.java。

通过阅读本章，读者可以掌握：

- 如何显示对话框；
- 如何为对话框添加一个或多个按钮；
- 如何自动关闭对话框；
- 如何定制对话框；
- 如何显示 Toast 信息框。

5.1 普通对话框

使用 CommonDialog 类可以实现一个对话框，该类有很多 API，我们可以定制不同风格的对话框，本节将详细介绍 CommonDialog 类的用法。

5.1.1 显示一个最简单的对话框

对话框包含的最基本的两个信息是标题（title）和内容（content），通过 CommonDialog 类的 setTitleText 方法和 setContentText 方法可以设置对话框的标题和内容。通过 show 方法可以显示对话框，实现代码如下：

```
CommonDialog commonDialog1 = new CommonDialog(this);
commonDialog1.setTitleText("信息");
commonDialog1.setContentText("任务已经完成！");
commonDialog1.show();
```

运行程序，单击“最简单的对话框”按钮，就会弹出图 5-1 所示的对话框。

图 5-1　最简单的对话框

对话框是模态的，也就是说，当对话框显示时，后面的所有组件都不可操作。这样的对话框并不符合我们印象中的对话框。因为没有“关闭”“确定”这样的按钮，所以关闭对话框的唯一方法是按手机的回退键（或任何等同于回退键的手势操作）。在下一节我们将学习如何为对话框添加“关闭”按钮。

5.1.2 为对话框添加“关闭”按钮

通过 CommonDialog 类的 setButton 方法可以为对话框添加一个或多个按钮，该方法的原型如下：

```
public CommonDialog setButton(int buttonNum, String text, ClickedListener listener);
```

下面是 setButton 方法的参数描述。

- buttonNum：按钮的索引。如果多个按钮共用同一个单击事件，该参数可以用来区分单击的到底是哪一个按钮。
- text：在按钮上显示的文本。
- listener：单击按钮响应的事件监听器。

为对话框添加的按钮，默认不会响应任何动作，所以需要用单击事件监听器来响应单击按钮的动作。例如，如果要关闭按钮，可以调用 CommonDialog 类的 destroy 方法。实现代码如下：

```
CommonDialog commonDialog1 = new CommonDialog(this);
commonDialog1.setTitleText("信息");
commonDialog1.setContentText("带关闭按钮的对话框！");
commonDialog1.setButton(1, "关闭", new IDialog.ClickedListener() {
    @Override
    public void onClick(IDialog iDialog, int i) {
        // 关闭对话框
        commonDialog1.destroy();
    }
});
commonDialog1.show();
```

运行程序，单击“带关闭按钮的对话框”按钮，会弹出图 5-2 所示的对话框，单击“关闭”按钮后，就会关闭对话框。

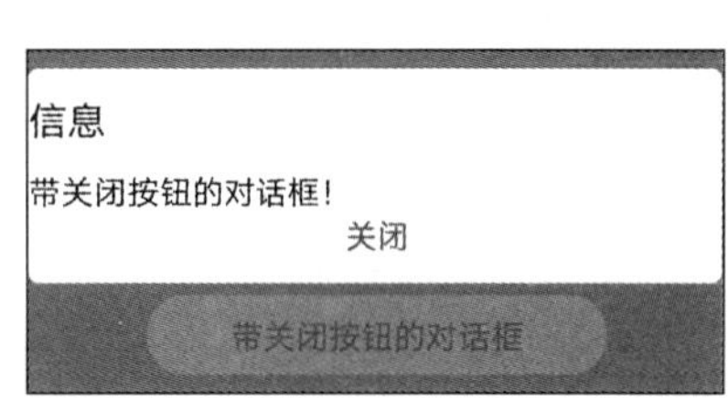

图 5-2 带关闭按钮的对话框

5.1.3 为对话框添加多个按钮

使用 setButton 方法理论上可以为对话框添加多个按钮，但目前最多只能为对话框添加两个按钮，超过两个按钮，多余的按钮将被忽略。下面的代码为对话框添加了“确定”按钮和“取消”按钮。

```
CommonDialog commonDialog1 = new CommonDialog(this);
commonDialog1.setTitleText("信息");
commonDialog1.setContentText("带 2 个按钮的对话框！");
// “确定”按钮和“取消”按钮共用一个单击事件监听器
IDialog.ClickedListener clickedListener = new IDialog.ClickedListener() {
    @Override
    public void onClick(IDialog iDialog, int i) {
        commonDialog1.destroy();
        // 这里的 i 就是 setButton 方法的第 1 个参数值
        switch (i) {
            case 1:   // “确定”按钮响应的动作
                button3.setText("单击了确定按钮");
```

```
                break;
            case 2:  // "取消"按钮响应的动作
                button3.setText("单击了取消按钮");
                break;
        }
    }
};
commonDialog1.setButton(1, "确定", clickedListener);
commonDialog1.setButton(2, "取消", clickedListener);
commonDialog1.show();
```

运行程序，单击“带 2 个按钮的对话框”按钮，会弹出图 5-3 所示的对话框。

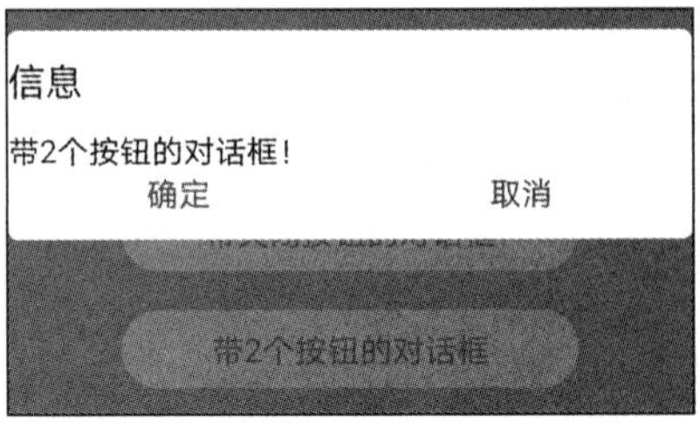

图 5-3 带 2 个按钮的对话框

在本例中，“确定”按钮和“取消”按钮共用了一个单击事件监听器，并且单击这两个按钮后，会为 button3 设置不同的文本。

注意，如果为对话框指定两个按钮，setButton 方法的第 1 个参数值必须是连续的，例如，不能使用下面的代码为对话框指定两个按钮。

```
commonDialog1.setButton(1, "确定", clickedListener);
commonDialog1.setButton(3, "取消", clickedListener);
```

在这段代码中，按钮索引分别为 1 和 3，所以第 2 个按钮会被忽略。

5.1.4 调整按钮的尺寸

前面弹出的按钮在水平方向上是充满整个屏幕的，使用 setSize 方法可以重新设置对话框的尺寸（宽度和高度）。该方法的原型如下：

```
public BaseDialog setSize(int width, int height);
```

这里的 width 和 height 指对话框的宽度和高度，不过还有一个问题，width 和 height 的单位是什么呢？

由于移动设备的屏幕千差万别，因此在设计 App 时通常不会采用物理像素，而是采用虚拟像素，例如，HarmonyOS 的虚拟像素单位是 vp。不管设备实际的分辨率是多少，使用虚拟像素的设备的分辨率都是一样的。当 App 运行在实际的设备上时，系统会自动根据屏幕密度将虚拟像素转换为物理像素。如果屏幕密度是 160dpi①，那么 1vp = 1px，这里 px 表示物理像素。所以如果 App 在拥有 160dpi 屏幕密度的设备上运行，那么 100vp 就相当于 100px 了。如果屏幕密度是 320dpi，那么 100vp 就会转换为 200px。这样可以保证组件的相对位置和相对尺寸不变。

现在回答前面提出的问题。setSize 方法的两个参数的单位是 px，但我们不知道 App 到底会运行在怎样的设备上，所以我们期望使用 vp 来指定宽度和高度。因此，就需要通过下面的代码获取当前设备的屏幕密度，然后与 160 相除，算出屏幕密度与标准屏幕密度的比值，再与 vp 相乘，将

① 160dpi 就是每英寸包含 160 个物理像素。

vp 转换为 px。

```
getResourceManager().getDeviceCapability().screenDensity;
```

完整的实现代码如下：

```
CommonDialog commonDialog1 = new CommonDialog(this);
commonDialog1.setTitleText("信息");
commonDialog1.setContentText("调整对话框的尺寸！");
// 计算屏幕密度与标准屏幕密度的比值
int density = getResourceManager().getDeviceCapability().screenDensity /
                                  DeviceCapability.SCREEN_MDPI;  // 该值是 160
// 将 300vp 转换为实际的像素
int width = 300 * density;
// 将 100vp 转换为实际的像素
int height = 100 * density;
// 重新调整对话框的尺寸
commonDialog1.setSize(width, height);
commonDialog1.setButton(1, "关闭", new IDialog.ClickedListener() {
   @Override
    public void onClick(IDialog iDialog, int i) {
       commonDialog1.destroy();
   }
});
commonDialog1.show();
```

运行程序，单击“调整按钮的尺寸”按钮，会显示图 5-4 所示的对话框。

图 5-4 调整按钮的尺寸

5.1.5 自动关闭对话框

除了单击按钮关闭对话框，CommonDialog 类还允许单击非对话框区域自动关闭对话框，这里需要使用 setAutoClosable 方法允许自动关闭对话框。完整实现代码如下：

```
CommonDialog commonDialog1 = new CommonDialog(this);
commonDialog1.setTitleText("信息");
commonDialog1.setContentText("自动关闭对话框！");
// 设置对话框自动关闭
commonDialog1.setAutoClosable(true);
commonDialog1.setButton(1, "关闭", null);
commonDialog1.show();
```

运行程序，单击“自动关闭对话框”按钮，会显示图 5-5 所示的对话框。因为没有为“关闭”按钮添加监听事件，所以单击“关闭”按钮没有任何反应，但如果单击对话框以外的区域，对话框就会自动关闭。

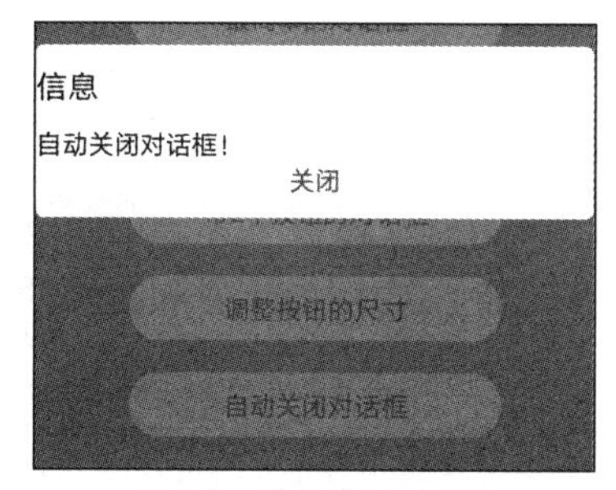

图 5-5 自动关闭对话框

5.2 定制对话框

原生对话框的样式单一，要想设计出更完美的对话框，需要使用 setContentCustomComponent

方法定制对话框，该方法可以指定一个视图（通常直接使用布局文件装载），可以用该视图完全替换标题、内容和按钮，也可以只替换标题和内容。下面的代码设置了标题和内容的颜色，并为对话框添加了图标。

```
Component directionalLayout =
LayoutScatter.getInstance(getContext()).parse(ResourceTable.Layout_custom_dialog,
null, false);
Text text = (Text)directionalLayout.findComponentById(ResourceTable.Id_content);
// 设置对话框的内容
text.setText("定制带图像的对话框！");
CommonDialog commonDialog = new CommonDialog(this);
commonDialog.setAlignment(LayoutAlignment.CENTER);
int density = getResourceManager().getDeviceCapability().screenDensity /
                            DeviceCapability.SCREEN_MDPI;
int width = 300 * density;
int height = 125 * density;
// 设置对话框的尺寸
commonDialog.setSize(width,height);
commonDialog.setAutoClosable(true);
commonDialog.setButton(1, "关闭", new IDialog.ClickedListener() {
   @Override
   public void onClick(IDialog iDialog, int i) {
      commonDialog.destroy();
   }
});
// 定制对话框
commonDialog.setContentCustomComponent(directionalLayout);
commonDialog.show();
```

这段代码使用了名为custom_dialog.xml的布局文件，该布局文件的代码如下：

```
<?xml version="1.0" encoding="utf-8"?>
<DirectionalLayout
       xmlns:ohos="http://schemas.huawei.com/res/ohos"
       ohos:width="match_parent"
       ohos:orientation="vertical"
       ohos:padding="5vp"
       ohos:height="match_content">
   <Text
       ohos:height="match_content"
       ohos:width="match_parent"
       ohos:text_color="#FF0000"
       ohos:text_size="25fp"
       ohos:text="这是标题"
       ohos:text_alignment="center"/>
   <DirectionalLayout
       ohos:orientation="horizontal"
       ohos:alignment="center"
       ohos:width="match_parent"
       ohos:height="match_content">
   <Image
       ohos:scale_mode="zoom_center"
       ohos:height="50vp"
       ohos:width="50vp"
```

```
        ohos:layout_alignment="center"
        ohos:left_margin="15vp"
        ohos:image_src="$media:components"/>
    <Text
        ohos:id="$+id:content"
        ohos:height="match_content"
        ohos:width="match_parent"
             ohos:text_color="#0000FF"
        ohos:text_size="18fp"
        ohos:text_alignment="vertical_center"/>
    </DirectionalLayout>
</DirectionalLayout>
```

运行程序，单击“定制对话框”按钮，会显示图 5-6 所示的对话框。

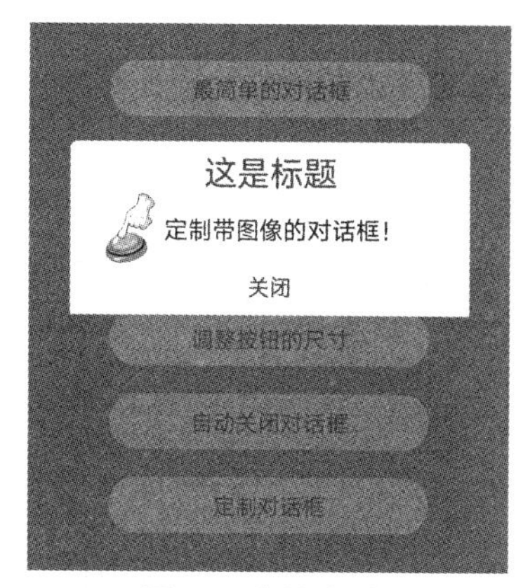

图 5-6 定制对话框

5.3 Toast 信息框

HarmonyOS 还提供了一种 Toast 信息框。Toast 信息框与前面讲的对话框的主要区别是前者不会获得焦点，而且是非模态显示。也就是说，在 Toast 信息框显示时，仍然可以操作后面的组件，而且 Toast 信息框在过一段时间（可由用户指定关闭时间）后会自动关闭。而对话框是模态显示的，一旦显示了对话框，除非关闭对话框，否则无法操作对话框后面的组件。Toast 信息框一般用于向用户反馈一些信息，而又不影响用户操作其他组件的场景。

ToastDialog 类用于显示 Toast 信息框，通过 setComponent 方法设置内容视图，通过 show 方法显示 Toast 信息框。实现代码如下：

```
Text text = new Text(this);
text.setWidth(MATCH_CONTENT);
text.setHeight(MATCH_CONTENT);
text.setTextSize(50);
text.setText("这是一个提示信息框");
text.setPadding(30, 20, 30, 20);
text.setMultipleLine(true);
text.setTextColor(Color.WHITE);
text.setTextAlignment(TextAlignment.CENTER);

ShapeElement style = new ShapeElement();
style.setShape(ShapeElement.RECTANGLE);
style.setRgbColor(new RgbColor(0x666666FF));
style.setCornerRadius(15);
text.setBackground(style);
DirectionalLayout mainLayout = new DirectionalLayout(this);
mainLayout.setWidth(MATCH_PARENT);

mainLayout.setHeight(MATCH_CONTENT);
mainLayout.setAlignment(LayoutAlignment.CENTER);
mainLayout.addComponent(text);
// 创建 ToastDialog 对象
ToastDialog toastDialog = new ToastDialog(this);
```

```
    toastDialog.setSize(MATCH_PARENT, MATCH_CONTENT);
    // 设置 Toast 信息框关闭时间（在 3 秒后自动关闭）
    toastDialog.setDuration(3000);
    // 设置 Toast 信息框为透明，否则会出现白色背景
    toastDialog.setTransparent(true);
    toastDialog.setAlignment(LayoutAlignment.CENTER);
    // 为 Toast 信息框设置内容视图
    toastDialog.setComponent(mainLayout);
    toastDialog.show();
```

运行程序，单击“提示信息框”按钮，会显示图 5-7 所示的 Toast 信息框，在 3 秒后，Toast 信息框会自动关闭。

图 5-7　Toast 信息框

5.4　总结与回顾

有了对话框，App 的交互性就更强了。如果只在窗口上给用户一些提示信息，用户有可能看不到。更糟糕的是，App 也不知道用户是否看到提示。如果这时用户正进行一些重要的操作，例如，要删除一些敏感信息，在删除之前，App 应该提示一下用户，是否需要真的删除这些信息？若用户没看到这条提示，而 App 并不知道用户是否看到提示，导致用户直接删除了信息，用户操作后发现删错了，若想恢复就困难了！

所以对于这些危险的操作，最好有可以阻止用户操作的提示，也就是本章介绍的对话框。一旦弹出对话框，除非用户关闭对话框，否则用户不能进行任何操作。

到现在为止，已经介绍了很多 HarmonyOS API，但我们会发现一个问题，如果 App 关闭，甚至卸载，再次启动或安装，以前的数据还会保留吗？答案是数据当然会保留。但这要用到一种技术，就是下一章要讲的数据管理。

第6章　数据管理

数据是 App 非常重要的组成部分，保存 App 的配置、读写用户数据、检索英文单词，都离不开数据的支撑。HarmonyOS 除了拥有操作本地数据的强大 API，还可以操作分布式数据和分布式文件，这是 HarmonyOS 独有的特性。本章将详细介绍 HarmonyOS 操作本地数据、分布式数据和分布式文件的 API 的使用方法。

通过阅读本章，读者可以掌握：

- 如何读写配置文件；
- 如何使用 SQL 操作 SQLite 数据库；
- 如何使用谓词操作 SQLite 数据库；
- 如何使用对象关系映射；
- 如何分布式文件；
- 如何分布式数据。

6.1 读写配置文件

大多数 App 都会或多或少使用配置文件来保存 App 的配置信息，如默认用户名、用户操作的最后状态等。配置文件通常使用文本格式，例如，JSON 和 XML 是最典型的文本格式。读写配置文件的方式很多，本节将详细介绍 HarmonyOS 提供的读写配置文件的 API，其中核心的类是 Preferences 类。

6.1.1 Preferences 类的基本用法

Preferences 类用于读写纯文本格式的配置文件。在纯文本格式的配置文件中，数据是以 key-value 对形式组织的，所以可以像操作 Map 一样操作配置文件中的数据。

如果要创建 Preferences 对象，需要先创建 DatabaseHelper 对象，然后通过 getPreferences 方法创建 Preferences 对象，代码如下：

```
DatabaseHelper databaseHelper = new DatabaseHelper(context);
Preferences preferences = databaseHelper.getPreferences("data.txt");
```

其中，data.txt 是配置文件名，这里不能指定路径，因为系统会将 data.txt 文件放在默认的存储路径下。使用 context.getPreferencesDir 方法可以获取 data.txt 文件的存储路径。

Preferences 类提供了一系列 putXxx 方法和 getXxx，用于写入和读取 key-value 对，其中，Xxx 表示 String、Int、Float 等。

【例 6.1】 使用 Preferences 对象将若干个 key-value 对保存在 data.cfg 文件中，并读取该文件中的 key-value 对，然后将其输出到 HiLog 视图中。

代码位置：src/main/java/com/unitymarvel/demo/data/PreferencesDemo.java

```
public class PreferencesDemo {
    // 在 data.cfg 文件中写入数据
    public static boolean writeData(Context context) {
        DatabaseHelper databaseHelper = new DatabaseHelper(context);
        // context 入参类型为 ohos.app.Context
        String fileName = "data.cfg";
        // 获取 Preferences 对象
        Preferences preferences = databaseHelper.getPreferences(fileName);
        // 开始写入 key-value 对
        preferences.putString("name", "李宁");
        preferences.putInt("age",18);
        preferences.putFloat("salary", 12345);
        Set<String> hobbies = new HashSet<>();
        hobbies.add("阅读");
        hobbies.add("旅游");
        preferences.putStringSet("hobby", hobbies);
        // preferences.flush(); // 异步保存 data.cfg 文件
        return preferences.flushSync();  // 同步保存 data.cfg 文件
    }
    // 从 data.cfg 文件读取数据
    public static void readData(Context context) {
        DatabaseHelper databaseHelper = new DatabaseHelper(context);
        // context 入参类型为 ohos.app.Context
        String fileName = "data.cfg";
        Preferences preferences = databaseHelper.getPreferences(fileName);
        // 开始读取数据，并将读取的数据输出到 HiLog 视图中
        Tools.print("readData:name:" + preferences.getString("name","未知"));
        Tools.print("readData:hobby:" + preferences.getStringSet("hobby",null));
        Tools.print("readData:all:" + preferences.getAll());
        // 输出配置文件的存储路径
        Tools.print("readData:path:" + context.getPreferencesDir().toString());
    }

}
```

最后，在按钮的单击事件中依次调用 writeData 方法和 readData 方法。

代码位置：src/main/java/com/unitymarvel/demo/data/DataManager.java

```
button1.setClickedListener(new Component.ClickedListener() {
    @Override
    public void onClick(Component component) {
        // 向配置文件写入数据
        PreferencesDemo.writeData(DataManager.this);
```

```
        // 从配置文件读取数据
        PreferencesDemo.readData(DataManager.this);
    }
});
```

运行程序，单击“读写配置文件”按钮，会在 HiLog 视图中输出图 6-1 所示的信息。

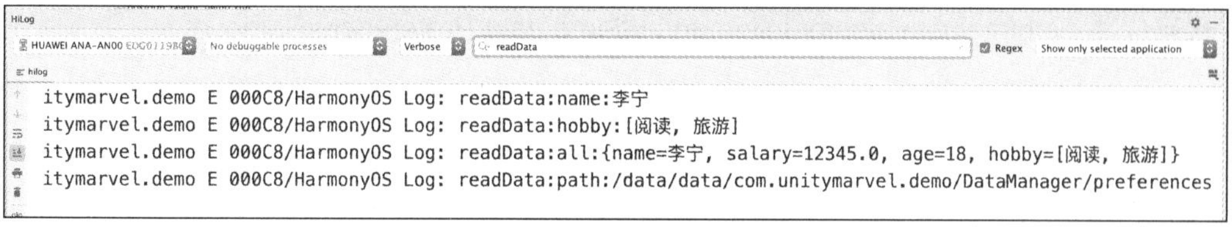

```
itymarvel.demo E 000C8/HarmonyOS Log: readData:name:李宁
itymarvel.demo E 000C8/HarmonyOS Log: readData:hobby:[阅读, 旅游]
itymarvel.demo E 000C8/HarmonyOS Log: readData:all:{name=李宁, salary=12345.0, age=18, hobby=[阅读, 旅游]}
itymarvel.demo E 000C8/HarmonyOS Log: readData:path:/data/data/com.unitymarvel.demo/DataManager/preferences
```

图 6-1 HiLog 视图中的输出

阅读前面的代码应注意如下几点。

- Preferences 类除了可以将简单类型（int、String、Float 等）的值写入配置文件，还可以使用 putStringSet 方法写入集合类型的值。
- 在调用 putInt、putFloat 等方法写入数据时，并没有将数据真正写到配置文件中，只是在内存中进行操作，所以还需要调用 flush 方法或 flushSync 方法完成写入文件的操作。其中，flush 是异步方法，flushSync 是同步方法，如果写入成功，则返回 true。
- 每一个 getXxx（getInt、getFloat 等）方法都需要被指定一个默认值，当指定的 key 不存在时返回这个默认值。
- Preferences 对象在操作配置文件时，首先会将整个配置文件都读取到内存，然后再读写配置文件中的内容，所以并不用担心 Preferences 对象会覆盖原来的配置文件。

为了了解配置文件的具体存储格式，读者可以进入配置文件的存储路径，并找到这个配置文件。最后执行 cat data.cfg 命令，会看到图 6-2 所示的内容。很明显，Preferences 对象读写的配置文件是 XML 格式的。

```
HWANA:/data/data/com.unitymarvel.demo/DataManager/preferences # cat data.cfg
<?xml version="1.0" encoding="UTF-8"?>
<preferences version="1.0">
  <string key="name">李宁</string>
  <float key="salary" value="12345.0"/>
  <int key="age" value="18"/>
  <set key="hobby">
    <string>阅读</string>
    <string>旅游</string>
  </set>
</preferences>
```

图 6-2 配置文件的内容

6.1.2 监控配置文件的写入动作

Preferences 对象允许对配置文件的写入动作进行监控，监控的工具就是观察者（Observer）对象。观察者对象对应的类必须实现 Preferences.PreferencesObserver 接口，该接口中有一个 onChange 方法，用于监听写入 key 的动作。当调用 flush 方法或 flushSync 方法成功将 key-value 对写入文件时后，就会调用 onChange 方法，该方法的原型如下：

```
public void onChange(Preferences preferences, String key);
```

其中，preferences 是 Preferences 对象，key 是已经写入的 key-value 对中的 key。

【例 6.2】 向 data.cfg 文件多次写入 key 为 salary 的 key-value 对，并通过观察者对象计算所有写入的 salary 的平均值。

代码位置：src/main/java/com/unitymarvel/demo/data/PreferencesDemo.java

```
public class PreferencesDemo {
   // 观察者对象
   private static class SalaryAverage implements Preferences.PreferencesObserver {
      // 写入 salary 的计数器
      private int count = 0;
      // 多次写入 salary 的总和
      private float salarySum = 0;
      @Override
      public void onChange(Preferences preferences, String key) {
         if ("salary".equals(key)) {
            // 写入 salary 的次数加 1
            count++;
            // salary 累加
            salarySum += preferences.getFloat(key,0);
            // 输出 salary 的平均值
            Tools.print("salary average:" + salarySum / count);
         }
      }
   }
   // 注册和注销观察者对象
   public static void salaryObserver(Context context) {
      DatabaseHelper databaseHelper = new DatabaseHelper(context);
      String fileName = "data.cfg";
      Preferences preferences = databaseHelper.getPreferences(fileName);
      // 创建观察者对象
      SalaryAverage salaryAverage = new SalaryAverage();
      // 注册观察者对象
      preferences.registerObserver(salaryAverage);
      preferences.putFloat("salary", 12345);
      preferences.flushSync();        // 调用了一次 onChange 方法
      preferences.putFloat("salary", 54321);
      preferences.flushSync();        // 调用了一次 onChange 方法
      preferences.putFloat("salary", 2345);
      preferences.flushSync();        // 调用了一次 onChange 方法
      // 注销观察者对象
      preferences.unregisterObserver(salaryAverage);
   }
}
```

最后，在按钮的单击事件中调用 salaryObserver 方法。

代码位置：src/main/java/com/unitymarvel/demo/data/DataManager.java

```
button2.setClickedListener(new Component.ClickedListener() {
   @Override
   public void onClick(Component component) {
```

```
            PreferencesDemo.salaryObserver(DataManager.this);
        }
    });
```

运行程序，单击“Preferences 观察者”按钮，会在 HiLog 视图中输出图 6-3 所示的内容。我们可以看到，本例输出了 3 次 salary average，每次都是由于调用 flushSync 方法后触发了观察者对象中的 onChange 方法才输出这些信息的。

```
HiLog
HUAWEI ANA-AN00  No debuggable processes  Verbose  salary average:  Regex  Show only selected application
hilog
58.360 16744-16744/com.unitymarvel.demo E 000C8/HarmonyOS Log: salary average:12345.0
58.371 16744-16744/com.unitymarvel.demo E 000C8/HarmonyOS Log: salary average:33333.0
58.375 16744-16744/com.unitymarvel.demo E 000C8/HarmonyOS Log: salary average:23003.666
```

图 6-3 HiLog 视图中的输出

6.1.3 移动和删除配置文件

Preferences 类提供了 movePreferences 方法和 deletePreferences 方法，分别用来移动和删除配置文件。其实根据配置文件的路径直接找到配置文件的方式也能完成同样的工作，只不过使用这两个方法操作更方便。movePreferences 方法的原型如下：

```
public boolean movePreferences(Context sourceContext, String sourceName, String targetName);
```

其中，sourceName 和 targetName 分别表示原来的文件名和移动后的文件名。这个功能相当于重命名配置文件。该方法如果成功移动配置文件，返回 true；否则，返回 false。

deletePreferences 方法的原型如下：

```
public boolean deletePreferences(String name) ;
```

name 参数表示要删除的文件名。如果成功删除了文件，返回 true；否则，返回 false。

【例 6.3】 将 data1.cfg 文件重命名为 data2.cfg，然后删除 data2.cfg 文件。

代码位置：src/main/java/com/unitymarvel/demo/data/PreferencesDemo.java

```
public class PreferencesDemo {
    // 移动配置文件（重命名配置文件）
    public static boolean move(Context context) {
        DatabaseHelper databaseHelper = new DatabaseHelper(context);
        String srcFile = "data1.cfg";
        Preferences preferences = databaseHelper.getPreferences(srcFile);
        preferences.putString("name", "Bill");
        preferences.flushSync();
        String targetFile = "data2.cfg";
        // 移动配置文件（重命名配置文件）
        return databaseHelper.movePreferences(context,srcFile,targetFile);
    }
    // 删除配置文件
    public static boolean delete(Context context) {
        DatabaseHelper databaseHelper = new DatabaseHelper(context);
```

```
        String filename = "data2.cfg";
        Preferences preferences = databaseHelper.getPreferences(filename);
        // 删除配置文件
        return databaseHelper.deletePreferences(filename);
    }
}
```

运行程序，单击“移动配置文件”按钮，会看到配置文件所在的目录下出现了一个 data2.cfg 文件，然后再单击“删除配置文件”按钮，会看到 data2.cfg 文件被删除了。

因为 Preferences 对象会将配置文件中的所有数据都读取到内存中进行操作，并且以文件名为 key，所以可以通过 removePreferencesFromCache 方法将读取到内存中的数据清除，这样该 Preferences 对象就不能再使用了。实现代码如下：

```
DatabaseHelper databaseHelper = new DatabaseHelper(context);
String fileName = "data.cfg";
// 将配置文件的数据从内存中清除
databaseHelper.removePreferencesFromCache(fileName);
```

6.2 操作 SQLite 数据库

SQLite 数据库是移动端 App 最常用的本地数据库。HarmonyOS 提供了丰富的 API 用来操作 SQLite 数据库，本节将详细讲解如何通过 SQL 和相应的谓词方法操作本地的 SQLite 数据库。

6.2.1 使用 SQL 操作 SQLite 数据库

SQL 是操作数据库最简单的方式，我们需要通过一系列 API 来执行 SQL 语句。使用 HarmonyOS 提供的 API 操作 SQLite 数据库的步骤如下。

（1）指定数据库文件的路径（通过 StoreConfig 对象指定）。

（2）定义用于创建和升级数据表的回调对象。

（3）创建 DatabaseHelper 对象。

（4）通过 DatabaseHelper 对象的 getRdbStore 方法获取 RdbStore 对象，该对象提供了若干方法用于操作 SQLite 数据库。

（5）开始使用 executeSql 方法、querySql 方法或其他方法来操作 SQLite 数据库，其中 executeSql 方法用于执行非 Select 的 SQL 语句，querySql 方法用来执行 Select 语句，并返回结果集（ResultSet）对象。

【例 6.4】 按照前面的步骤创建一个名为 data.sqlite 的数据库，该数据库包含一个名为 t_users 的表，依次对该表插入数据以及查询数据，最后将查询结果输出到 HiLog 视图中。

代码位置：src/main/java/com/unitymarvel/demo/data/SQLiteDemo.java

```
public class SQLiteDemo {
    public static void sqliteWithSQL(Context context) {
```

```
        RdbStore store;
        // 第 1 步：指定数据库文件的路径，data.sqlite 为数据库文件名，不需要指定路径
        StoreConfig config = StoreConfig.newDefaultConfig("data.sqlite");
        // 第 2 步：定义用于创建和升级数据表的回调对象
        RdbOpenCallback callback = new RdbOpenCallback() {
            // 创建表时调用
            @Override
            public void onCreate(RdbStore store) {
                // 创建 t_users 表
                store.executeSql("CREATE TABLE IF NOT EXISTS t_users (id INTEGER PRIMARY
                KEY, name VARCHAR(20), msg TEXT)");
                Tools.print("成功创建 t_users 表");
            }
            // 升级表时调用
            @Override
            public void onUpgrade(RdbStore store, int oldVersion, int newVersion) {
            }
        };
        // 第 3 步：创建 DatabaseHelper 对象
        DatabaseHelper helper = new DatabaseHelper(context);
        // 第 4 步：获取 RdbStore 对象
        store = helper.getRdbStore(config,
                1,
                callback,
                null);
        String deleteSQL = "delete from t_users;";
        // 第 5 步：执行 SQL 语句

        // 删除 t_users 表中的数据
        store.executeSql(deleteSQL,null);

        String insertSQL = "insert into t_users(id, name, msg) values(?,?,?);";
        // 向 t_users 表中插入 3 条记录
        store.executeSql(insertSQL, new Object[]{1, "Bill", "msg1"});
        store.executeSql(insertSQL, new Object[]{2, "Mike", "msg2"});
        store.executeSql(insertSQL, new Object[]{3, "John", "msg3"});

        String selectSQL = "select * from t_users where name=?";
        // 查询 t_users 表中 name 字段值为 Mike 的记录
        ResultSet resultSet = store.querySql(selectSQL,
                                  new String[]{String.valueOf("Mike")});
        // 将查询结果集的当前指针指向第 1 条记录
        if(resultSet.goToFirstRow()) {
            // 输出查询结果
            Tools.print("sqliteWithSQL:id:" + resultSet.getInt(0));
            Tools.print("sqliteWithSQL:name:" + resultSet.getString(1));
            Tools.print("sqliteWithSQL:msg:" + resultSet.getString(2));
        } else {
            Tools.print("sqliteWithSQL:未搜索到结果" );
        }
        store.close();    // 关闭数据库

    }
}
```

单击“使用 SQL 操作 SQLite 数据库”按钮，会看到 HiLog 视图中输出图 6-4 所示的查询结果。

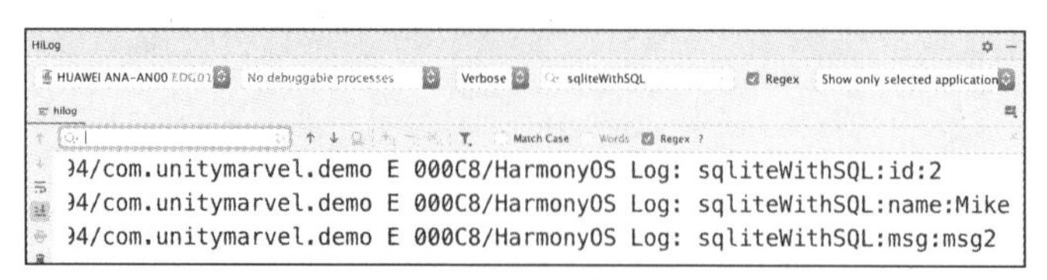

图 6-4 输出查询结果

阅读这段代码，需要注意如下几点。

- 用 StoreConfig 对象指定数据库文件名时，不需要指定路径，这是因为路径是固定的。
- RdbOpenCallback 对象中的 onCreate 方法会在第一次创建数据库时调用，通常在 onCreate 方法中为数据库创建表和视图。
- RdbOpenCallback 对象中的 onUpgrade 方法在数据库版本升级后被调用，通常用于升级表结构，修改视图，或升级表中的数据。在使用 getRdbStore 方法获取 RdbStore 对象时会指定一个数据库版本号（本例是 1），系统会将这个版本号保存到数据库中。如果下一次指定的版本号大于数据库中保存的版本号，那么 onUpgrade 方法就会被调用。
- 本例在执行 insert 语句之前删除了 t_users 表中的所有数据，这是因为 t_users 表的 id 字段是主键，不允许重复。所以需要在插入之前删除所有的数据，否则第 2 次执行这段代码时程序会崩溃。
- 使用 querySql 方法返回的 ResultSet 对象的指针指向第 1 条记录的前面，并不是第 1 条记录本身，所以为了获取 ResultSet 对象的第 1 条记录，需要使用 goToFirstRow 方法将指针移动到第 1 条记录上。
- ResultSet 对象提供了一系列 getXxx 方法，用于获取当前记录中特定字段的值，其中 Xxx 表示 Int、String 等。getXxx 方法的参数是字段索引，从 0 开始。

使用 context.getDatabaseDir 方法可以获取数据库的存储目录，不过获取的是 databases 目录，其实 SQLite 数据库文件存放在 databases 下的 db 子目录中，所以 data.sqlite 文件的完整路径如下：

```
/data/data/com.unitymarvel.demo/DataManager/databases/db/data.sqlite
```

将 data.sqlite 文件下载到 PC 上，用 DB Browser for SQLite 打开该文件，会看到 t_users 表中的数据，如图 6-5 所示。

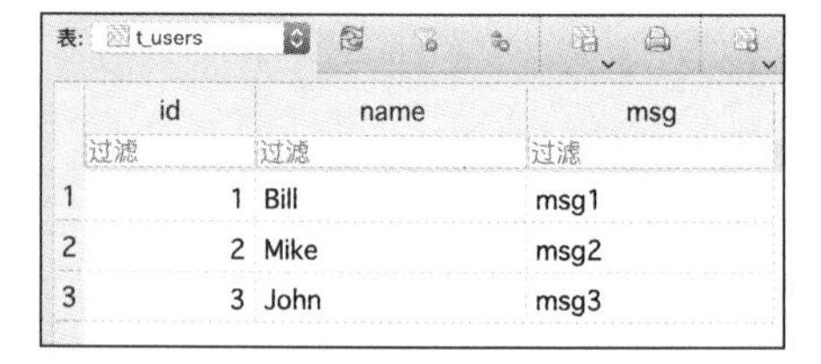

表: t_users

	id	name	msg
	过滤	过滤	过滤
1	1	Bill	msg1
2	2	Mike	msg2
3	3	John	msg3

图 6-5 t_users 表中的数据

6.2.2 使用谓词操作 SQLite 数据库

RdbStore 对象提供了 4 个方法（insert、delete、update 和 query），分别用来对数据表进行增删改查操作。使用这 4 个方法操作数据库，不需要使用 SQL 语句。这 4 个方法的原型如下：

```
long insert(String var1, ValuesBucket var2);
int delete(AbsRdbPredicates var1);
int update(ValuesBucket var1, AbsRdbPredicates var2);
ResultSet query(AbsRdbPredicates var1, String[] var2);
```

我们可以看到，这 4 个方法中并没有参数用于传递 SQL 语句，例如，insert 方法的第 1 个参数（var1）用于传递要插入数据的表名，第 2 个参数（var2）用于传递要插入的 key-value 对。所有 ValuesBucket 类型的参数值传递的都是 key-value 对，例如，update 方法的第 1 个参数。

在这 4 个方法中还有一类参数——AbsRdbPredicates 类型的参数，表示谓词，指通过方法来描述值的大小和值的关系。例如，greaterThan 方法表示比指定的字段值大。

query 方法用于使用谓词查询数据表，它的参数 var2 表示要查询的数据表的列名集合。

【例 6.5】 使用谓词重新实现例 6.4，并且最后查询多条记录。

代码位置：src/main/java/com/unitymarvel/demo/data/SQLiteDemo.java

```
public class SQLiteDemo {
   // 使用谓词操作 SQLite 数据库
   public static void sqliteWithPredicates(Context context) {
      RdbStore store;
      StoreConfig config = StoreConfig.newDefaultConfig("products.sqlite");

      RdbOpenCallback callback = new RdbOpenCallback() {
         @Override
         public void onCreate(RdbStore store) {
            store.executeSql("CREATE TABLE IF NOT EXISTS t_users (id INTEGER PRIMARY
            KEY, name VARCHAR(20), msg TEXT)");
            Tools.print("成功创建 t_users 表");
         }
         @Override
         public void onUpgrade(RdbStore store, int oldVersion, int newVersion) {}
      };
      DatabaseHelper helper = new DatabaseHelper(context);

      store = helper.getRdbStore(config, 1, callback, null);
      try {
         RdbPredicates rdbPredicates = new RdbPredicates("t_users");
         // 删除 t_users 表中的所有记录
         store.delete(rdbPredicates);

         ValuesBucket values = new ValuesBucket();
         values.putInteger("id", 1);
         values.putString("name", "Bill");
         values.putString("msg", "Msg1");
         // 向 t_users 表中插入一条记录
         store.insert("t_users", values);

         values.clear();
         values.putInteger("id", 2);
         values.putString("name", "Mike");
         values.putString("msg", "Msg2");
         // 向 t_users 表中插入一条记录
         store.insert("t_users", values);

         values.clear();
         values.putInteger("id", 3);
         values.putString("name", "John");
```

```
            values.putString("msg", "Msg3");
            // 向 t_users 表中插入一条记录
            store.insert("t_users", values);

            // 定义要查询的字段
            String[] columns = new String[] {"id", "name", "msg"};
            rdbPredicates =
                new RdbPredicates("t_users").greaterThan("id",1).orderByDesc("msg");
            // 搜索 t_users 表中 id 字段值大于 1 的记录，并且按 msg 字段值倒序排列
            ResultSet resultSet = store.query(rdbPredicates, columns);
            // 对查询结果进行迭代，并输出查询结果中的所有数据
            while(resultSet.goToNextRow()) {
               Tools.print("sqliteWithPredicates:id:" + resultSet.getInt(0));
               Tools.print("sqliteWithPredicates:name:" + resultSet.getString(1));
               Tools.print("sqliteWithPredicates:msg:" + resultSet.getString(2));
            }
            store.close();          // 关闭数据库
        }catch (Exception e) {
            Tools.print("sqliteWithPredicates:error:" + e.getMessage());
        }
    }
}
```

运行程序，并单击“使用谓词操作 SQLite 数据库”按钮，会在 HiLog 视图中输出图 6-6 所示的查询结果。

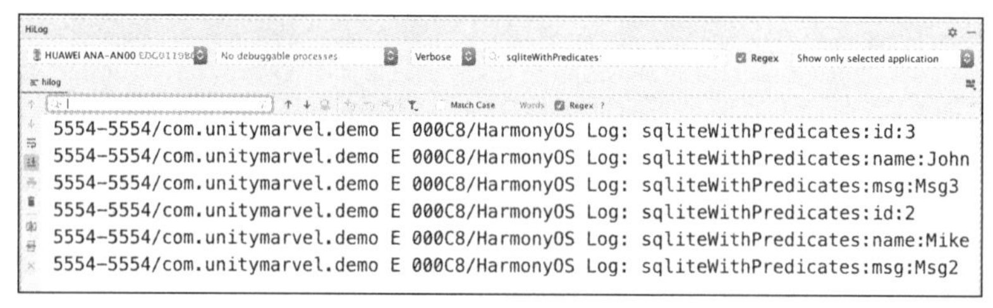

图 6-6　使用谓词查询数据表的结果

6.2.3　使用事务

操作数据库的关键之一就是保持数据的一致性。如果只执行一条 SQL 语句，例如，insert、update 等，是可以保证数据一致性的。一旦 SQL 语句执行出错，系统就会取消所有的数据修改。也就是说，单条 SQL 语句是原子操作。但如果执行多条 SQL 语句，就不是原子操作了。例如，连续执行两条 insert 语句，由于主键冲突，执行第 2 条 insert 语句时抛出异常，但我们的要求是要么都插入成功，要么都失败，在这种情况下，第 1 条 insert 语句执行成功，第 2 条 insert 语句执行失败，是不允许的。

为了解决这个问题，HarmonyOS 提供了事务操作 API，通过调用 RdbStore.beginTransaction 方法开启事务，然后使用 RdbStore.endTransaction 方法结束事务。如果成功结束事务，那么就会成功保存所有数据，否则夹在 RdbStore.beginTransaction 和 RdbStore.endTransaction 之间的所有修改数据库的操作都会回滚，也就是回到调用 RdbStore.beginTransaction 方法之前的状态。HarmonyOS 事务

默认是回滚的，所以并没有提供回滚方法，而是提供了一个 RdbStore.markAsCommit 方法。当所有的数据库操作执行成功后，通过该方法标记事务成功，这时才会真正提交数据到数据库。如果该方法未被调用，那么事务就会回滚。

如果要监控事务的执行过程，可以使用 RdbStore.beginTransactionWithObserver 方法开始事务。该方法需要传入一个事务观察者对象。观察者对象对应的类必须实现 TransactionObserver 接口，该接口有 3 个方法——onBegin、onCommit 和 onRollback，分别用于监听事务开始、事务提交和事务回滚。

【例 6.6】 向 transaction.sqlite 数据库中插入两条记录，因为主键冲突，所以在执行第 2 条 insert 语句时会抛出异常。并且因为对这两条 insert 语句使用了事务，所以记录并不会被真正插入数据表中。最后通过事务观察者对象监听事务的执行过程。

代码位置：src/main/java/com/unitymarvel/demo/data/SQLiteDemo.java

```
public class SQLiteDemo {
    // 事务观察者
    private static  class MyTransactionObserver implements TransactionObserver {
        @Override
        public void onBegin() {
            Tools.print("MyTransactionObserver:开始事务" );
        }

        @Override
        public void onCommit() {
            Tools.print("MyTransactionObserver:事务提交" );
        }

        @Override
        public void onRollback() {
            Tools.print("MyTransactionObserver:开始回滚" );
        }
    }
    public static void withTransaction(Context context) {
        RdbStore store;
        StoreConfig config = StoreConfig.newDefaultConfig("transaction.sqlite");

        RdbOpenCallback callback = new RdbOpenCallback() {
            @Override
            public void onCreate(RdbStore store) {
                store.executeSql("CREATE TABLE IF NOT EXISTS t_users (id INTEGER PRIMARY KEY,
                name VARCHAR(20), msg TEXT)");
                Tools.print("成功创建 t_users 表");
            }
            @Override
            public void onUpgrade(RdbStore store, int oldVersion, int newVersion) {
            }
        };
        DatabaseHelper helper = new DatabaseHelper(context);
        store = helper.getRdbStore(config,1,callback,null);
```

```
        try {
            String deleteSQL = "delete from t_users;";
            store.executeSql(deleteSQL, null);

            String insertSQL = "insert into t_users(id, name, msg) values(?,?,?);";

            //store.beginTransaction();
            // 开始事务，并通过观察者监听事务
            store.beginTransactionWithObserver(new MyTransactionObserver());

            store.executeSql(insertSQL, new Object[]{1, "Bill", "msg1"});
            // 这条语句会抛出异常
            store.executeSql(insertSQL, new Object[]{1, "Mike", "msg2"});
            // 如果成功执行所有的 insert 语句，做一个成功提交的标记
            store.markAsCommit();
        }catch (Exception e) {
            Tools.print("withTransaction:error:" + e.getMessage());
        }finally {
            // 如果没有调用 markAsCommit 方法就结束事务，那么事务会回滚
            store.endTransaction();
            store.close();
        }
    }

}
```

现在运行程序，然后单击“使用事务”按钮，在 HiLog 视图中会输出图 6-7 所示的信息。查看数据库，发现 t_users 表中并没有任何数据，说明事务成功回滚了。读者可以将事务去掉，再单击“使用事务”按钮，会发现 t_users 表中只有 1 条记录，这说明第 2 条 insert 语句执行失败，而且第 1 条 insert 语句对数据表的修改并未回滚。

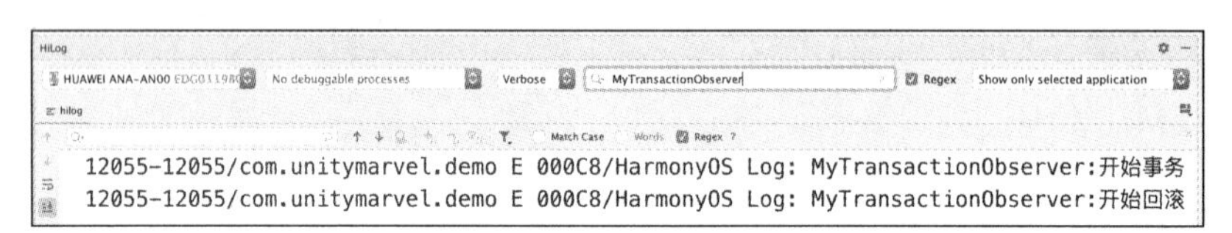

图 6-7 监听事务

6.3 对象关系映射

对象关系映射（Object Relational Mapping，ORM），也就是将对象的操作映射成关系型数据库的操作（SQL 操作），这样就可以通过操作 Java 对象的方式来操作数据库了。

因为 HarmonyOS 中 ORM 使用注解来描述实体类，所以在使用 ORM 操作数据库之前，要在 HarmonyOS SDK 中找到如下 3 个 jar 文件：

- orm_annotations_processor_java.jar；
- orm_annotations_java.jar；
- javapoet_java.jar。

这 3 个 jar 文件通常在<HarmonyOS SDK 根目录>/java/3.0.1.93/build-tools/lib 目录下，其中 3.0.1.93 由于读者使用的 SDK 版本不同会有所不同。

要想在 DevEco Studio 中使用这 3 个 jar 文件，需要将这 3 个 jar 文件复制到<HarmonyOS 工程根目录>/entry/libs 目录下。因为只有 libs 目录下的 jar 文件才可以直接在 Java 源代码文件中引用。然后需要在<HarmonyOS 工程根目录>/entry/build.gradle 文件中的 dependencies 部分添加如下内容：

```
dependencies {
   compile files("libs/orm_annotations_java.jar","libs/orm_annotations_processor_java.
   jar","libs/javapoet_java.jar")
   annotationProcessor files("libs/orm_annotations_java.jar","libs/orm_annotations_
   processor_java.jar","libs/javapoet_java.jar")
}
```

做完这些准备工作后，就可以使用注解来修改实体类了。目前支持的注解如表 6-1 所示。

表 6-1 ORM 注解

注解名称	描述
@Database	被@Database 修饰的类必须继承 OrmDatabase 类，表示数据库
@Entity	被@Entity 修饰的类必须继承 OrmObject 类，表示数据表
@Column	被@Column 修饰的变量对应数据表中的字段
@PrimaryKey	被@PrimaryKey 修饰的变量对应数据表中的主键
@ForeignKey	被@ForeignKey 修饰的变量对应数据表中的外键
@Index	被@Index 修饰的变量对应数据表索引的属性

【例 6.7】 使用 ORM 创建 products.db 数据库和 t_users 表，并添加 2 条记录、修改 1 条记录中的 name 字段值。注意，将这些操作放在事务中。

首先，创建 t_users 表的实体类。

代码位置：src/main/java/com/unitymarvel/demo/data/User.java

```
package com.unitymarvel.demo.data;
import ohos.data.orm.OrmObject;
import ohos.data.orm.annotation.Entity;
import ohos.data.orm.annotation.PrimaryKey;
// 指定 User 对应的表是 t_users
@Entity(tableName = "t_users")
public class User extends OrmObject {
   // 标识 id 变量是自增的字段
   @PrimaryKey(autoGenerate = true)
   private Integer id;
   private String name;
   private String msg;

   public Integer getId() {
      return id;
   }

   public String getName() {
```

```
        return name;
    }

    public String getMsg() {
        return msg;
    }

    public void setId(Integer id) {
        this.id = id;
    }

    public void setName(String name) {
        this.name = name;
    }

    public void setMsg(String msg) {
        this.msg = msg;
    }
}
```

然后，创建 products.db 对应的实体类。

代码位置：src/main/java/com/unitymarvel/demo/data/Products.java

```
package com.unitymarvel.demo.data;
import ohos.data.orm.OrmDatabase;
import ohos.data.orm.annotation.*;
// 指定数据库中有一个 t_users 表
@Database(entities = {User.class}, version = 1)
public abstract class Products extends OrmDatabase {
}
```

最后，使用前面创建的 2 个实体类和 ORM 相关 API 操作 SQLite 数据库。

代码位置：src/main/java/com/unitymarvel/demo/data/OrmDemo.java

```
package com.unitymarvel.demo.data;

import com.unitymarvel.demo.Tools;
import ohos.app.Context;
import ohos.data.DatabaseHelper;
import ohos.data.orm.OrmContext;
import ohos.data.orm.OrmPredicates;
import java.util.List;

public class OrmDemo {
    public static void testOrm(Context context) {
        DatabaseHelper helper = new DatabaseHelper(context);
        // 打开或创建 products.db 数据库，并将该数据库与 Products 类绑定
        OrmContext ormContext = helper.getOrmContext("Products", "products.db",
        Products.class);
        try{
            // 开启事务
            ormContext.beginTransaction();
            // 删除所有数据
```

```
            OrmPredicates predicates = ormContext.where (User.class);
            List<User> users = ormContext.query(predicates);
            for(User user: users) {
                ormContext.delete(user);
                ormContext.flush();
            }

            // 插入记录
            User user1 = new User();
            user1.setName("Bill");
            user1.setMsg("Msg1");
            boolean isSuccessed1 = ormContext.insert(user1);

            User user2 = new User();
            user2.setName("Mike");
            user2.setMsg("Msg2");
            boolean isSuccessed2 = ormContext.insert(user2);
            // 调用 flush 方法会将内存中的数据一次性写入数据库
            ormContext.flush();

            // 更新数据
            predicates = ormContext.where(User.class);
            predicates.equalTo("name","Mike");
            users = ormContext.query(predicates);
            User user = users.get(0);
            user.setName("李宁");
            ormContext.update(user);
            ormContext.flush();
            // 提交事务
            ormContext.commit();
        }catch (Exception e) {
            Tools.print("testOrm:error:" + e.getMessage());
            // 如果抛出异常，回滚事务
            ormContext.rollback();
        }finally {
             ormContext.close();
        }
    }
}
```

运行程序，单击“对象关系映射（ORM）”按钮，会生成一个 products.db 文件，将该文件下载到 PC，然后查看 t_users 表，会看到图 6-8 所示的数据。

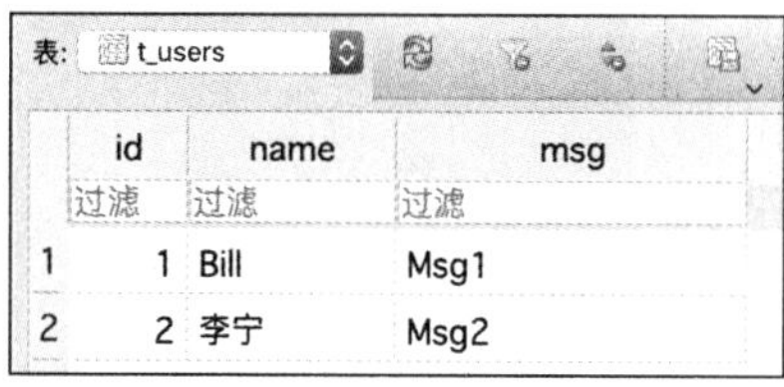

	id	name	msg
	过滤	过滤	过滤
1	1	Bill	Msg1
2	2	李宁	Msg2

图 6-8 用 ORM API 操作 SQLite 数据库

6.4 分布式文件

HarmonyOS 的优势之一就是分布式，其中最容易使用的功能就是分布式文件。通过分布式文件，我们可以在使用同一个华为账号登录的多个设备之间共享同一个文件，而且操作过程是透明的。分布式文件的同步工作完全由 HarmonyOS 负责，App 只需要获取用于存储分布式文件的目录，然后将文件写入该目录，HarmonyOS 就会自动将该文件同步到其他 HarmonyOS 设备上。同一个 App 在多个设备上的分布式目录是相同的。

分布式文件的关键是使用 context.getDistributedDir 方法获取分布式目录，但在读写该目录中的文件之前，先要在 config.json 文件中添加如下权限：

```
"reqPermissions": [
  {
    "name": "ohos.permission.DISTRIBUTED_DATASYNC"
  },
  {
    "name": "ohos.permission.servicebus.DISTRIBUTED_DEVICE_STATE_CHANGE"
  },
  {
    "name": "com.huawei.permission.MANAGE_DISTRIBUTED_PERMISSION"
  },
  {
    "name": "ohos.permission.INTERNET"
  },
  {
    "name": "ohos.permission.GET_NETWORK_INFO"
  },
  {
    "name": "com.huawei.hwddmp.servicebus.BIND_SERVICE"
  }
]
```

然后还需要用 Java 代码对分布式特性进行授权，所以需要编写下面的授权方法。

```
private void requestPermission() {
   String[] permission = {"ohos.permission.DISTRIBUTED_DATASYNC"};
   List<String> applyPermissions = new ArrayList<>();
   for (String element : permission) {

      if (verifySelfPermission(element) != 0) {
         if (canRequestPermission(element)) {
            applyPermissions.add(element);
         } else {
         }
      } else {

      }
   }
   requestPermissionsFromUser(applyPermissions.toArray(new String[0]), 0);
}
```

最后在 Ability 的 onStart 方法的开始部分调用 requestPermission 方法进行授权。

【例 6.8】 将 data.txt 文件写入一部 HarmonyOS 手机的分布式文件存储目录，然后用另一部 HarmonyOS 手机读取 data.txt 文件的内容，并将该文件的内容输出到 HiLog 视图中。

代码位置：src/main/java/com/unitymarvel/demo/data/DistributedFile.java

```
package com.unitymarvel.demo.data;

import com.unitymarvel.demo.Tools;
```

```
import ohos.app.Context;
import java.io.BufferedReader;
import java.io.File;
import java.io.FileReader;
import java.io.FileWriter;

public class DistributedFile {
    // 写分布式文件
    public static void writeDistributedFile(Context context) {
        // 获取分布式文件存储目录
        File distDir = context.getDistributedDir();
        // 组合生成分布式文件的路径
        String distributedFileName = distDir + File.separator + "data.txt";
        Tools.print("distributedFileName:" + distributedFileName);
        // 下面的代码开始向 data.txt 文件写数据
        File sharedNoteFile = new File(distributedFileName);
        if(sharedNoteFile.exists()) {
            sharedNoteFile.delete();
        }
        try {
            FileWriter fileWriter = new FileWriter(distributedFileName, false);
            fileWriter.write("世界，您好！");
            fileWriter.close();
        }
        catch (Exception e) {
        }
    }
    // 读分布式文件
    public static void readDistributedFile(Context context) {
        // 获取分布式文件存储目录
        File distDir = context.getDistributedDir();
        // 组合生成分布式文件的路径
        String distributedFileName = distDir + File.separator + "data.txt";
        File distributedFile = new File(distributedFileName);
        try {
            //读取 data.txt 文件的内容
            if(distributedFile.exists()) {
                FileReader fileReader = new FileReader(distributedFileName);
                BufferedReader br = new BufferedReader(fileReader);
                Tools.print("readDistributedFile:" + br.readLine());
                fileReader.close();
            }
        }
        catch (Exception e) {
        }
    }
}
```

运行本例至少需要两部 HarmonyOS 手机（或其他 HarmonyOS 设备），假设这两部 HarmonyOS 手机是 A 和 B。首先在 A 上运行程序，然后单击“写分布式文件”按钮，接下来在 B 上运行程序，然后单击“读分布式文件”按钮，会看到 HiLog 视图中输出图 6-9 所示的信息。

同时我们会发现在 A 和 B 的分布式文件存储目录下，都有一个名为 data.txt 的文件，该文件是 HarmonyOS 自动同步的。

注意，当第一次在 HarmonyOS 设备上运行本例时，会弹出图 6-10 所示的授权对话框，单击“始终允许”按钮，下次运行 App 就不会弹出这个对话框了（如果重新安装 App，第一次运行本例时仍然会弹出这个对话框）。

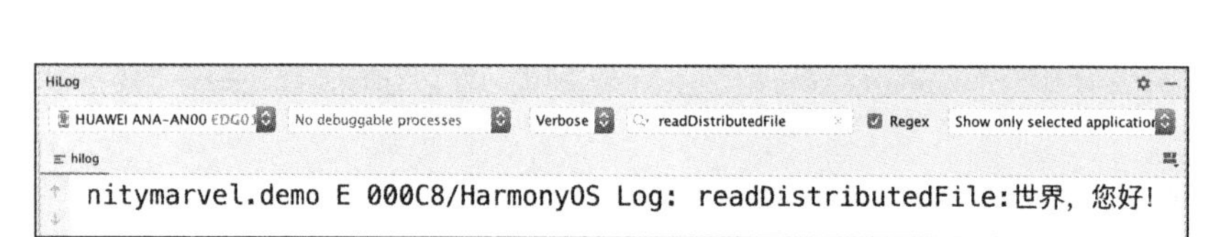

图 6-9 输出分布式文件的内容

图 6-10 授权对话框

6.5 分布式数据

分布式数据与分布式文件类似，也可以在 HarmonyOS 设备之间同步数据。只不过分布式数据更强大，主要特点如下。

- 分布式数据可以根据设备 ID 向特定的设备同步数据，而分布式文件不能选择特定的设备，会向所有可利用的设备同步文件。
- 分布式数据向其他设备同步的是数据，而不是文件，所以效率会更高。
- 分布式数据可以使用谓词对数据进行查询。

6.5.1 同步数据

分布式数据可以在两部或多部设备之间同步 key-value 格式的数据。下面是使用分布式数据在多个设备之间同步数据的步骤。

（1）创建 KvManager 对象。

```
KvManagerConfig config = new KvManagerConfig(context);
KvManager kvManage = KvManagerFactory.getInstance().createKvManager(config);
```

（2）创建 SingleKvStore 对象。

```
Options CREATE = new Options();
```

```
    CREATE.setCreateIfMissing(true).setEncrypt(false).setKvStoreType(KvStoreType.SINGLE_
VERSION);
    String storeID = "testApp";
    SingleKvStore singleKvStore = kvManager.getKvStore(CREATE, storeID);
```

（3）对于接收数据方，需要实现 KvStoreObserver 接口，如果接收到数据，该接口的 onChange 方法会被调用，可以在该方法中通过 SingleKvStore 对象提供的相应 API 读取接收到的数据。

```
public class DistributedData implements KvStoreObserver{
   @Override
   public void onChange(ChangeNotification changeNotification) {
       String name = singleKvStore.getString("name");
   }
}
```

（4）对于接收数据方，需要使用下面的代码订阅分布式数据。其中，subscribe 方法的第 2 个参数就是实现了 KvStoreObserver 接口的类的实例。

```
singleKvStore.subscribe(SubscribeType.SUBSCRIBE_TYPE_ALL, new DistributedData());
```

（5）在发送数据方，需要使用下面的代码写入数据。

```
String key = "name";
String value = "Bill";
singleKvStore.putString(key, value);
```

（6）在发送数据方，当写入数据后，需要使用下面的代码根据设备 ID 同步分布式数据。

```
List<DeviceInfo> deviceInfoList = getAvailableDeviceIds();
List<String> deviceIdList = new ArrayList<>();
for (DeviceInfo deviceInfo : deviceInfoList) {
   deviceIdList.add(deviceInfo.getDeviceId());
}
singleKvStore.sync(deviceIdList, SyncMode.PUSH_ONLY);
```

【例 6.9】 使用分布式数据在 A 和 B 两个设备上同步数据。在 A 设备上同步数据，B 设备就会接收到这些数据，并输出到 HiLog 视图中。在 B 设备上同步数据，A 设备也会接受到这些数据，并输出到 HiLog 视图中。

代码位置：src/main/java/com/unitymarvel/demo/data/DistributedData.java

```
package com.unitymarvel.demo.data;

import com.unitymarvel.demo.Tools;
import ohos.app.Context;
import ohos.data.distributed.common.*;
import ohos.data.distributed.user.SingleKvStore;
import ohos.distributedschedule.interwork.DeviceInfo;
import ohos.distributedschedule.interwork.DeviceManager;
import java.util.ArrayList;
import java.util.List;
// 该类同时用于发送数据端和接收数据端
public class DistributedData implements KvStoreObserver{
   private SingleKvStore singleKvStore;
   private KvManager kvManager;
```

```
    public DistributedData(Context context) {
        KvManagerConfig config = new KvManagerConfig(context);
        kvManager = KvManagerFactory.getInstance().createKvManager(config);

        Options CREATE = new Options();

        CREATE.setCreateIfMissing(true).setEncrypt(false).setKvStoreType(KvStoreType.
        SINGLE_VERSION);
        String storeID = "testApp";
        singleKvStore = kvManager.getKvStore(CREATE, storeID);
        // 订阅分布式数据
        singleKvStore.subscribe(SubscribeType.SUBSCRIBE_TYPE_ALL, this);
    }
    // 但接收到数据后，该方法会被调用
    @Override
    public void onChange(ChangeNotification changeNotification) {
        // 读取接收到的数据，并将这些数据输出到 HiLog 视图中
        Tools.print("onChange:" + singleKvStore.getString("name"));
    }
    // 获取所有可用设备的信息
    public List<DeviceInfo> getAvailableDeviceIds() {

        List<DeviceInfo> deviceInfoList =
             DeviceManager.getDeviceList(DeviceInfo.FLAG_GET_ONLINE_DEVICE);
        if (deviceInfoList == null) {
            return new ArrayList<>();
        }

        if (deviceInfoList.size() == 0) {
            return new ArrayList<>();
        }

        return deviceInfoList;
    }
    // 写入数据，并将这些数据同步到所有可用的设备
    public void writeData() {
        String key = "name";
        String value = "李宁";
        // 写入 key-value 对
        singleKvStore.putString(key, value);

        // 获取所有可用设备的 ID
        List<DeviceInfo> deviceInfoList = getAvailableDeviceIds();
        List<String> deviceIdList = new ArrayList<>();
        for (DeviceInfo deviceInfo : deviceInfoList) {
            deviceIdList.add(deviceInfo.getDeviceId());
        }
        // 将数据同步到所有可用设备，或者称为同步数据
        singleKvStore.sync(deviceIdList, SyncMode.PUSH_ONLY);
    }
}
```

运行本例至少需要两部 HarmonyOS 设备，假设是设备 A 和设备 B。在这两个设备上都运行程序，然后在 A 设备上单击“同步分布式数据”按钮，在 B 设备的 HiLog 视图就会看到图 6-11 所示

的信息。在 B 设备上单击“同步分布式数据”按钮，在 A 设备的 HiLog 视图中也会输出同样的信息。

图 6-11 同步分布式数据

注意，分布式数据也同样需要加 6.4 节的权限，否则无法同步分布式数据。

6.5.2 用谓词查询分布式数据

HarmonyOS 还提供了一组谓词 API（类似于 ORM 的谓词 API），用于查询分布式数据。假设有两个设备——A 和 B，现在将数据集从 A 同步到 B，并在 B 中使用谓词搜索特定的记录，步骤如下。

（1）创建记录集的字段。记录集的字段通常包含字段名、字段类型以及其他属性。字段由 FieldNode 对象表示，一个 FieldNode 对象表示一个字段，创建字段的基本方式如下：

```
FieldNode idNode = new FieldNode("id");
idNode.setType(FieldValueType.INTEGER);
idNode.setNullable(false);
```

（2）创建 Schema。Schema 相当于数据库中的表，所以为数据集创建一个表（Schema）的代码如下：

```
Schema schema = new Schema();
schema.setSchemaMode(SchemaMode.COMPATIBLE);
```

（3）为数据集添加字段。这一步需要将第（1）步创建的字段添加到第（2）步创建的 Schema 中，代码如下：

```
schema.getRootFieldNode().appendChild(idNode);
```

（4）创建索引列表。为了让查询更有效率，通常会在经常查询的字段上创建索引，一个记录集可以有 1 个或多个索引，由一个 List<String>对象表示，代码如下：

```
List<String> indexes = new ArrayList<>();
indexes.add("$.id");
```

要注意，这里的 id 是字段名，在添加索引时，要在前面加“$.”，表示 JSON 数据的一个隐含根节点。

（5）为 Schema 设置索引。这一步需要将第（4）步创建的索引添加到第（2）步创建的 Schema 中，代码如下：

```
schema.setIndexes(indexes);
```

（6）设置 Schema 选项。通过 Options 对象可以对 Schema 进行配置，通常的配置如下：

```
Options options = new Options();
options.setCreateIfMissing(true).setAutoSync(false).setEncrypt(false).setKvStoreType
(KvStoreType.SINGLE_VERSION);
options.setSchema(schema);
```

（7）创建数据库（SingleKvStore 对象）。这里说的数据库相当于内存数据库，一个数据库需要用一个字符串作为唯一标识，如“mydata”。

```
KvManagerConfig config = new KvManagerConfig(this);
KvManger kvManager = KvManagerFactory.getInstance().createKvManager(config);
SingleKvStore singleKvStore = kvManager.getKvStore(options, "mydata");
```

这里的 SingleKvStore 对象就表示一个数据库，mydata 相当于数据库名。不能重复创建数据库，否则会抛出异常。

（8）创建观察者类。如果数据需要从设备 A 同步到设备 B，那么在设备 B 中需要使用观察者类来截获同步过来的数据，观察者类必须实现 KvStoreObserver 接口，实现代码如下：

```
public class DistributedDataObserver implements KvStoreObserver {
    // 当设备 B 接收到同步过来的数据时，onChange 方法会被调用
    @Override
    public void onChange(ChangeNotification changeNotification) {
        // 处理业务的代码
    }
}
```

（9）订阅观察者对象。在设备 B 中需要使用下面的代码订阅第（8）步创建的观察者类的实例，这样才能截获同步过来的数据。

```
singleKvStore.subscribe(SubscribeType.SUBSCRIBE_TYPE_ALL,
                                            new DistributedDataObserver());
```

（10）写入数据。不管是普通的数据，还是数据集类型的数据，都需要使用 putXxx 方式写入数据（Xxx 表示 String、Boolean 等），只不过数据集中的记录使用 JSON 描述，所以这里只能通过 putString 方法写入数据，代码如下：

```
singleKvStore.putString("key1","{\"id\":1}");
```

这里的 key1 可以任意指定，只要不和其他的 key 重复即可。后面通过 Object 形式的 JSON 描述一条记录。因为第（1）步只创建了 1 个字段，而且是整数类型，所以 JSON 对象只包含一个名为 id 的属性。JSON 对象中的属性名、属性个数和类型必须与第（1）步创建的字段名、字段个数和类型相匹配，否则会抛出异常。

（11）同步数据。通过 SingleKvStore.sync 方法可以将数据同步到其他 HarmonyOS 设备上，同步需要获取其他 HarmonyOS 设备的 ID，代码如下：

```
// 获取所有可用 HarmonyOS 设备的信息
List<DeviceInfo> deviceInfoList = getAvailableDeviceIds();
List<String> deviceIdList = new ArrayList<>();
for (DeviceInfo deviceInfo : deviceInfoList) {
    // 将设备 ID 添加到 deviceIdList 列表中
    deviceIdList.add(deviceInfo.getDeviceId());
}
// 向 deviceIdList 列表中包含的所有设备同步数据
singleKvStore.sync(deviceIdList, SyncMode.PUSH_ONLY);
```

（12）在设备 B 用谓词查询数据。设备 B 通过观察者对象的 onChange 方法接收到数据后，可以通过下面的代码查询数据。

```
Query query = Query.select();
query.equalTo("$.id", "1");
//使用谓词查询，entries 包含了所有符合条件的结果
List<Entry> entries = singleKvStore.getEntries(query);
for (Entry entry : entries) {
    // 在 HiLog 视图中输出查询到的结果（JSON 字符串）
    Tools.print("onChange:entries:" + entry.getValue().getString());
}
```

【例 6.10】 完整地演示用谓词查询分布式数据的全过程。按照以上的步骤，添加 3 个字段（id、name 和 age），并且搜到 name 字段值等于“Mike”的记录，然后在设备 B 中将搜索结果输出到 HiLog 视图中，最后将 JSON 格式的数据转换为 JSON 对象，并显示搜索结果中的 name 字段和 age 字段的值。

代码位置：src/main/java/com/unitymarvel/demo/data/MyDistributedData.java

```
package com.unitymarvel.demo.data;

import com.unitymarvel.demo.ResourceTable;
import com.unitymarvel.demo.Tools;
import ohos.aafwk.ability.Ability;
import ohos.aafwk.content.Intent;
import ohos.agp.components.Button;
import ohos.agp.components.Component;
import ohos.data.distributed.common.*;
import ohos.data.distributed.user.SingleKvStore;
import ohos.distributedschedule.interwork.DeviceInfo;
import ohos.distributedschedule.interwork.DeviceManager;
import ohos.utils.zson.ZSONObject;
import java.util.ArrayList;
import java.util.List;

public class MyDistributedData extends Ability {
   // 定义数据库名
   private final String STORE_ID = "DataDemo" ;
   private SingleKvStore singleKvStore;
   private KvManager kvManager;
   // 观察者类
   private class DistributedDataObserver implements KvStoreObserver {
      @Override
      public void onChange(ChangeNotification changeNotification) {
         // 因为 onChange 方法在非 UI 线程中运行，所以显示 ToastDialog 信息框必须让其
         // 在 UI 线程中运行，因此要使用 getUITaskDispatcher 方法来运行代码
         getUITaskDispatcher().asyncDispatch(new Runnable() {
            @Override
            public void run() {
               // 构造谓词查询
               Query query = Query.select();
               // name 字段值等于 Mike，这里指定字段，需要前面加"$."
               query.equalTo("$.name", "Mike");

               // 使用谓词查询（方式 1）
```

```
                List<Entry> entries = singleKvStore.getEntries(query);
                // 使用谓词查询（方式 2）
                KvStoreResultSet resultset = singleKvStore.getResultSet(query);
                // 在 HiLog 视图中输出所有的查询结果
                for (Entry entry : entries) {
                    Tools.print("onChange:entries:" + entry.getValue().getString());
                }
                // 在 HiLog 视图中输出所有的查询结果
                while (resultset.goToNextRow()) {
                    Tools.print("onChange:resultset:" +
                            resultset.getEntry().getValue().getString());
                    // 获取查询结果中的 JSON 字符串
                    String json = resultset.getEntry().getValue().getString();
                    // 将 JSON 字符串转换为 JSON 对象
                    ZSONObject object =  ZSONObject.stringToZSON(json);
                    // 获取 name 属性值
                    String name = object.getString("name");
                    // 获取 age 属性值
                    int age = object.getInteger("age");
                    Tools.showTip(MyDistributedData.this, "name:" + name + " age:" + age);
                    Tools.print("onChange:resultset:" + "name:" + name + " age:" + age);
                }
            }
        });
    }
}
// 主要负责创建数据库（SingleKvStore）、表（Schema）和字段（FieldNode）
public void init() {
    KvManagerConfig config = new KvManagerConfig(this);
    kvManager = KvManagerFactory.getInstance().createKvManager(config);
    // 下面的代码创建了 3 个字段
    FieldNode idNode = new FieldNode("id");
    idNode.setType(FieldValueType.INTEGER);
    idNode.setNullable(false);

    FieldNode nameNode = new FieldNode("name");
    nameNode.setType(FieldValueType.STRING);

    FieldNode ageNode = new FieldNode("age");
    ageNode.setType(FieldValueType.INTEGER);
    // 创建表
    Schema schema = new Schema();
    schema.setSchemaMode(SchemaMode.COMPATIBLE);
    List<String> indexes = new ArrayList<>();
    indexes.add("$.id");
    // 设置索引
    schema.setIndexes(indexes);
    // 为 Schema 添加字段
    schema.getRootFieldNode().appendChild(idNode);
    schema.getRootFieldNode().appendChild(nameNode);
    schema.getRootFieldNode().appendChild(ageNode);
    // 设置 Schema
```

```
        Options options = new Options();
        options.setCreateIfMissing(true).setAutoSync(false).setEncrypt(false).setKvStoreType
        (KvStoreType.SINGLE_VERSION);
        options.setSchema(schema);

        // 创建数据库
        singleKvStore = kvManager.getKvStore(options, STORE_ID);
        // 订阅观察者对象
        singleKvStore.subscribe(SubscribeType.SUBSCRIBE_TYPE_ALL,
                              new DistributedDataObserver());
        Tools.showTip (this,"DistributedData:" + STORE_ID);
    }
    // 获取所有可用的设备
    public List<DeviceInfo> getAvailableDeviceIds() {
        List<DeviceInfo> deviceInfoList =
                DeviceManager.getDeviceList(DeviceInfo.FLAG_GET_ONLINE_DEVICE);
        if (deviceInfoList == null) {
            return new ArrayList<>();
        }
        if (deviceInfoList.size() == 0) {
            return new ArrayList<>();
        }
        return deviceInfoList;
    }

    @Override
    public void onStart(Intent intent) {
        super.onStart(intent);
        super.setUIContent(com.unitymarvel.demo.ResourceTable.Layout_my_distributed_data);
        // 完成初始化工作
        init();
        Button button1 = (Button) findComponentById(ResourceTable.Id_button1);
        if (button1 != null) {
            button1.setClickedListener(new Component.ClickedListener() {
                // 写入数据
                @Override
                public void onClick(Component component) {
                    try {
                        //写数据
                        singleKvStore.putString("key1",
                        "{\"id\":1,\"name\":\"Bill\",\"age\":20}");
                        singleKvStore.putString("key2",
                        "{\"id\":2,\"name\":\"Mike\",\"age\":25}");
                        singleKvStore.putString("key3",
                        "{\"id\":3,\"name\":\"John\",\"age\":50}");
                        Tools.showTip(MyDistributedData.this,
                                   "MyDistributedData:已经成功写入数据");
                    }catch (Exception e) {
                        Tools.showTip(MyDistributedData.this,"DistributedFile:error:" +
                                                        e.getMessage());
                    }
                }
            });
```

```
}
Button button2 = (Button) findComponentById(ResourceTable.Id_button2);
if (button2 != null) {
   button2.setClickedListener(new Component.ClickedListener() {
      @Override
      public void onClick(Component component) {
         try {
             List<DeviceInfo> deviceInfoList = getAvailableDeviceIds();
             List<String> deviceIdList = new ArrayList<>();
             for (DeviceInfo deviceInfo : deviceInfoList) {
                 deviceIdList.add(deviceInfo.getDeviceId());
             }
             // 同步数据
             singleKvStore.sync(deviceIdList, SyncMode.PUSH_ONLY);
             Tools.showTip(MyDistributedData.this, "数据同步成功! ");
         } catch (Exception e) {
             Tools.showTip(MyDistributedData.this,"DistributedFile:error:" +
             e.getMessage());
         }
      }
   });
}
Button button3 = (Button) findComponentById(ResourceTable.Id_button3);
if (button3 != null) {
   button3.setClickedListener(new Component.ClickedListener() {
      // 从本地读数据
      @Override
      public void onClick(Component component) {
         try {
             Query query = Query.select();
             query.equalTo("$.id", 1);
             List<Entry> entries = singleKvStore.getEntries(query);
             if (entries.size() > 0) {
                Tools.showTip(MyDistributedData.this,
                entries.get(0).getValue().getString());
             } else {
                Tools.showTip(MyDistributedData.this, "未找到数据");
             }
          } catch (Exception e) {
             Tools.showTip(MyDistributedData.this,"DistributedFile:error:" +
             e.getMessage());
          }
      }
   });
}
Button button4 = (Button) findComponentById(ResourceTable.Id_button4);
if (button4 != null) {
   button4.setClickedListener(new Component.ClickedListener() {
      @Override
      public void onClick(Component component) {
         try {
             // 关闭数据库
             kvManager.closeKvStore(singleKvStore);
```

```
                    Tools.showTip(MyDistributedData.this, "成功关闭数据库");
                }catch (Exception e) {
                    Tools.showTip(MyDistributedData.this,"DistributedFile:error:" +
                    e.getMessage());
                }
            }
        });
    }
    Button button5 = (Button) findComponentById(ResourceTable.Id_button5);
    if (button5 != null) {
        button5.setClickedListener(new Component.ClickedListener() {
            @Override
            public void onClick(Component component) {
                try{
                    // 删除数据库
                    kvManager.deleteKvStore(STORE_ID);
                    Tools.showTip(MyDistributedData.this, "成功删除数据库");
                 }catch (Exception e) {
                    Tools.showTip(MyDistributedData.this,"DistributedFile:error:" +
                    e.getMessage());
                 }

            }
        });
    }

  }
}
```

阅读这段代码，需要了解如下几点。

- 分布式数据的记录使用 JSON 格式表示，如果要想获取具体字段值，需要使用 JSON 库解析 JSON 字符串。
- 因为观察者对象中的 onChange 方法在非 UI 线程中运行，所以如果想在该方法中访问 UI 组件，或显示 ToastDialog 信息框，需要使用 getUITaskDispatcher().asyncDispatch 方法运行，否则会抛出异常。
- 在为 Schema 添加索引或使用谓词查询时，指定字段需要在字段名前面加“$.”前缀，表示 JSON 中隐含的根节点。
- 在同步完数据后，需要使用 KvManager.closeKvStore 方法关闭数据库，然后使用 KvManager.deleteKvStore 方法删除数据库，否则不能再创建同名的数据库，即使重新运行 App，甚至重新安装 App 也不行。
- 查询方法有两个——getEntries 和 getResultSet，前者以列表形式返回查询结果，后者以 KvStoreResultSet 形式返回查询结果。

现在分别在设备 A 和设备 B 上运行程序，在两个设备上显示的效果均如图 6-12 所示。

然后依次单击设备 A 上的“写入数据”按钮和“同步数据”按钮，这时设备 B 会在 ToastDialog 信息框中显示 name 字段值和 age 字段值，如图 6-13 所示。

在设备 B 的 HiLog 视图中会显示图 6-14 所示的输出信息。

图 6-12　用谓词查询分布式数据

图 6-13　设备 B 的查询结果

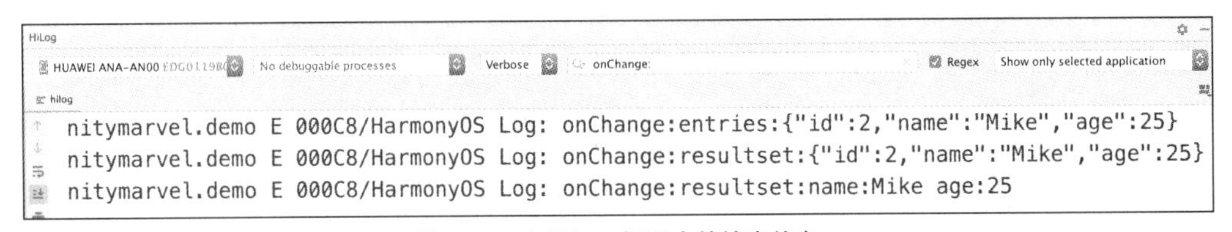

图 6-14　在 HiLog 视图中的输出信息

在设备 B 上依次单击“写入数据”按钮和“同步数据”按钮，会在设备 A 中显示同样的查询结果。在完成以上操作后，要依次单击“关闭数据库”和“删除数据库”按钮来关闭和删除数据库，否则再次运行将抛出异常。如果忘记了单击这两个按钮，可以将数据库（STORE_ID）换一个名字。

6.6 总结与回顾

古人云：人过留名，雁过留声。对 App 来说，保存数据很重要。

App 的价值就是产生和管理数据，任何一类 App，都会产生大量的数据，这些数据通常都要求永久留存。留存数据所使用的技术就是本章讲的数据管理。在 HarmonyOS 中可以使用 Preferences 保存数据，Preferences 将数据以 XML 格式保存。

如果需要查询数据以及对数据进行增、删、改操作，可以使用 SQLite 数据库。HarmonyOS 提供了丰富的 API 来操作数据库。它不仅支持直接用 SQL 来操作 SQLite，还支持 ORM，可以像操作 Java 对象一样操作 SQLite 数据库。

除了这些传统的文件和数据库操作，HarmonyOS 还有一项特殊的功能，就是让文件和数据在不同 HarmonyOS 设备间同步。这一过程对于开发人员和用户都是透明的。开发人员不需要关心通信问题，只要告诉系统，需要同步哪些文件或数据即可。这就是本章最后介绍的分布式文件和分布式数据。

第 7 章 Data Ability

Data Ability 让 App 拥有向外提供数据的能力，这些数据可能来源于关系型数据库、文本文件、二进制文件等。HarmonyOS 允许访问本机的 Data Ability，也允许通过 deviceId 访问其他 HarmonyOS 设备的 Data Ability。本章将详细讲解 App 与 Data Ability 交互的过程。

通过阅读本章，读者可以掌握：

- 什么是 Data Ability；
- Data Ability URI 的结构；
- 如何创建 Data Ability；
- 如何用 Data Ability 操作数据库；
- 如何用 Data Ability 操作文件；
- 如何访问本地的 Data Ability；
- 如何访问其他 HarmonyOS 设备的 Data Ability。

7.1 Data Ability 概述

顾名思义，Data Ability 与数据有关，其实 Data Ability 的确与数据有关，Data Ability 的主要目的就是将数据抽象化。也就是说，使用 Data Ability 时，只需要知道获取数据的内容，而不需要知道数据的来源。

Data Ability 提供的数据可能来自本地的一个文本文件，或者是一个 SQLite 数据库，或者干脆就是数组、列表、映射中的数据，甚至可能来自网络，例如，通过 HTTP(S)或 Socket 从服务端获取数据，这一切对调用者来说是透明的。这么做至少有如下几个好处。

- 因为数据源对于调用者是透明的，所以在调用端并不会涉及与数据源强耦合的代码。因此，如果要更换数据源非常方便，只需要修改 Data Ability 内部的代码即可。只要保持返回数据的结构和接收参数不变，在调用端就不需要修改任何代码。
- 因为很多数据源非常复杂，例如，获取某些信息可能需要同时搜索多个数据表，或者用非常复杂的复合查询。如果在业务逻辑端（Data Ability 的调用端）充斥着大量这样的代码，会让代码维护非常困难，所以需要将这些复杂代码封装在 Data Ability 中，而返回的结果集只是一个简单的二维表，这样一来，就将复杂性隐藏在 Data Ability 内部。因为 Data Ability 可能会被多次调用，所以这种方式会让代码更容易维护（因为复杂的代码都集中在了同一个地方）。

调用Data Ability需要使用一个统一资源标记符（Uniform Resource Identifier，URI），一个URI唯一标识一个Data Ability。在HarmonyOS中，既可以调用本机的Data Ability，也可以跨设备调用Data Ability。

理论上来说，Data Ability的数据源可以是任何形式，不过Data Ability提供了两组方便的API，可以调用文件形式的数据源和SQLite数据库形式的数据源，本章后面的内容会详细介绍这些API。

7.2 Data Ability中的URI

不管是访问本机的Data Ability，还是其他设备中的Data Ability，都需要使用URI。URI由6个部分组成，结构如下；

```
dataability://[deviceid]/<authority>/<path>[?query][#fragment]
```

这6个部分的描述如表7-1所示。

表7-1 Data Ability URI组成部分描述

组成部分	描述
dataability	这一部分是固定的，表示访问的是Data Ability
deviceid	deviceid只有在跨设备访问Data Ability时才有用，访问本地Data Ability时，忽略deviceid
authority	唯一标识Data Ability的字符串，通常是Data Ability类的全名（包括包名），相当于网址中的域名或IP
path	特定资源的路径信息，可以任意指定，用于区分Data Ability中的具体资源
query	用于查询参数
fragment	可用于指定要访问的子资源（更细粒度区分Data Ability中的资源）

其中，authority和path用尖括号括起来了，表示这两部分是必要的，其他部分都用方括号括起来了，表示其他部分都是可选的。

URI示例如下。

- 跨设备URI：dataability://deviceid/com.unitymarvel.data/product/20。
- 本地URI：dataability:///com.unitymarvel.data/product/20。

注意，本地URI尽管省略了deviceid，但deviceid后面的斜杠（/）并没有省略，所以dataability:和authority之间是3个斜杠。

7.3 创建Data Ability

Data Ability可以手动创建，也可以自动创建。本节会讲解手工创建Data Ability和自动创建Data Ability这两种方式。

1. 手动创建Data Ability

Data Ability类是一个普通的Java类，只是该Java类必须继承ohos.aafwk.ability.Ability类。其

实 Ability 类中并没有任何方法必须在子类中实现，所以一个 Data Ability 类中可以不添加任何代码。

```
package com.unitymarvel.data;
// 一个最简单的 Data Ability
public class SimpleDataAbility extends Ability {
}
```

如果要让 Data Ability 拥有某些功能，那么还是需要在 Data Ability 类中添加一些方法的，本节只是教大家如何创建一个最简单的 Data Ability，所以就不添加任何代码了。

接下来在 config.json 文件中配置 SimpleDataAbility，代码如下：

```
{
   "permissions": [
      "com.unitymarvel.data.DataAbilityShellProvider.PROVIDER"
    ],
    "name": "com.unitymarvel.data.SimpleDataAbility",
    "type": "data",
    "uri": "dataability://com.unitymarvel.data.SimpleDataAbility"
}
```

在配置 SimpleDataAbility 时，下面的几个信息必须指定。

- SimpleDataAbility 的全名，使用 name 属性指定。
- Ability 的类型，使用 type 属性指定。如果是 Data Ability，type 属性值应该是 data。
- 定义 Data Ability 的 URI，使用 uri 属性指定。该属性只需要指定 dataability://authority 即可，不管这个 Data Ability 是被本地 App 访问，还是被其他设备上的 App 访问，都不要指定 deviceid、query 等部分，而且 dataability:和 authority 之间是两个斜杠（/），这一点要和访问 Data Ability 时使用的 URI 区分开。
- 为 Data Ability 自定义一个权限，使用 permissions 属性指定。在使用这个 Data Ability 的 App 的 config.json 文件中，要申请一个权限。这个权限可以自定义，如果使用自动生成 Data Ability 的方式，IDE 会自动生成一个默认的权限。

2. 自动创建 Data Ability

DevEco Studio 提供了自动创建 Data Ability 的功能。选中某个 Package，单击鼠标右键，在弹出的菜单中选择 New→Ability→Empty Data Ability 菜单项，如图 7-1 所示。

选择该菜单项后，会弹出图 7-2 所示的 New Ability 窗口。在 Data Name 文本框中输入 Data Ability 的名字，在 Package Name 文本框中输入 Data Ability 的包名，然后单击 Finish 按钮创建 Data Ability。

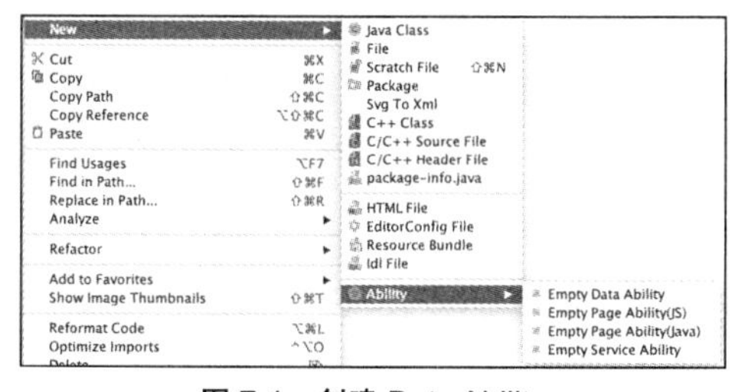

图 7-1 创建 Data Ability

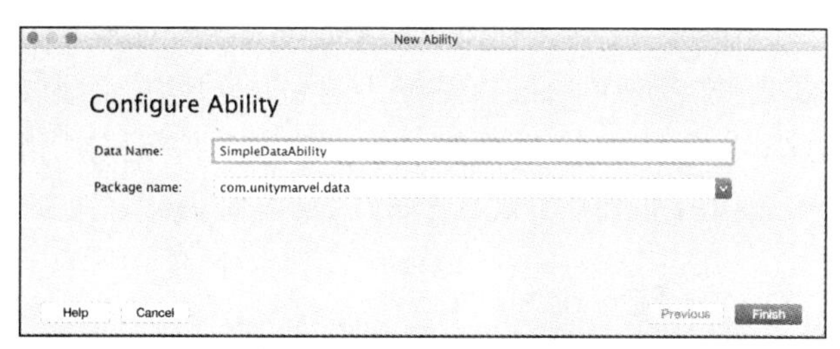

图 7-2 New Ability 窗口

创建完 Data Ability 后，IDE 会自动创建 SimpleDataAbility 类、添加前面所描述的两类方法，以及在 config.json 文件中添加该 Data Ability 的配置代码。

7.4 访问本地数据库

通过 Data Ability，我们可以将本机的 SQLite 数据库共享给其他设备。并且，我们可以利用 Data Ability 对数据库进行增、删、改和查操作，这些操作分别对应 insert、delete、update 和 query 这 4 个方法。这 4 个方法不需要都实现，通常根据需要的操作选择实现的方法。这 4 个方法的原型如下：

```
public int insert(Uri uri, ValuesBucket value);
public int delete(Uri uri, DataAbilityPredicates predicates);
public int update(Uri uri, ValuesBucket value, DataAbilityPredicates predicates);
public ResultSet query(Uri uri, String[] columns, DataAbilityPredicates predicates);
```

我们可以看到，这 4 个方法的第 1 个参数都是 uri，这个 uri 就是前面讲的 URI。根据操作的不同，这 4 个方法会使用不同类型的参数。下面是这些参数的含义。

- uri：用于指定 Data Ability 的 URI。
- value：用于指定 key-value 对，其中 key 表示插入值对应的字段名，value 表示插入值。
- predicates：用谓词描述的删除条件。
- columns：查询时返回的字段名，如果为 null，表示返回记录集的所有字段。

在 Data Ability 中操作数据库，可以直接使用 SQL 来处理，也可以使用 ORM 来处理。但推荐使用 ORM，这是因为上述 4 个方法的条件都是通过 DataAbilityPredicates 描述的，可以非常容易地将 DataAbilityPredicates 对象转换为 ORM 支持的谓词对象，操作起来非常方便。如果直接使用 SQL 操作，则需要用大量代码将 DataAbilityPredicates 对象中的条件转换为 SQL 语句，操作起来非常麻烦。

例如，在 query 方法中根据传入的参数值查询数据，只需要使用下面的代码即可：

```
OrmPredicates ormPredicates = DataAbilityUtils.createOrmPredicates(predicates,
                                                                  Product.class);
ResultSet resultSet = ormContext.query(ormPredicates, columns);
```

其中，predicates 是 query 方法的参数，通过 DataAbilityUtils.createOrmPredicates 方法可以直接将 DataAbilityPredicates 对象转换为 ORM 支持的 OrmPredicates 对象。然后就可以正常使用 ORM 的相关 API 操作数据库了。

在客户端，需要使用 DataAbilityHelper.query 方法通过 Data Ability 查询数据，调用该方法，其实就相当于调用 Data Ability 中的 query 方法。因为 query 方法需要指定查询条件，所以需要创建 DataAbilityPredicates 对象，并通过相应的谓词方法指定查询条件。其他 3 个方法（insert、delete 和 update）的使用方式与 query 方法类似。

```
DataAbilityHelper helper = DataAbilityHelper.creator(this);
// 通过谓词指定查询条件
DataAbilityPredicates predicates = new DataAbilityPredicates();
predicates.greaterThan ("id",2);
```

```
// 查询数据
ResultSet resultSet = helper.query(uri,null,predicates);
```

【例 7.1】 创建一个名为 DBDataAbility 的 Data Ability，并且在初始化时（使用 onStart 方法）创建一个名为 data.sqlite 的数据库，以及一个 t_products 表，然后通过 SQL 语句插入 2 条记录。接下来实现 query、insert、update 和 delete 这 4 个方法，在这 4 个方法中使用 ORM 操作数据库。在客户端通过相应的 API 访问 Data Ability，在对数据库进行 insert、delete 和 update 操作后，使用 query 方法查询数据，并用 Toast 信息框显示查询结果。

首先看一下 DBDataAbility 类的实现。

代码位置：src\main\java\com\unitymarvel\demo\ability\DBDataAbility.java

```
package com.unitymarvel.demo.ability;

import com.unitymarvel.demo.Tools;
import ohos.aafwk.ability.Ability;
import ohos.aafwk.ability.DataAbilityHelper;
import ohos.aafwk.content.Intent;
import ohos.app.Context;
import ohos.data.DatabaseHelper;
import ohos.data.dataability.DataAbilityUtils;
import ohos.data.orm.OrmContext;
import ohos.data.orm.OrmPredicates;
import ohos.data.rdb.RdbOpenCallback;
import ohos.data.rdb.RdbStore;
import ohos.data.rdb.StoreConfig;
import ohos.data.resultset.ResultSet;
import ohos.data.rdb.ValuesBucket;
import ohos.data.dataability.DataAbilityPredicates;
import ohos.utils.net.Uri;
// Data Ability 类
public class DBDataAbility extends Ability {
   // 定义要操作的数据库名
   private static final String DATABASE_NAME ="data.sqlite";
   private static final String DATABASE_NAME_ALIAS = "DBDataAbility";
   private OrmContext ormContext = null;
   private RdbStore store;
   public void initDB(Context context) {
      StoreConfig config = StoreConfig.newDefaultConfig("data.sqlite");

      RdbOpenCallback callback = new RdbOpenCallback() {
         @Override
         public void onCreate(RdbStore store) {
            // 用 SQL 语句创建 t_products 表
            store.executeSql("CREATE TABLE IF NOT EXISTS t_products (id INTEGER
            PRIMARY KEY, name VARCHAR(20), msg TEXT)");
            Tools.print("DBDataAbility","成功创建 t_products 表");
         }
         @Override
         public void onUpgrade(RdbStore store, int oldVersion, int newVersion) {
```

```
        }
    };
    DatabaseHelper helper = new DatabaseHelper(context);
    store = helper.getRdbStore(config,1,callback,null);

    String deleteSQL = "delete from t_products";
    // 删除t_products中的所有数据，否则再次运行程序会抛出异常（主键冲突）
    store.executeSql(deleteSQL);
    // 下面的代码使用SQL插入2条记录
    String insertSQL = "insert into t_products(id, name, msg) values(?,?,?);";
    store.executeSql(insertSQL, new Object[]{1, "harmonyos20", "超级手机"});
    insertSQL = "insert into t_products(id, name, msg) values(?,?,?);";
    store.executeSql(insertSQL, new Object[]{10, "进取号", "星际联盟战舰"});
    store.close();
    // 获得OrmContext对象
    ormContext = helper.getOrmContext(DATABASE_NAME_ALIAS, DATABASE_NAME,
                                                    Products.class);
}

@Override
public void onStart(Intent intent) {
    super.onStart(intent);
    // 在初始化Data Ability时初始化数据库
    initDB(this);
}
// 通用的校验方法，主要校验uri是否合法，path必须为data才可以操作数据库
// 校验成功，返回true，校验失败，返回false
public boolean verify(Uri uri) {
    if (ormContext == null) {
        Tools.print(this.getClass().getSimpleName(), "查询失败, ormContext为null! ");
        return false;
    }
    // 获取URI中的path，必须是data才可以
    String path = uri.getDecodedPathList().get(1);
    if(!path.equals("data")) {
        Tools.print(this.getClass().getSimpleName(), "<" + path + "> 数据库未找到");
        return false;
    }
    return true;
}
// 查询数据
@Override
public ResultSet query(Uri uri, String[] columns, DataAbilityPredicates predicates) {

    if(verify(uri)) {
        // 查询数据库
        OrmPredicates ormPredicates = DataAbilityUtils.createOrmPredicates(predicates,
        Product.class);
        // 返回查询结果集
        ResultSet resultSet = ormContext.query(ormPredicates, columns);
        if (resultSet == null) {
            Tools.print(this.getClass().getSimpleName(), "未查看任何结果");
```

```
        }
        // 返回结果集
        return resultSet;
    } else {
        return null;
    }
}
// 插入单条记录，成功插入，则返回插入对象的 row id，否则返回-1
@Override
public int insert(Uri uri, ValuesBucket value) {

    if(verify(uri)) {
        // 创建 Product 对象，对应 t_products 表
        Product product = new Product();
        product.setId(value.getInteger("id"));
        product.setName(value.getString("name"));
        product.setMsg(value.getString("msg"));
        boolean isSuccessed = true;
        try {
            // 插入数据
            isSuccessed = ormContext.insert(product);
        } catch (Exception e) {
            Tools.print(this.getClass().getSimpleName(), "insert fail: " + e.getMessage());
            return -1;
        }

        isSuccessed = ormContext.flush();
        if (!isSuccessed) {
            Tools.print(this.getClass().getSimpleName(), "failed to insert flush");
            return -1;
        }
        ormContext.flush();
        DataAbilityHelper.creator(this, uri).notifyChange(uri);
        int id = Math.toIntExact(product.getRowId());
        return id;
    } else {
        return -1;
    }
}
// 删除记录，成功删除，则返回删除对象的 id，否则返回-1
@Override
public int delete(Uri uri, DataAbilityPredicates predicates) {
    if(verify(uri)) {
        OrmPredicates ormPredicates = DataAbilityUtils.createOrmPredicates
        (predicates, Product.class);
        // 删除记录
        int value = ormContext.delete(ormPredicates);
        ormContext.flush();
        DataAbilityHelper.creator(this, uri).notifyChange(uri);
        return value;
    } else {
        return -1;
    }
}
```

```
    // 更新记录，成功更新，则返回更新对象的 id，否则返回-1
    @Override
    public int update(Uri uri, ValuesBucket value, DataAbilityPredicates predicates) {
        if(verify(uri)) {
            OrmPredicates ormPredicates = DataAbilityUtils.createOrmPredicates(predicates,
            Product.class);
            int result = ormContext.update(ormPredicates, value);
            ormContext.flush();
            DataAbilityHelper.creator(this, uri).notifyChange(uri);
            return result;
        } else {
            return -1;
        }
    }
}
```

DBDataAbility 类的配置代码如下：

```
{
    "permissions": [
      "com.unitymarvel.demo.ability.DataAbilityShellProvider.PROVIDER"
    ],
    "name": "com.unitymarvel.demo.ability.DBDataAbility",
    "icon": "$media:icon",
    "description": "$string:dbdataability_description",
    "type": "data",
    "uri": "dataability://com.unitymarvel.demo.ability.DBDataAbility"
}
```

在 DBDataAbility 类中涉及两个实体类——Products 和 Product。这两个类分别对应数据库 data.sqlite 和表 t_products。下面是这两个类的代码。

代码位置：src\main\java\com\unitymarvel\demo\ability\Products.java

```
package com.unitymarvel.demo.ability;
import ohos.data.orm.OrmDatabase;
import ohos.data.orm.annotation.Database;

@Database(entities = {Product.class}, version = 1)
public abstract class Products extends OrmDatabase {
}
```

代码位置：src\main\java\com\unitymarvel\demo\ability\Product.java

```
package com.unitymarvel.demo.ability;

import ohos.data.orm.OrmObject;
import ohos.data.orm.annotation.Entity;
import ohos.data.orm.annotation.PrimaryKey;

@Entity(tableName = "t_products")
public class Product extends OrmObject {
    @PrimaryKey(autoGenerate = true)
    private Integer id;
```

```
    private String name;
    private String msg;

    public Integer getId() {
        return id;
    }
    public String getName() {
        return name;
    }
    public String getMsg() {
        return msg;
    }
    public void setId(Integer id) {
        this.id = id;
    }
    public void setName(String name) {
        this.name = name;
    }
    public void setMsg(String msg) {
        this.msg = msg;
    }
}
```

最后看一下调用 DBDataAbility 的代码（省略了无关的代码）。

代码位置：src\main\java\com\unitymarvel\demo\ability\DataAbilityDemo.java

```
public class DataAbilityDemo extends Ability {
    private Uri dbUri =
        Uri.parse("dataability:///com.unitymarvel.demo.ability.DBDataAbility/data");
    ...
    @Override
    public void onStart(Intent intent) {
        super.onStart(intent);
        super.setUIContent(ResourceTable.Layout_data_ability_demo);
        ...
        Button button1 = (Button)findComponentById(ResourceTable.Id_button1);
        if(button1 != null) {
            // 访问本地数据库
            button1.setClickedListener(new Component.ClickedListener() {
                @Override
                public void onClick(Component component) {
                    DataAbilityHelper helper =
                        DataAbilityHelper.creator(DataAbilityDemo.this);
                    try {
                        try {
                            // 插入 1 条记录
                            ValuesBucket valuesBucket = new ValuesBucket();
                            valuesBucket.putInteger("id", 2);
                            valuesBucket.putString("name", "iPhone");
                            valuesBucket.putString("msg", "Apple 公司的手机");
                            helper.insert(dbUri, valuesBucket);

                            // 批量插入记录
```

```
            ValuesBucket[] values = new ValuesBucket[2];
            values[0] = new ValuesBucket();
            values[0].putInteger("id", 3);
            values[0].putString("name", "Android");
            values[0].putString("msg", "开源手机操作系统");
            values[1] = new ValuesBucket();
            values[1].putInteger("id", 4);
            values[1].putString("name", "特斯拉");
            values[1].putString("msg", "电动汽车");
            helper.batchInsert(dbUri, values);

        } catch (Exception e) {

        }

        // 删除1条记录
        DataAbilityPredicates predicates = new DataAbilityPredicates();
        predicates.equalTo("id",3);
        helper.delete(dbUri,predicates);

        // 更新id等于4的记录
        predicates = new DataAbilityPredicates();
        predicates.equalTo("id",4);

        ValuesBucket valuesBucket = new ValuesBucket();
        valuesBucket.putString("name", "Redmi 9A");
        valuesBucket.putString("msg", "红米手机");
        helper.update(dbUri, valuesBucket, predicates);

        // 查询数据
        predicates = new DataAbilityPredicates();
        predicates.between ("id",1,4);

        ResultSet resultSet = helper.query(dbUri,null,predicates);
        String content = "";

        while(resultSet.goToNextRow())
        {
            String result = resultSet.getInt(0) + "\r\n" +
                    resultSet.getString(1) + "\r\n" +
                    resultSet.getString(2) + "\r\n";
            content += result;
        }
        // 用Toast信息框显示查询结果
        Tools.showTip(DataAbilityDemo.this, content.trim());
        resultSet.close();

    } catch (Exception e) {
        Tools.showTip(DataAbilityDemo.this, "error: " + e.getMessage());
    }
}
```

```
            });
        }
    }
    ...
}
```

插入记录有以下两个方法可用。

- insert：对应 Data Ability 中的 insert 方法，插入单条记录。
- batchInsert：对应 Data Ability 中的 batchInsert 方法，插入多条记录。但该方法被声明为 final，即不能在子类中重写该方法。这是因为 batchInsert 方法已经实现了通过多次调用 insert 方法插入记录，所以用户并不需要自己实现 batchInsert 方法，只需要在客户端调用 batchInsert 即可。

运行程序，然后单击“访问本地数据库”按钮，会显示图 7-3 所示的 Toast 信息框。

图 7-3 通过 Data Ability 访问本地数据库

7.5 访问本地文件

Data Ability 可以用于访问本地文件，为了实现这个功能，需要实现 Ability 的 openFile 方法，该方法的原型如下：

```
public FileDescriptor openFile(Uri uri, String mode);
```

其中，uri 参数值是访问 Data Ability 的 URI，可以通过 uri 参数获取 URI 的每一部分。mode 参数值是打开文件的模式，例如，r 表示只读，w 表示只写，rw 表示可读写。openFile 方法可以返回一个 FileDescriptor 对象，但不能直接返回 FileDescriptor 对象，而要使用 MessageParcel 对象的 dupFileDescriptor 方法复制待操作文件流的文件描述符，并将其返回。也就是说，openFile 方法实际上返回的是 dupFileDescriptor 方法的返回值。

如果要调用 Data Ability，需要创建 DataAbilityHelper 对象，然后使用该对象的相关方法访问 Data Ability。例如，使用 DataAbilityHelper. openFile 方法可以打开文件，其实就是通过 URI 调用 Data Ability 类中的 openFile 方法。

【例 7.2】 实现一个用于读取本地文件的 Data Ability（名为 FileDataAbility）。在程序启动时，会自动在 App 的 data 目录下生成一个名为 data.txt 的文件，并写入两行字符串。然后 Data Ability 中的 openFile 方法会读取 data.txt 文件，并返回对应的 FileDescriptor 对象。在调用端，会利用 FileDescriptor 对象读取 data.txt 文件的内容，并显示在 Toast 信息框中。

首先看一下 FileDataAbility 类的实现。

代码位置：src\main\java\com\unitymarvel\demo\ability\FileDataAbility.java

```
package com.unitymarvel.demo.ability;
```

```
import ohos.aafwk.ability.Ability;
import ohos.rpc.MessageParcel;
import ohos.utils.net.Uri;
import java.io.File;
import java.io.FileDescriptor;
import java.io.FileInputStream;

public class FileDataAbility extends Ability {
    // 打开本地文件
    @Override
    public FileDescriptor openFile(Uri uri, String mode) {
        // 获取 App 的 data 目录
        File dataDirFile = getDataDir();
        // 创建 MessageParcel 对象 （用于打包数据）
        MessageParcel messageParcel = MessageParcel.obtain();
        // 根据 uri 中的路径组合成要读取的文件名，也就是 data.txt 文件的完整路径
        // uri.getDecodedPathList().get(1)获得了 URI 的 path 部分，也就是 data.txt
        File file = new File(dataDirFile.getAbsolutePath() + "/" +
                                        uri.getDecodedPathList().get(1));
        // 如果未指定 mode 或指定为只读模式，则将文件设为只读状态
        if (mode == null || "r".equals(mode)) {
            file.setReadOnly();
        }
        try {
            // 打开文件
            FileInputStream fis = new FileInputStream(file);
            // 获取 FileDescriptor 对象
            FileDescriptor fd = fis.getFD();
            // 复制 FileDescriptor 对象，并返回该对象
            return messageParcel.dupFileDescriptor(fd);
        } catch (Exception e) {
        }
        return null;
    }
}
```

FileDataAbility 类在 config.json 文件中的配置代码如下：

```
{
    "permissions": [
      "com.unitymarvel.demo.ability.DataAbilityShellProvider.PROVIDER"
    ],
    "name": "com.unitymarvel.demo.ability.FileDataAbility",
    "icon": "$media:icon",
    "description": "$string:filedataability_description",
    "type": "data",
    "visible": true,
    "uri": "dataability://com.unitymarvel.demo.ability.FileDataAbility"
}
```

从 FileDataAbility 的配置代码可以得到自定义的权限以及访问 FileDataAbility 的 URI，接下来编写调用 FileDataAbility 的代码。这些代码在 DataAbilityDemo 中，DataAbilityDemo 是一个 Page

Ability，在运行 DataAbilityDemo 时会在 data 目录下创建一个 data.txt 文件，并写入两行文本。

代码位置：src\main\java\com\unitymarvel\demo\ability\DataAbilityDemo.java

```
package com.unitymarvel.demo.ability;

import com.unitymarvel.demo.ResourceTable;
import com.unitymarvel.demo.Tools;
import ohos.aafwk.ability.Ability;
import ohos.aafwk.ability.DataAbilityHelper;
import ohos.aafwk.content.Intent;
import ohos.agp.components.Button;
import ohos.agp.components.Component;
import ohos.utils.net.Uri;
import java.io.*;

public class DataAbilityDemo extends Ability {
   // 指定定义要写入的文件
   private String filename = "data.txt";
   // 指定 data.txt 文件的完整路径，在 dataability:后面必须指定 3 个斜杠（/）
   private Uri uri =
        Uri.parse("dataability:///com.unitymarvel.demo.ability.FileDataAbility/" +
        filename);
   // 在 data 目录下创建 data.txt 文件，并写入两行字符串
   private void writeFile() {
      // 获取 data 目录
      File dataDirFile = getDataDir();
      try {
          // 以写模式打开或创建 data.txt 文件
          FileWriter fw = new FileWriter(dataDirFile.getAbsoluteFile() + "/" + filename);
          // 写入两行字符串
          fw.write("hello world\r\n 世界，您好！");
          // 关闭文件
          fw.close();
      }catch (Exception e) {
      }
   }
   @Override
   public void onStart(Intent intent) {
      super.onStart(intent);
      super.setUIContent(ResourceTable.Layout_data_ability_demo);
      // 在 data 目录下创建 data.txt 文件
      writeFile();
      // 这里省略了与本例无关的代码
       ...
      Button button2 = (Button)findComponentById(ResourceTable.Id_button2);
      if(button2 != null) {
         button2.setClickedListener(new Component.ClickedListener() {
            @Override
            public void onClick(Component component) {
               // 创建 DataAbilityHelper 对象
               DataAbilityHelper helper =
             DataAbilityHelper.creator(DataAbilityDemo.this);
               try {
```

```
                    // 以只读的方式打开文件，服务端会以只读模式返回 FileDescriptor 对象
                    FileDescriptor fd = helper.openFile(uri, "r");
                    if(fd == null) {
                        throw new Exception("文件不存在");
                    }
                    // 使用 FileDescriptor 对象打开文件
                    FileInputStream fis = new FileInputStream(fd);
                    // 根据 FileInputStream 对象创建 FileReader 对象
                    FileReader fr = new FileReader(fd);

                    BufferedReader br = new BufferedReader(fr);
                    String line = "";
                    String content = "";
                    // 读取 data.txt 文件的每一行信息
                    while((line = br.readLine()) != null) {
                    content += line + "\r\n";
                    }
                    // 关闭文件
                    br.close();
                    // 用 Toast 信息框显示 data.txt 文件内容
                    Tools.showTip(DataAbilityDemo.this, content.trim());

                } catch (Exception e) {
                    Tools.showTip(DataAbilityDemo.this, "error: " + e.getMessage());
                }
            }
        });
    }
  }
}
```

为了方便，本例使用同一个 App 调用自身的 Data Ability，其实也可以将 Data Ability 和调用代码分开，作为两个独立的 App 来运行，方式都是一样的。

现在运行程序，单击“访问本地文件”按钮，会通过 Toast 信息框显示 data.txt 文件中的内容，如图 7-4 所示。

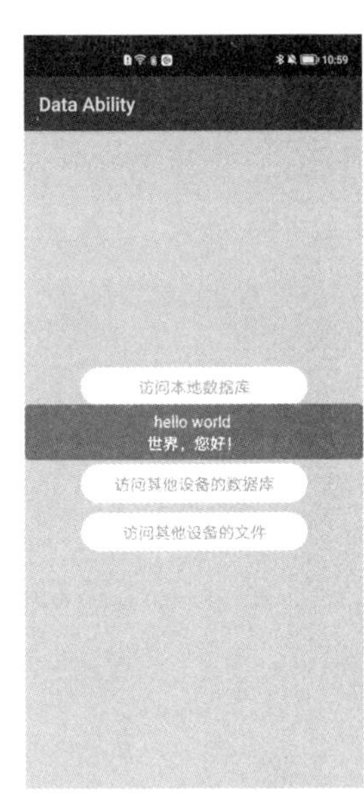

图 7-4 访问本地文件

读者可以用 adb shell 命令进到 HarmonyOS 手机的 Shell，然后进入下面的目录：

```
/data/data/com.unitymarvel.demo
```

这个目录就是 App 的 data 目录，在该目录下会有一个 data.txt 文件，使用 cat 命令查看该文件的内容，会得到图 7-5 所示的结果。这就是前面通过 Data Ability 读取的文件的内容。

```
HWANA:/data/data/com.unitymarvel.demo # ls
cache code_cache data.txt files
HWANA:/data/data/com.unitymarvel.demo # cat data.txt
hello world
世界，您好！ HWANA:/data/data/com.unitymarvel.demo #
```

图 7-5 查看 data.txt 文件的内容

注意，在指定文件路径时，要指定 HarmonyOS 设备上的路径，不要指定开发时 PC 的路径，否则无法找到对应的文件。

7.6 跨设备访问数据库

想访问其他 HarmonyOS 设备上的 Data Ability，只需在 URI 中指定 deviceId 即可。不过，需要在 config.json 文件中按如下方式修改 DBDataAbility 的两处配置。

（1）将 visible 属性值设为 true。在默认情况下，DBDataAbility 对非本机的 App 是不可见的，所以需要在 DBDataAbility 的配置代码中将 visible 属性值设置为 true，设置完后的配置代码如下：

```
{
   "permissions": [
     "com.unitymarvel.demo.ability.DataAbilityShellProvider.PROVIDER"
   ],
   "name": "com.unitymarvel.demo.ability.DBDataAbility",
   "icon": "$media:icon",
   "description": "$string:dbdataability_description",
   "type": "data",
   "visible": true,
   "uri": "dataability://com.unitymarvel.demo.ability.DBDataAbility"
}
```

（2）将 DBDataAbility 的权限设为系统级，配置代码如下：

```
"defPermissions": {
   "name": "com.unitymarvel.demo.ability.DataAbilityShellProvider.PROVIDER",
   "grantMode": "system_grant"
}
```

【例 7.3】 使用例 7.1 中实现的 DBDataAbility，只是在另一部 HarmonyOS 手机中调用本机的 DBDataAbility。所以，要完成本例，至少需要两部 HarmonyOS 手机。

下面是调用另一部 HarmonyOS 手机中 DBDataAbility 的代码（省略了无关的代码）。

代码位置：src\main\java\com\unitymarvel\demo\ability\DataAbilityDemo.java

```
public class DataAbilityDemo extends Ability {
   private void requestPermission() {
       String[] permission = {
               "ohos.permission.DISTRIBUTED_DATASYNC",
       };
       List<String> applyPermissions = new ArrayList<>();
       for (String element : permission) {

           if (verifySelfPermission(element) != 0) {
               if (canRequestPermission(element)) {
                   applyPermissions.add(element);
               } else {
               }
           } else {
```

```
            }
        }
        requestPermissionsFromUser(applyPermissions.toArray(new String[0]), 0);
    }
    @Override
    protected void onAbilityResult(int requestCode, int resultCode, Intent resultData) {
        String deviceId = resultData.getStringParam("deviceId");
        // 跨设备访问 DBDataAbility
        if(resultCode == 100 && requestCode == 98) {
            // 生成带 deviceId 的 URI
            Uri uri = Uri.parse("dataability://" + deviceId +
                       "/com.unitymarvel.demo.ability.DBDataAbility/data");
            DataAbilityHelper helper = DataAbilityHelper.creator(DataAbilityDemo.this);
            try {
                // 插入 1 条记录
                try {
                    ValuesBucket valuesBucket = new ValuesBucket();
                    valuesBucket.putInteger("id", 12);
                    valuesBucket.putString("name", "猎禽舰");
                    valuesBucket.putString("msg", "克林贡帝国战舰");
                    helper.insert(uri, valuesBucket);
                }catch (Exception e) {

                }
                // 查询数据
                DataAbilityPredicates predicates = new DataAbilityPredicates();
                predicates.greaterThan ("id",2);

                ResultSet resultSet = helper.query(uri,null,predicates);
                String content = "";

                while(resultSet.goToNextRow())
                {
                    String result = resultSet.getInt(0) + "\r\n" +
                           resultSet.getString(1) + "\r\n" +
                           resultSet.getString(2) + "\r\n";
                    content += result;

                }

                Tools.showTip(DataAbilityDemo.this, content.trim());

                resultSet.close();

            } catch (Exception e) {
                Tools.showTip(DataAbilityDemo.this, "error: " + e.getMessage());
            }
        }
        ...
    }
    @Override
    public void onStart(Intent intent) {
        super.onStart(intent);
        super.setUIContent(ResourceTable.Layout_data_ability_demo);
```

```
        requestPermission();
        ...
        // 跨设备访问 DBDataAbility
        Button button3 = (Button)findComponentById(ResourceTable.Id_button3);
        if(button3 != null) {
            button3.setClickedListener(new Component.ClickedListener() {
                @Override
                public void onClick(Component component) {
                    Intent intentPageAbility = new Intent();
                    Operation operation = new Intent.OperationBuilder()
                            .withBundleName("com.unitymarvel.demo")
                            .withAbilityName("com.unitymarvel.demo.ability.DeviceIdsAbility")
                             .build();
                    intentPageAbility.setOperation(operation);
                    startAbilityForResult(intentPageAbility,98);
                }
            });
        }
        ...
    }
}
```

这段代码使用了在 2.5.2 节实现的用于显示设备列表的 Page Ability。

运行程序，单击“访问其他设备的数据库”按钮，会弹出选择设备窗口，选择一个同样安装了当前 App 的设备，然后会弹出图 7-6 所示的 Toast 信息框，单击另一部设备中的“访问本地数据库”按钮，在本地数据库中插入、删除和更新一些数据，然后再次单击本机的“访问其他设备的数据库”按钮，会弹出图 7-7 所示的 Toast 信息框。

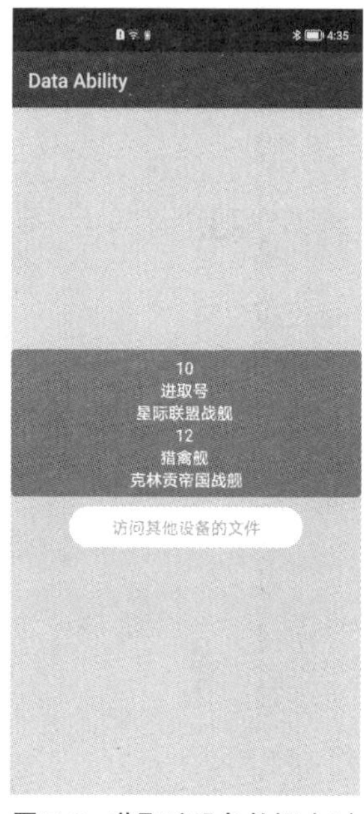

图 7-6 获取跨设备数据（1）

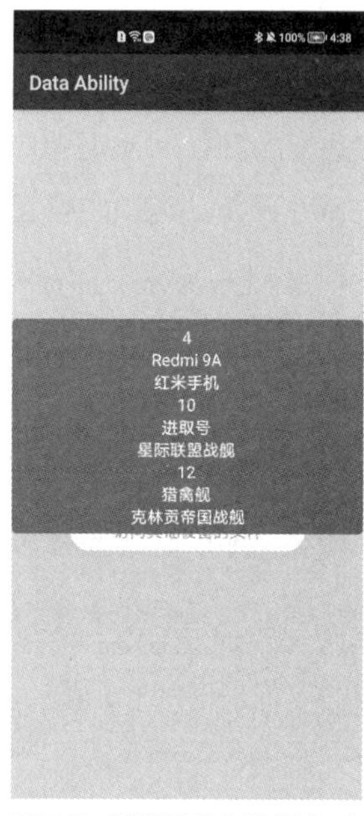

图 7-7 获取跨设备数据（2）

7.7 跨设备访问文件

与跨设备访问数据库类似，跨设备访问文件也需要指定 deviceId，而且 config.json 中的配置与上一节类似。

【例 7.4】 使用例 7.2 中实现的 FileDataAbility，只是在另一部 HarmonyOS 手机中调用本机的 FileDataAbility。所以，要完成本例，至少需要两部 HarmonyOS 手机。

调用其他 HarmonyOS 设备中 FileDataAbility 的代码如下：

代码位置：src\main\java\com\unitymarvel\demo\ability\DataAbilityDemo.java

```
Intent intentPageAbility = new Intent();
Operation operation = new Intent.OperationBuilder()
       .withBundleName("com.unitymarvel.demo")
       .withAbilityName("com.unitymarvel.demo.ability.DeviceIdsAbility")
       .build();
intentPageAbility.setOperation(operation);
// 显示选择设备 ID 的窗口
startAbilityForResult(intentPageAbility,99);
```

代码位置：src\main\java\com\unitymarvel\demo\ability\DataAbilityDemo.java

```
// 选择某一个特定设备后，会回调该方法
@Override
protected void onAbilityResult(int requestCode, int resultCode, Intent resultData) {
   String deviceId = resultData.getStringParam("deviceId");
   if(resultCode == 100 && requestCode == 98) {
      ...
   } else if(resultCode == 100 && requestCode == 99) {   // 跨设备访问文件
      Uri uri = Uri.parse("dataability://" + deviceId +
              "/com.unitymarvel.demo.ability.FileDataAbility/" + filename);
      DataAbilityHelper helper = DataAbilityHelper.creator(DataAbilityDemo.this);
      try {
         // 这个 URI 带 deviceId
         FileDescriptor fd = helper.openFile(uri, "r");
         helper.query(uri,null,null);
         if(fd == null) {
            throw new Exception("文件不存在");
         }
         FileInputStream fis = new FileInputStream(fd);
         FileReader fr = new FileReader(fd);

         BufferedReader br = new BufferedReader(fr);
         String line = "";
         String content = "";
         while((line = br.readLine()) != null) {
            content += line + "\r\n";
         }
         br.close();
```

```
            Tools.showTip(DataAbilityDemo.this, content.trim());

        } catch (Exception e) {
            Tools.showTip(DataAbilityDemo.this, "error: " + e.getMessage());
          }
      }
  }
```

首先在两部 HarmonyOS 手机上运行 App，然后在其中一部 HarmonyOS 手机上单击“访问其他设备的文件”按钮，会弹出一个 deviceId 选择窗口，选择已经运行 App 的设备后，会弹出与图 7-4 相同的 Toast 信息框。

7.8 总结与回顾

在第 2 章中我们介绍了 HarmonyOS 的一项功能，即用 Page Ability 创建窗口。本章讲解了 HarmonyOS 的另外一项功能——Data Ability。Data Ability 是用来获取数据的，那么可能有人会问，Data Ability 和上一章讲的数据管理有什么区别呢？数据管理中的技术，是直接将数据暴露给用户，相当于“直销模式”。而 Data Ability 其实是一种代理的方式，用户并不知道返回的数据来源，只需要使用它们。这些数据可能来自文本文件、XML 文件、JSON 文件、SQLite 数据库、网络。也就是说，Data Ability 就是对数据的抽象，相当于“代销模式”。对于“直销模式”，用户看到的是原始数据，也就是数据出厂时的状态，而对于“代销模式”，用户获得的数据有可能被加工过。就像商品的代销一样，代销商有可能在商品上添加自己的包装，或者附带其他的产品，总之，通过“代销模式”获得的数据，有可能不是数据的原始状态。

综上所述，Data Ability 有两个作用——数据抽象和数据加工。

另外，Data Ability 也支持 HarmonyOS 的分布式特性，也就是说，一台 HarmonyOS 设备上的 Data Ability 可以被任何其他 HarmonyOS 设备访问，当然，这些 HarmonyOS 设备需要得到授权。

第 8 章　Service Ability

Service Ability 是 HarmonyOS 中一种非常重要的 Ability，是 PA 中的一种。顾名思义，Service Ability 与服务有关。Service Ability 允许在后台运行，也允许向外暴露一些 API，供其他 App 调用。Service Ability 与 HarmonyOS 中其他 Ability 一样，既允许在本机调用，也允许跨设备调用。

通过阅读本章，读者可以掌握：

- 如何启动和停止本机的 Service Ability；
- 如何跨设备启动和停止 Service Ability；
- 如何调用 Service Ability 的 API。

8.1 Service Ability 的生命周期

Service Ability 有两个生命周期，一个是在后台运行的生命周期，另一个是远程调用的生命周期。这两个生命周期有重合的方法，也有独立的方法。

1. 在后台运行的 Service Ability 的生命周期

在后台运行的 Service Ability，由以下 4 个生命周期方法组成。

- onStart：Service Ability 初始化时调用，在 Service Ability 的创建过程中只调用一次。
- onCommand：每次启动 Service Ability 时被调用。如果是第一次启动 Service Ability，首先会调用 onStart 方法，然后调用 onCommand 方法。以后再启动 Service Ability，就只会调用 onCommand 方法。
- onBackground：停止 Service Ability 时第一个调用的方法，调用该方法后，Service Ability 会切换到后台运行，HarmonyOS 会在系统资源紧张时回收这个 Service Ability。
- onStop：停止 Service Ability 时最后一个调用的方法。

在后台运行的 Service Ability 生命周期示意如图 8-1 所示。

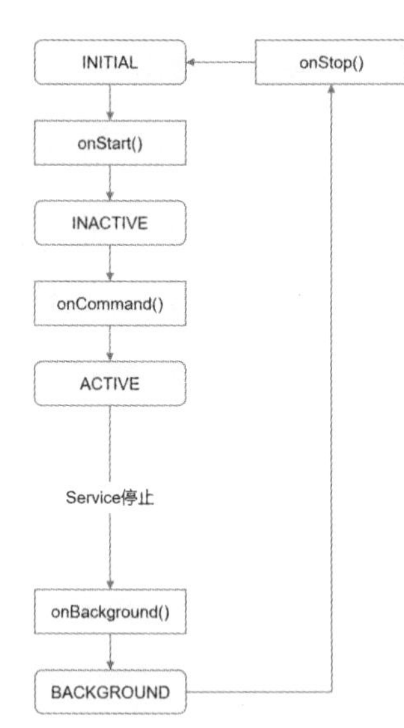

图 8-1　在后台运行的 Service Ability 生命周期

2. 作为远程服务的 Service Ability 的生命周期

Service Ability 的另外一个功能就是向外暴露一些方法（API），以便其他 App 调用，与这一功能相关的生命周期方法有如下 5 个。

- onStart：成功连接 Service Ability 后，会调用该方法，一般用于初始化 Service Ability。
- onConnect：需要返回远程对象时，会调用该方法。onConnect 方法返回的对象就是客户端要使用的对象。
- onDisconnect：客户端与 Service Ability 断开连接时被调用。
- onBackground：断开连接后，调用完 onDisconnect 方法，会调用该方法。
- onStop：断开连接最后调用的方法，表示 Service Ability 彻底停止。

作为远程服务的 Service Ability 的生命周期如图 8-2 所示。

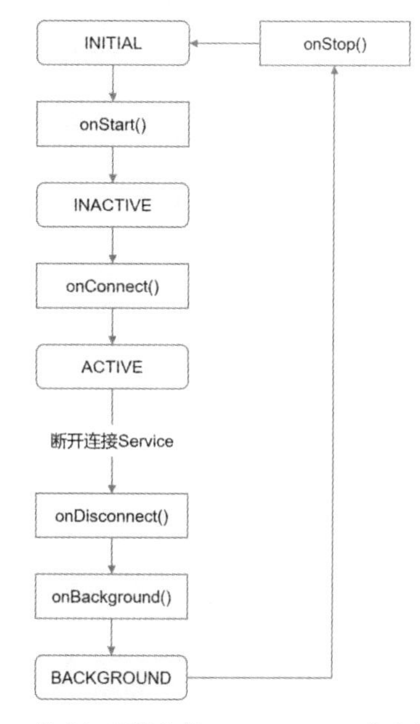

图 8-2 作为远程服务的 Service Ability 的生命周期

8.2 后台运行 Service Ability

Service Ability 允许在后台常驻内存，并在不打扰用户的情况下执行一些任务，如下载文件、播放音乐等。本节将深入介绍从创建 Service Ability，到执行 Service Ability，以及停止 Service Ability 的完整步骤。

8.2.1 操作本地的 Service Ability

Service Ability 类与其他 Ability 类相同，都必须从 Ability 类继承，只是需要在 config.json 文件的配置代码中将 type 属性值设置为 service。为了可以跨设备调用，需要将 visible 属性值设置为 true。

如果只是让 Service Ability 在后台运行，最多只需要实现图 8-1 所示的 4 个生命周期方法。这 4 个生命周期方法的原型如下：

```
public void onStart(Intent intent) ;
public void onBackground();
public void onStop();
public void onCommand(Intent intent, boolean restart, int startId);
```

在这 4 个方法中，onBackground 方法和 onStop 方法都没有参数，这两个方法只是用来拦截状态的。onStart 方法与 Page Ability 中的 onStart 方法相同，用于初始化。intent 参数主要用来传递数据。onCommand 方法有 3 个参数，其中第 1 个参数（intent）与 onStart 方法的 intent 参数含义相同，后两个参数的含义如下。

- restart：设置 Service Ability 的启动模式。如果该参数的值为 false，表示 Service Ability 停止后不会再启动（除非再次显式启动 Service Ability）；如果该参数的值为 true，表示 Service Ability 停止后会自动重启。
- startId：Service Ability 启动的次数。例如，Service Ability 被启动了 6 次，那么该参数的值为 6。

【例 8.1】　创建一个名为 MyServiceAbility 的 Service Ability，在该 Service Ability 中实现 onStart、onCommand、onBackground 和 onStop 这 4 个方法，并用 startAbility 方法和 stopAbility 方法启动和停止 MyServiceAbility。

首先看一下 MyServiceAbility 类的实现。

代码位置：src\main\java\com\unitymarvel\demo\ability\MyServiceAbility.java

```
package com.unitymarvel.demo.ability;

import com.unitymarvel.demo.Tools;
import ohos.aafwk.ability.Ability;
import ohos.aafwk.content.Intent;
import ohos.event.notification.NotificationRequest;

public class MyServiceAbility extends Ability {
   @Override
   public void onStart(Intent intent) {
      super.onStart(intent);
      Tools.print("MyServiceAbility", "onStart");
   }
   @Override
   public void onBackground() {
      super.onBackground();
      Tools.print("MyServiceAbility", "onBackground");
   }
   @Override
   public void onStop() {
      super.onStop();
      Tools.print("MyServiceAbility", "onStop");
   }
   @Override
   public void onCommand(Intent intent, boolean restart, int startId) {
      Tools.print("MyServiceAbility", "onCommand");
   }
}
```

在 MyServiceAbility 类的 4 个方法中只是输出相应的信息，这样在 HiLog 视图中可以观察生命周期方法调用的情况。

MyServiceAbility 类在 config.json 文件中的配置代码如下：

```
{
  "skills": [
    {
```

```
            "actions": [
                "action.service.background.MyServiceAbility"
            ]
        }
    ],
    "name": "com.unitymarvel.demo.ability.MyServiceAbility",
    "icon": "$media:icon",
    "visible": true,
    "description": "$string:myserviceability_description",
    "type": "service"
}
```

我们可以看到，在配置 MyServiceAbility 时指定了 action，所以在启动和停止 MyServiceAbility 时既可以采用隐式的方式，也可以采用显式的方式。本例采用了隐式的方式来启动和停止 MyServiceAbility。

下面是启动和停止 MyServiceAbility 的代码。

代码位置：src\main\java\com\unitymarvel\demo\ability\BackgroundServiceAbility.java

```
package com.unitymarvel.demo.ability;

import com.unitymarvel.demo.ResourceTable;
import com.unitymarvel.demo.Tools;
import ohos.aafwk.ability.Ability;
import ohos.aafwk.content.Intent;
import ohos.agp.components.Button;
import ohos.agp.components.Component;
import java.util.ArrayList;
import java.util.List;

public class BackgroundServiceAbility extends Ability {
   @Override
   public void onStart(Intent intent) {
      super.onStart(intent);
      super.setUIContent(com.unitymarvel.demo.ResourceTable.Layout_background_service);
      requestPermission();
      Button button1 = (Button)findComponentById(ResourceTable.Id_button1);
      if(button1 != null) {
         button1.setClickedListener(new Component.ClickedListener() {
            @Override
            public void onClick(Component component) {
               Intent intent = new Intent();
               // 指定服务的 action
               intent.setAction("action.service.background.MyServiceAbility");
               // 启动服务
               startAbility(intent);
               Tools.showTip(BackgroundServiceAbility.this, "服务已经启动");
            }
         });
      }
      Button button2 = (Button)findComponentById(ResourceTable.Id_button2);
      if(button2 != null) {
```

```
            button2.setClickedListener(new Component.ClickedListener() {
                @Override
                public void onClick(Component component) {
                    Intent intent = new Intent();
                    // 指定服务的 action
                    intent.setAction("action.service.background.MyServiceAbility");
                    // 停止服务
                    stopAbility(intent);
                    Tools.showTip(BackgroundServiceAbility.this, "服务已经停止");
                }
            });
        }

    }
}
```

运行程序，第 1 次单击“启动服务”按钮，在 HiLog 视图中会依次输出如下信息：

```
onStart
onCommand
```

这说明第 1 次启动服务，服务需要初始化，所以执行了 onStart 方法。再次单击“启动服务”按钮，会发现，这次只输出了 onCommand，所以再次启动服务，其实只调用 onCommand 方法。现在单击“停止服务”按钮，在 HiLog 视图中会输出如下信息。这说明停止服务时会依次调用 onBackground 方法和 onStop 方法。

```
onBackground
onStop
```

8.2.2 跨设备操作 Service Ability

跨设备启动和停止 Service Ability 需要指定 deviceId，所以需要使用在 2.5.2 节创建的设备列表窗口，使用方式已经多次给出，这里不赘述。从设备列表窗口选择一个已经运行了这个 App 的设备，然后会通过 onAbilityResult 方法来处理跨设备启动和停止 Service Ability 的工作，下面是该方法的代码。

代码位置：src\main\java\com\unitymarvel\demo\ability\BackgroundServiceAbility.java

```
protected void onAbilityResult(int requestCode, int resultCode, Intent resultData) {
    // 跨设备启动服务
    if(requestCode == 99) {
        // 获取设备 ID
        String deviceId = resultData.getStringParam("deviceId");
        Intent intent = new Intent();
        ElementName elementName = new
        ElementName(deviceId,"com.unitymarvel.demo","com.unitymarvel.demo.ability.
        MyServiceAbility");
        intent.setElement(elementName);
        startAbility(intent);
        Tools.showTip(BackgroundServiceAbility.this, "服务已经启动");
```

```
        } else if(requestCode == 98) {
            // 获取设备 ID
            String deviceId = resultData.getStringParam("deviceId");
            Intent intent = new Intent();
            ElementName elementName = new
            ElementName(deviceId,"com.unitymarvel.demo","com.unitymarvel.demo.ability.
            MyServiceAbility");
            intent.setElement(elementName);
            stopAbility(intent);
            Tools.showTip(BackgroundServiceAbility.this, "服务已经停止");
        }
    }
```

如果要测试本例，至少需要准备两部 HarmonyOS 手机，然后在这两部 HarmonyOS 手机上都安装本例的 App，最后在其中一部 HarmonyOS 手机上单击"跨设备启动服务"按钮和"跨设备停止服务"按钮，接下来在另一部 HarmonyOS 手机的 HiLog 视图中观察输出结果，会发现与启动和停止本地服务的输出结果完全相同。

8.3 跨设备调用 Service Ability 中的 API

Service Ability 的另外一个功能就是向外部暴露一些 API，这些 API 可以供本机的其他 App 调用，也可以供其他 HarmonyOS 设备的 App 调用。本机调用和跨设备调用的主要区别就是本地调用时的 deviceId 为空串，而跨设备调用时需要指定被调用设备的 deviceId。因为本机调用与跨设备调用的差别不大，所以本节主要介绍跨设备调用 Service Ability 中的 API。

【例 8.2】 创建一个名为 TestServiceAbility 的 Service Ability，并向外暴露两个方法——add 和 greet，分别用来计算两个整数之和以及返回问候语。然后在另一个 HarmonyOS 手机上调用 TestServiceAbility 中暴露出来的两个方法，最后在 Toast 信息框中显示这两个方法返回的结果。

要完成本例，需要按如下的步骤操作。

1. 创建接口定义文件

接口定义文件的扩展名是 idl，用于描述要暴露的 API，代码格式与 Java 接口类似，只是不需要使用 package 指定包。为了方便，后面的章节统一将接口定义文件统称为 idl 文件。

idl 文件通常会被放到工程目录的 src/main/idl 子目录下，例如，本例的包名是 com.unitymarvel.demo，所以 idl 文件通常会在图 8-3 所示的位置。

idl 文件可以手动创建，也可以自动创建，作者推荐使用自动的方式创建 idl 文件。选中 idl 目录，单击鼠标右键，在弹出的菜单中选择 New→Idl File 菜单项，会弹出图 8-4 所示的 Create IDL File 窗口，在窗口中的文本框内输入 IRMIServiceInterface，然后单击 OK 按钮创建 IRMIServiceInterface.idl 文件。

使用自动方式创建 idl 文件，会自动将 idl 文件放到 App 的包中（即自动配置 config.json 文件中的 bundleName 属性值）。

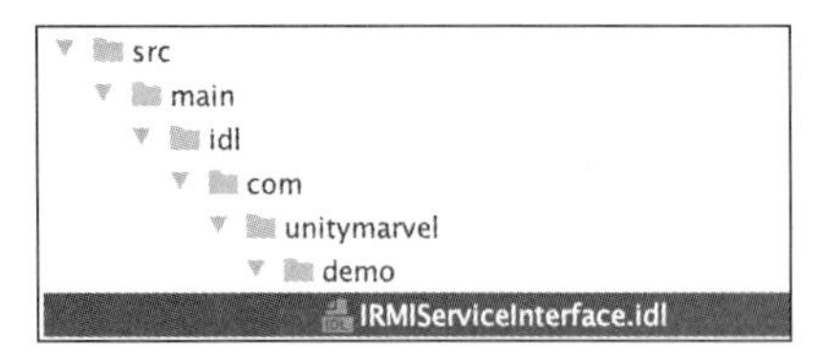

图 8-3 idl 文件的位置

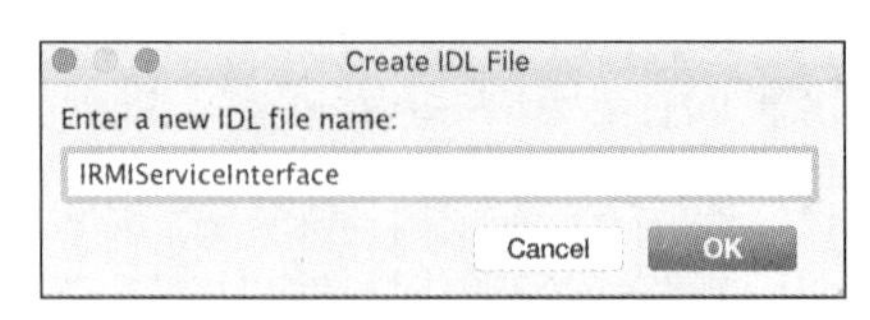

图 8-4 Create IDL File 对话框

2. 在 idl 文件中编写接口代码

idl 文件的代码格式与 Java 接口类似，只是不需要使用 package 指定包，而是将包名加到接口名前面，本例的 idl 文件代码如下：

```
interface com.unitymarvel.demo.IRMIServiceInterface {
   int add([in] int a, [in] int b);
   String greet([in] String name);
}
```

在 IRMIServiceInterface 接口中有两个方法——add 和 greet，分别用来求两个整数之和以及返回问候语。该方法总体上与在 Java 接口中定义的方法类似，只是在参数类型前多了一个[in]，这个标记是 idl 文件中规定的，任何方法的参数前都必须加类似的标记。[in]表示参数只作为输入，[out]表示参数只作为输出，[inout]表示参数既可作为输入，也可以作为输出。

3. 自动生成 Java 文件

idl 文件是不能直接用的，所以需要利用在上一步创建的 idl 文件自动生成相关的 Java 文件。现在编译整个工程，IDE 就会自动利用 idl 文件在工程目录中 entry/build/generated/source/idl 目录下生成图 8-5 所示的 3 个文件。

本节后面的内容会详细描述这 3 个 Java 文件的情况。

注意，如果 idl 文件自动生成的 3 个 Java 文件没有在 IDE 中出现，读者可以关闭 IDE，然后重新打开 IDE，就会加载这 3 个 Java 文件了。

```
entry
  build
    generated
      source
        annotation
        idl
          com.unitymarvel.demo
            IRMIServiceInterface
            RMIServiceInterfaceProxy
            RMIServiceInterfaceStub
```

图 8-5 利用 idl 文件自动生成的 3 个 Java 文件

4. 创建 Service Ability 类

在本例中创建的 Service Ability 类是 TestServiceAbility 类，因为这个 Service Ability 类的目的是作为 RMI（远程方法调用），而且不需要初始化，所以只需要实现 onConnect 方法和 onDisconnect 方法。对于本例的功能，onDisconnect 方法并不是必须的。因此，完全可以只实现 onConnect 方法。

代码位置： src\main\java\com\unitymarvel\demo\ability\TestServiceAbility.java

```
package com.unitymarvel.demo.ability;

import com.unitymarvel.demo.RMIServiceInterfaceStub;
import com.unitymarvel.demo.Tools;
```

```
import ohos.aafwk.ability.Ability;
import ohos.aafwk.content.Intent;
import ohos.rpc.IRemoteObject;
import ohos.rpc.RemoteException;

public class TestServiceAbility extends Ability {
   // 当客户端连接 TestServiceAbility 时会调用该方法
   @Override
   protected IRemoteObject onConnect(Intent intent) {
      Tools.print("MyServiceAbility", "onConnect");
      return null;
   }
}
```

从这段代码可以看出，onConnect 方法返回 null。其实该方法应该返回一个远程对象，也就是客户端需要获得的对象，该对象对应的类会在下一步实现。

5. 编写远程类

在这一步我们要用到利用 idl 文件自动生成的 3 个 Java 文件中的：RMIServiceInterfaceStub.java，该文件可以称为桩（Stub）类，主要作用是定义了 idl 文件中的方法，而且这些方法都是抽象的，所以如果一个类从 RMIServiceInterfaceStub 类继承，那么该类必须实现 idl 文件中定义的方法。

从 RMIServiceInterfaceStub 类继承的子类的实例就是远程对象，该对象需要通过 onConnect 方法被返回。本例中远程对象对应的类是 RMIRemoteObject。

代码位置：src\main\java\com\unitymarvel\demo\ability\TestServiceAbility.java

```
private class RMIRemoteObject extends RMIServiceInterfaceStub {
   public RMIRemoteObject(String descriptor) {
      super(descriptor);
   }
   @Override
   public int add(int a, int b) throws RemoteException {
      return a + b;
   }
   @Override
   public String greet(String name) throws RemoteException {
      return "hello " + name;
   }
}
```

RMIServiceInterfaceStub 类还有另外一个作用，就是接收远程调用请求，然后解析请求中的数据，将其转换为远程对象方法的调用。在 RMIServiceInterfaceStub 类中有一个 onRemoteRequest 方法，主要功能就是将远程调用数据反序列化，解析出里面的数据，然后传递给远程对象中的 add 方法和 greet 方法。onRemoteRequest 方法的代码如下：

```
public boolean onRemoteRequest(
   /* [in] */ int code,
   /* [in] */ MessageParcel data,
   /* [out] */ MessageParcel reply,
   /* [in] */ MessageOption option) throws RemoteException {
```

```
    String token = data.readInterfaceToken();
    if (!DESCRIPTOR.equals(token)) {
        return false;
    }
    switch (code) {
        case COMMAND_ADD: {   // 处理请求 add 方法的数据
            int a = data.readInt();
            int b = data.readInt();
            int result;
            result = add(a, b);
            reply.writeNoException();
            reply.writeInt(result);
            return true;
        }
        case COMMAND_GREET: {   // 处理请求 greet 方法的数据
            String name = data.readString();
            String result;
            result = greet(name);
            reply.writeNoException();
            reply.writeString(result);
            return true;
        }
        default:
            return super.onRemoteRequest(code, data, reply, option);
    }
}
```

RMIServiceInterfaceStub 类实现了 IRMIServiceInterface 接口，IRMIServiceInterface.java 文件也是 IRMIServiceInterface.idl 生成的 3 个 Java 文件之一，该接口的代码如下：

```
package com.unitymarvel.demo;

import ohos.rpc.IRemoteBroker;
import ohos.rpc.RemoteException;

public interface IRMIServiceInterface extends IRemoteBroker {
    int add(
        /* [in] */ int a,
        /* [in] */ int b) throws RemoteException;
    String greet(
        /* [in] */ String name) throws RemoteException;
};
```

从 IRMIServiceInterface 接口的代码可以看出，该接口是根据 IRMIServiceInterface.idl 文件自动生成的。所以 IRMIServiceInterface.java 文件和 RMIServiceInterfaceStub.java 文件是用于服务端（包含 Service Ability 的一端）的文件。在发布服务端时，需要带这两个 Java 文件。而 RMIServiceInterfaceProxy.java 文件则用于客户端（调用 Service Ability 的一端）。后面会详细讲解该文件的用法。

6. 返回远程对象

这一步需要重新实现 onConnect 方法，该方法需要返回上一步实现的 RMIRemoteObject 类的实例。

```
protected IRemoteObject onConnect(Intent intent) {
   Tools.print("MyServiceAbility", "onConnect");
   // 返回远程对象
   return new RMIRemoteObject("rmi");
}
```

7. 配置 Service Ability

到现在为止，Service Ability 的编码工作已经全部完成，服务端还剩下最后一项工作，就是在 config.json 文件中配置这个 Service Ability，代码如下：

```
{
   "visible": true,
   "name": "com.unitymarvel.demo.ability.TestServiceAbility",
   "icon": "$media:icon",
   "type": "service"
}
```

8. 编写 Service Ability 的回调类

从这一步开始编写客户端的代码，因为连接服务端是异步的，所以首先需要在 RMIServiceAbility（一个 Page Ability）中编写一个用于接收连接状态的回调对象。

代码位置：src\main\java\com\unitymarvel\demo\ability\RMIServiceAbility.java

```
public class RMIServiceAbility extends Ability {
   private static RMIServiceInterfaceProxy serviceProxy;
   private static IAbilityConnection abilityConnection = new IAbilityConnection() {
      @Override
      public void onAbilityConnectDone(ElementName elementName, IRemoteObject
      iRemoteObject, int i) {
         // 如果成功连接 Service Ability，会将 iRemoteObject 转换为 RMIServiceInterfaceProxy 对象
         serviceProxy = new RMIServiceInterfaceProxy(iRemoteObject);
      }
      @Override
      public void onAbilityDisconnectDone(ElementName elementName, int i) {
         serviceProxy = null;
      }
   };
   ...
}
```

在这段代码中，通过 IAbilityConnection 接口创建了一个回调对象。该接口有两个方法——onAbilityConnectDone 和 onAbilityDisconnectDone。其中，onAbilityConnectDone 方法在成功连接 Service Ability 后会被调用，该方法的 iRemoteObject 参数值是在第（6）步 onConnect 方法返回的 RMIRemoteObject 对象。而通过 idl 文件生成的 RMIServiceInterfaceProxy.java 文件同样实现了 IRMIServiceInterface 接口，所以可以将 iRemoteObject 转换为 RMIServiceInterfaceProxy 对象。

目前我们已经了解了利用 idl 生成的 3 个 Java 文件的含义。这 3 个 Java 文件中包含一个接口文件（IRMIServiceInterface.java）和两个 Java 类文件（RMIServiceInterfaceProxy.java 和 RMIService-

InterfaceStub.java)。其中 RMIServiceInterfaceProxy.java 用于客户端，RMIServiceInterfaceStub. java 用于服务端。在服务端需要有一个类从 RMIServiceInterfaceStub 类继承，并且实现 IRMIServiceInterface 接口中的方法，该类的实例被称为远程对象。因为 RMIServiceInterfaceProxy 类同样实现了 IRMIServiceInterface 接口，所以可以将 iRemoteObject 转换为 RMIServiceInterfaceProxy 对象。图 8-6 所示是这 3 个类和接口的关系。

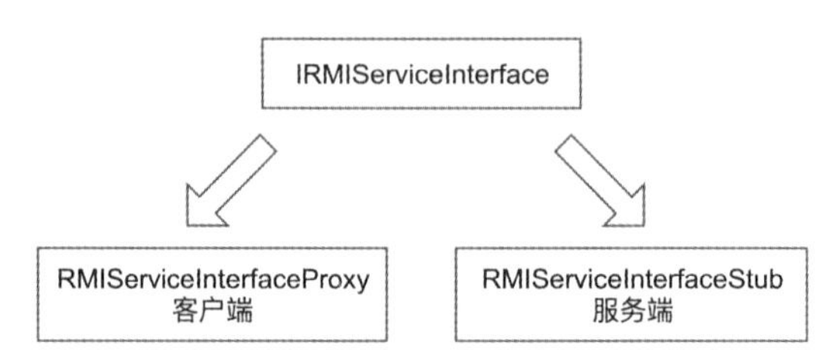

图 8-6 利用 idl 文件生成的接口和类的关系

9. 选择设备 ID

因为在本例中需要连接远程的设备，所以在连接之前，需要先获取要连接设备的 ID，本例仍然使用在 2.5.2 节实现的获取设备列表的 Page Ability。在单击"连接远程 Service Ability"按钮时显示这个 Page Ability。

代码位置：src\main\java\com\unitymarvel\demo\ability\RMIServiceAbility.java

```
Button button2 = (Button)findComponentById(ResourceTable.Id_button2);
if(button2 != null) {
   button2.setClickedListener(new Component.ClickedListener() {
      @Override
      public void onClick(Component component) {
         Intent intentPageAbility = new Intent();
         Operation operation = new Intent.OperationBuilder()
               .withBundleName("com.unitymarvel.demo")
               .withAbilityName("com.unitymarvel.demo.ability.DeviceIdsAbility")
               .build();
         intentPageAbility.setOperation(operation);
         // 显示设备列表 Page Ability
         startAbilityForResult(intentPageAbility,99);
      }
   });
}
```

10. 连接远程设备的 Service Ability

从设备列表中选择一个设备后，系统会回调 onAbilityResult 方法，在该方法中利用获取的 deviceId 连接远程设备，下面是该方法的代码。

代码位置：src\main\java\com\unitymarvel\demo\ability\RMIServiceAbility.java

```
protected void onAbilityResult(int requestCode, int resultCode, Intent resultData) {
   if(requestCode == 99) {
```

```
        // 获取 deviceId
        String deviceId = resultData.getStringParam("deviceId");
        // 在 Toast 信息框中显示 deviceId
        Tools.showTip(RMIServiceAbility.this, deviceId);
        String bundleName = "com.harmonyos.demo";
        String abilityName = "com.harmonyos.demo.ability.TestServiceAbility";
        Intent intentData = new Intent();
        ElementName elementName = new ElementName(deviceId, bundleName, abilityName);
        intentData.setElement(elementName);
        // 该标志必须设置，否则无法连接远程设备
        intentData.setFlags(Intent.FLAG_ABILITYSLICE_MULTI_DEVICE);
        // 连接远程设备
        boolean success = connectAbility(intentData,abilityConnection);
        if(success) {
           Tools.showTip(RMIServiceAbility.this, "远程 Service Ability 已经连接");
        } else {
           Tools.showTip(RMIServiceAbility.this, "无法连接远程 Service Ability");
        }
    }
}
```

单击“连接远程 Service Ability”按钮，如果显示“远程 Service Ability 已经连接”Toast 信息框，说明已经连接成功。

11. 调用 Service Ability 中的 API

成功连接远程 Service Ability 后，单击“调用 Service Ability API”按钮调用 add 方法和 greet 方法。

代码位置：src\main\java\com\unitymarvel\demo\ability\RMIServiceAbility.java

```
Button button3 = (Button)findComponentById(ResourceTable.Id_button3);
if(button3 != null) {
   button3.setClickedListener(new Component.ClickedListener() {
      // 调用 Service Ability 中的 API
      @Override
      public void onClick(Component component) {
         try {
            if (serviceProxy != null) {
               // 调用 add 方法
               int sum = serviceProxy.add(20, 30);
               // 调用 greet 方法
               String greet = serviceProxy.greet("李宁");
               // 用 Toast 信息框显示 add 方法和 greet 方法的返回结果
               Tools.showTip(RMIServiceAbility.this, "20 + 30 = " + sum + "\r\n" + greet);
            }
         } catch (Exception e) {
            Tools.showTip(RMIServiceAbility.this, e.getMessage());
         }
      }
   });
}
```

12. 关闭 Service Ability

如果不使用Service Ability,可以单击“关闭 Service Ability”按钮关闭 Service Ability，代码如下：

```
Button button4 = (Button)findComponentById(ResourceTable.Id_button4);
if(button4 != null) {
    button4.setClickedListener(new Component.ClickedListener() {
        @Override
        public void onClick(Component component) {
            // 关闭 Service Ability'
            disconnectAbility(abilityConnection);
            Tools.showTip(RMIServiceAbility.this, "服务已经断开");
        }
    });
}
```

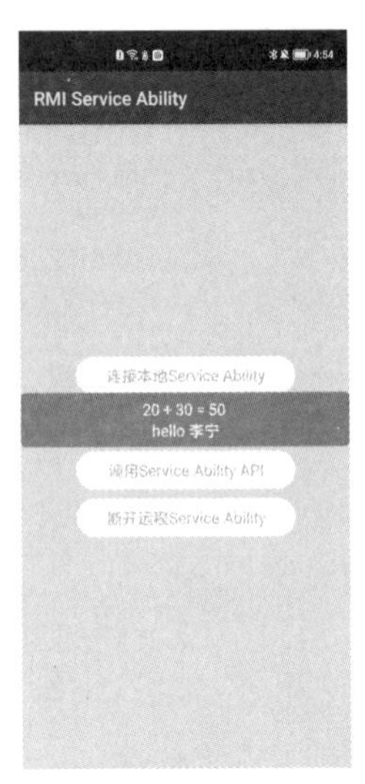

图 8-7 调用远程 Service Ability API

运行程序，单击“连接远程 Service Ability”按钮，如果连接成功，再单击“调用 Service Ability API”按钮，如果显示图 8-7 所示的 Toast 信息框，表示已经成功调用了 Service Ability 中的 API，并返回了正确的结果。

8.4 总结与回顾

第 2 章和第 7 章分别讲了 Page Ability 和 Data Ability，这是 HarmonyOS 的两项非常重要的功能，不过这两项功能使用上都非常单一，Page Ability 只负责创建窗口，Data Ability 仅负责提供数据。而本章讲的 Service Ability 则具有服务与方法调用的功能。

Service Ability 中的服务实现了 App 在后台运行的功能。在很多场景中，如后台播放音乐、后台下载文件，都需要有一个程序在后台一直运行，而服务就是用于实现这个的。

另外一项功能是方法调用，就是一个 App 想要调用另一个 App 中的某个方法，即跨进程调用，也可以称为 RMI（远程方法调用）。

用于后台运行的服务，与 Page Ability 一样，也有一个生命周期，只是没有 Page Ability 的生命周期复杂。通常分为启动服务、执行服务和停止服务 3 个阶段。启动服务只进行一次，然后每次调用服务时，就执行一次服务，如果服务不需要了，可以停止服务。

Service Ability 也同样支持 HarmonyOS 的分布式特性，也就是说，一部 HarmonyOS 设备可以启动另外一部 HarmonyOS 设备中的服务，或调用另外一部 HarmonyOS 设备中某个 App 中的某个方法。

到现在为止，HarmonyOS 中的 3 个核心功能都已经讲完了，分别是 Page Ability、Data Ability 和 Service Ability。它们都支持跨设备访问，如果与 UI 有关，就用 Page Ability；如果与数据有关，就用 Data Ability；如果涉及频繁的交互，就用 Service Ability 中的 RMI。

第 9 章　多媒体

HarmonyOS 中的多媒体功能主要包括音频、视频处理和相机。本章将详细讲解这 3 个部分的核心 API 的使用方法。在本章我们使用的音频文件格式是 mp3，使用的视频文件格式是 mp4。

本章的源代码在 multimedia 目录下。

通过阅读本章，读者可以掌握：

- 如何播放本地音频文件；
- 如何播放在线音频文件；
- 如何播放本地视频文件；
- 如何暂停播放和继续播放音频和视频文件；
- 如何停止播放音频和视频文件；
- 如何拍照和保存拍摄的照片。

9.1 音频

HarmonyOS 支持 mp3、wav 等常用的音频格式，本节就详细介绍一下如何播放本地音频文件和在线音频文件。

9.1.1 准备本地音频文件

要想播放本地音频文件，需要先准备一个本地音频文件。对于 HarmonyOS，我们可以将音频文件放在 sdcard 目录下，也可以放在 App 的 data 目录及其子目录下。为了方便，在本节示例中我们将音频文件放在 App 的 data 目录下，读写该目录下的文件并不需要设置任何权限。

现在的问题是，一旦重新安装 App，data 目录就会被初始化，以前复制到该目录下的所有文件就会消失，而且如果没有 root 权限，从 shell 中是不能进入 data 目录的，当然也不能查看和编辑该目录下的任何文件。所以如何将音频文件放在 data 目录下是我们首先要解决的问题。

当然，获取音频文件的方法很多，可以从网络上下载，也可以自己生成，不过这些方法可能需要各种权限，会特别麻烦。本节将采用一种屡试不爽的方法，将嵌入 hap 文件中的 mp3 文件提取出来，然后复制到 data 目录下。要完成这项工作，首先要准备一个 mp3 文件（文件不要太大，建议 1 MB 以内），本节示例中是 eng.mp3。

首先要做的是将 eng.mp3 文件放在 src/main/resources/rawfile 目录下，如图 9-1 所示。

HarmonyOS 工程中资源目录下的任何文件都可以通过相应的 API 获取，通过 ResourceManager.getRawFileEntry 方法可以获取 rawfile 目录下文件的 RawFileEntry 对象，通过 RawFileEntry.openRawFile 方法可以获取资源文件对应的 Resource 对象，通过 Resource.read 方法可以读取 eng.mp3 文件的字节流，然后利用 FileOutputStream 对象将读取的字节流写入 data 目录下的同名的文件中。下面是完整的实现代码。

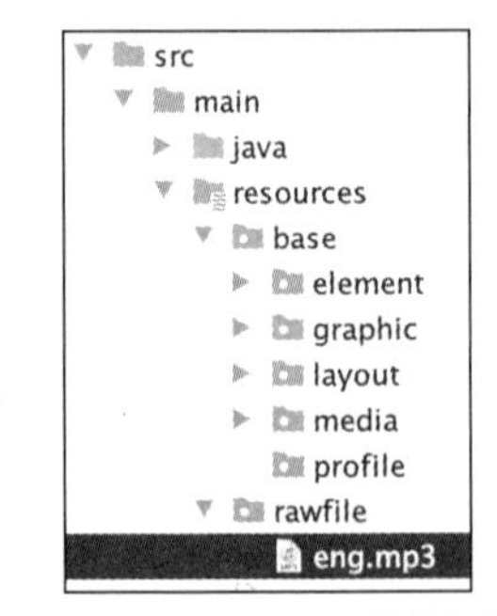

图 9-1 eng.mp3 文件的位置

代码位置：src/main/java/com/unitymarvel/demo/Tools.java

```
package com.unitymarvel.demo;
...
public class Tools {
    ...
    // 将 rawfile 目录下任意文件复制到 data 目录的通用方法
    public static String extractFile(Context context, String filename) {
        try {
            // 打开 rawfile 目录下的文件，并获取与该文件对应的 Resource 对象
            Resource resource = context.getResourceManager().
                    getRawFileEntry("resources/rawfile/" + filename).
                    openRawFile();
            // 定义 data 目录下同名文件的路径
            File path = new File(Paths.get(context.getDataDir().toString(),
                                                  filename).toString());
            // 如果文件存在，先删除该文件
            if (path.exists()) {
                path.delete();
            }
            // 根据路径创建文件输出流对象
            FileOutputStream fos = new FileOutputStream(path);
            // 每次要复制的字节（4KB）
            byte[] buffer = new byte[4096];
            // 每次复制的字节数，如果没有读取到文件末尾，通常是 4096 字节
            // 如果不够 4096 字节，那么返回实际读取的字节数
            int count = 0;
            // 循环读取 rawfile/eng.mp3 文件中的数据，直到读取到文件的末尾
            while ((count = resource.read(buffer)) >= 0) {
                // 一边读，一边写
                fos.write(buffer, 0, count);
            }
            // 关闭输入流
            resource.close();
            // 关闭输出流
            fos.close();
            // 返回 data/eng.mp3 文件的完整路径
            return path.toString();
        } catch (Exception e) {
            // 如果发生异常，返回空串
            return "";
        }
    }
}
```

阅读这段代码，需要了解如下几点。

- Tools 是一个通用类，里面包含若干个方法，例如，输出日志的方法、显示 Toast 信息框的方法，以及 extractFile 方法。
- extractFile 是一个通用的方法，可以将 rawfile 目录下任何一个文件以同名的形式复制到 data 目录下，包括 rawfile 子目录下的文件。
- 尽管 getRawFileEntry 方法的参数值是 rawfile 目录下文件的路径，但根目录是 resources 目录，所以在指定路径时，应该加上 resources。
- 因为要复制的文件的尺寸不定，所以 extractFile 方法采用了循环复制的方式，也就是说，每次复制固定的字节数（本例是 4 KB），这样不管文件有多大，都可以完成复制工作。

注意，extractFile 方法在本章介绍视频处理时还会被用来复制 mp4 文件到 data 目录。

9.1.2 播放本地音频文件

因为播放音频文件需要使用 ohos.media.player.Player 对象，所以先要声明一个 Player 类型的变量，然后初始化这个变量，代码如下：

```
Player player;
player = new Player(this);
```

下面是使用 Player 对象播放音频文件的具体步骤。

（1）复制音频文件到 data 目录。在上一节我们已经实现了用于复制 rawfile 目录下文件到 data 目录的通用方法 extractFile，这一步就使用 extractFile 方法将 rawfile/eng.mp3 文件复制到 data 目录下，代码如下：

```
String path = Tools.extractFile(this, "eng.mp3");
```

在调用 extractFile 方法时只需要直接指定文件名，如果文件在 rawfile 目录的子目录下，需要指定子目录。path 变量值就是 eng.mp3 文件复制到 data 目录后的完整路径，在后面的步骤中需要使用该变量。

（2）打开本地音频文件。在上一步我们已经实现了将 rawfile 目录下的 eng.mp3 文件复制到了 data 目录下，这一步我们需要使用 FileInputStream 对象打开该文件，代码如下：

```
FileInputStream fis = new FileInputStream(new File(path));
```

这里的 path 就是 data 目录下 eng.mp3 文件的绝对路径。

（3）获取文件描述符（FileDescriptor）。在指定音频源时需要使用 FileDescriptor 对象，因此需要先获取该对象，代码如下：

```
FileDescriptor fd = fis.getFD();
```

（4）创建 Source 对象。Player 对象没有直接指定音频文件路径的方法，因此需要通过 Source 对象指定本地音频源。Source 类的构造方法需要指定一个 FileDescriptor 对象，我们已经在上一步创建了该对象，这一步就可以直接利用 FileDescriptor 对象创建 Source 对象，代码如下：

```
Source source = new Source(fd);
```

其中，fd就是在上一步创建的FileDescriptor对象。

（5）为Player对象设置音频源。创建完Source对象后，可以用Source对象为Player对象指定音频源，代码如下：

```
player.setSource(source);
```

（6）准备必要的资源。为Player对象指定音频源后，并不能立刻播放，还需要调用prepare方法完成一些初始化工作，代码如下：

```
player.prepare();
```

（7）播放音频文件。一切准备就绪后，就可以调用play方法播放音频文件了，代码如下：

```
player.play();
```

9.1.3 暂停和继续播放音频

想要暂停播放音频，可以调用Player对象的pause方法，代码如下：

```
player.pause();
```

暂停音频播放后，Player对象并没有释放，所以可以直接再次调用play方法继续播放音频，代码如下：

```
player.play();
```

9.1.4 停止播放音频

调用Player对象的stop方法，将会停止播放音频，代码如下：

```
player.stop();
```

在停止播放音频后，通常需要调用release方法释放Player对象装载的资源，即调用prepare方法装载的资源，代码如下：

```
player.release();
```

为了让JVM尽快释放Player对象，需要将player变量设为null，代码如下：

```
player = null;
```

这是因为JVM只会回收没有任何变量指向的内存空间。

9.1.5 播放在线音频文件

播放在线音频文件相对简单，只需通过Source类的构造方法指定一个HTTP或HTTPS格式的音频文件即可，代码如下：

```
Source source = new Source("https://网址/eng.mp3");
player.setSource(source);
player.prepare();
player.play();
```

读者需要将 https://网址/eng.mp3 换成可用的 URL。

尽管在线播放音频文件没有直接使用网络，但毫无疑问，在内部肯定是需要使用网络从服务器端下载音频文件的，所以需要在 config.json 文件中申请 Internet 访问权限。

```
"reqPermissions": [
  {
    "name": "ohos.permission.INTERNET"
  }
]
```

9.1.6 播放音频的完整案例

尽管前面几节已经详细介绍了播放本地音频文件和在线音频文件的完整步骤，但读者对具体如何实现可能还是有一些模糊，为了让读者更完整地了解播放音频的具体实现，本节将给出一个具体的例子。运行这个例子，会看到图 9-2 所示的界面。通过这个界面中的 4 个按钮，可以演示前面的各种操作。

图 9-2 音频播放演示

【例 9.1】 实现一个完整的示例，演示播放本地音频文件，播放在线音频文件，暂停、继续和停止播放音频文件。

代码位置：src/main/java/com/unitymarvel/demo/AudioAbility.java

```
package com.unitymarvel.demo;

import ohos.aafwk.ability.Ability;
import ohos.aafwk.content.Intent;
import ohos.agp.components.Button;
import ohos.agp.components.Component;
import ohos.media.common.Source;
import ohos.media.player.Player;
import java.io.*;
import java.util.concurrent.atomic.AtomicBoolean;

public class AudioAbility extends Ability {
   private Player player;
   // 用于表示正在播放的变量（原子操作）
   private AtomicBoolean isPlaying = new AtomicBoolean(false);
   // 用于锁定对象
   private static final Object LOCK = new Object();
   // 检测 player 变量是否初始化，如果已经初始化，并且正在播放音频，则停止播放
   // 如果未初始化，则创建新的 Player 对象
   private void checkPlayer() {
      if (player == null) {
         player = new Player(this);
      } else {
         if (isPlaying.get()) {
            player.stop();
         }
      }
```

```
}
// 播放本地音频文件，filePath参数值是本地音频文件的文件名（不包括路径）
public boolean play(String filePath) {
    // 锁定对象，让play方法线程同步
    synchronized (LOCK) {
        checkPlayer();
        boolean isSuccess = false;
        FileDescriptor fd = null;
        // 打开本地音频文件
        try (FileInputStream fis = new FileInputStream(new File(filePath))) {
            // 获取文件描述符
            fd = fis.getFD();
            // 创建音频源对象
            Source source = new Source(fd);
            // 为Player对象指定音频源
            player.setSource(source);
            // 装载播放音频需要的资源，如果成功，返回true
            isSuccess = player.prepare();
            if (!isSuccess) {
                return false;
            }
            // 开始播放音频
            isSuccess = player.play();
            // 设置正在播放的状态
            isPlaying.set(isSuccess);
        } catch (IOException ignored) {
            return false;
        }
        return isSuccess;
    }
}
// 播放在线音频文件，url是HTTP或HTTPS格式的音频文件
public boolean playOnlineAudio(String url) {
    synchronized (LOCK) {
        checkPlayer();
        boolean isSuccess = false;
        try  {
            // 创建音频源对象
            Source source = new Source(url);
            // 为Source对象指定在线音频源
            player.setSource(source);
            // 装载播放音频需要的资源
            isSuccess = player.prepare();
            if (!isSuccess) {
                return false;
            }
            // 开始播放在线音频
            isSuccess = player.play();
            // 设置正在播放在线音频的状态
            isPlaying.set(isSuccess);
        } catch (Exception ignored) {
            return false;
        }
        return isSuccess;
```

```
        }
    }
    // 暂停播放音频
    public boolean pause() {
        synchronized (LOCK) {
            if (player == null) {
                return false;
            }
            boolean success = player.pause();
            // 设置未播放状态
            isPlaying.set(false);
            return success;
        }
    }
    // 停止播放音频
    public boolean stop() {
        synchronized (LOCK) {
            if (player == null) {
                return false;
            }
            // 停止播放
            boolean success = player.stop();
            // 设置未播放状态
            isPlaying.set(false);
            // 释放资源
            success = player.release();
            // 将 player 变量值设置为 null，等待 JVM 回收内存空间
            player = null;
            return success;
        }
    }
    @Override
    public void onStart(Intent intent) {
        super.onStart(intent);
        super.setUIContent(ResourceTable.Layout_audio_ability);
        // 处理“播放本地音频文件”按钮
        Button button1 = (Button)findComponentById(ResourceTable.Id_button_audio1);
        if(button1 != null) {
            button1.setClickedListener(new Component.ClickedListener() {
                // 播放本地音频文件
                @Override
                public void onClick(Component component) {
                    // 将 rawfile/eng.mp3 文件复制到 data 目录下，并返回 eng.mp3 文件的完整路径
                    String path = Tools.extractFile(AudioAbility.this, "eng.mp3");
                    // 开始播放本地音频文件
                    boolean success = play(path);
                    if(success) {
                        Tools.showTip(AudioAbility.this, "成功播放");
                    } else {
                        Tools.showTip(AudioAbility.this, "播放失败");
                    }
                }
            });
        }
```

```
// 处理“暂停/继续播放”按钮
Button button2 = (Button)findComponentById(ResourceTable.Id_button_audio2);
if(button2 != null) {
   button2.setClickedListener(new Component.ClickedListener() {
      // 暂停/继续播放
      @Override
      public void onClick(Component component) {
         if(button2.getText().equals("暂停播放")) {
            button2.setText("继续播放");
            // 暂停播放音频
            boolean success = pause();
            if(success) {
               Tools.showTip(AudioAbility.this, "成功暂停");
            } else {
               Tools.showTip(AudioAbility.this, "暂停失败");
            }
         } else {
            button2.setText("暂停播放");
            if(player != null) {
               // 继续播放音频
               boolean success = player.play();
               if(success) {
                  Tools.showTip(AudioAbility.this, "成功继续播放");
               } else {
                  Tools.showTip(AudioAbility.this, "继续播放失败");
               }
            }
         }
      }
   });
}
// 处理“停止播放”按钮
Button button3 = (Button)findComponentById(ResourceTable.Id_button_audio3);
if(button3 != null) {
   button3.setClickedListener(new Component.ClickedListener() {
      // 停止播放
      @Override
      public void onClick(Component component) {
         // 停止播放音频
         boolean success = stop();
         if(success) {
            Tools.showTip(AudioAbility.this, "成功停止播放");
         } else {
            Tools.showTip(AudioAbility.this, "停止播放失败");
         }
      }
   });
}
// 处理“播放在线音乐”按钮
Button button4 = (Button)findComponentById(ResourceTable.Id_button_audio4);
if(button4 != null) {
   button4.setClickedListener(new Component.ClickedListener() {
      // 播放在线音乐
      @Override
```

```
            public void onClick(Component component) {
               // 播放在线音乐，读者需要将 URL 换成可用的 mp3 文件
               boolean success =
                      playOnlineAudio("https://网址/music.mp3");
               if(success) {
                  Tools.showTip(AudioAbility.this, "成功播放在线音频");
               } else {
                  Tools.showTip(AudioAbility.this, "播放在线音频失败");
               }
            }
         });
      }
   }
}
```

运行程序，就会看到图 9-2 所示的界面。单击“播放本地音频文件”按钮，会立刻播放 eng.mp3 文件。然后单击“暂停播放”按钮，会暂时停止音频播放，并且“暂停播放”按钮会变成“继续播放”按钮。再次单击“继续播放”按钮，eng.mp3 文件又开始播放了。接下来单击“停止播放”按钮，会彻底停止 eng.mp3 的播放。最后单击“播放在线音乐”按钮，会播放代码中指定的 music.mp3 文件，但要注意，根据不同的网络环境，播放会有不同程度的延迟，因为首先需要从网络上下载一部分 music.mp3 文件。

9.2 视频

播放视频与播放音频一样，也需要使用 Player 对象。但与播放音频不同的是，因为视频需要 UI 显示画面，所以在播放视频时，需要将 Player 对象与 SurfaceProvider 对象绑定。SurfaceProvider 是一个可视化组件，通常用来显示图像和画面。Player 每播放一帧画面，就会在 SurfaceProvider 上显示当前播放的帧画面。当然，声音仍然是通过 Player 播放的。

Player 与 SurfaceProvider 绑定需要在 SurfaceOps.Callback 接口的 surfaceCreated 方法中完成，代码如下：

```
SurfaceProvider surfaceView;
surfaceView = new SurfaceProvider(this);
surfaceView.pinToZTop(true);
surfaceView.getSurfaceOps().get().addCallback(new SurfaceOps.Callback() {
   @Override
   public void surfaceCreated(SurfaceOps surfaceOps) {
      // Player 与 SurfaceProvier 绑定
      player.setSurfaceOps(surfaceOps);
      // 播放视频
      player.play();
   }
   @Override
   public void surfaceChanged(SurfaceOps surfaceOps, int i, int i1, int i2) {
   }
   @Override
   public void surfaceDestroyed(SurfaceOps surfaceOps) {
   }
});
```

其他操作基本上与播放音频类似。例 9.2 完整地演示了使用 Player 播放视频的过程。

【例 9.2】 实现一个完整的案例，用于演示播放本地视频文件（video.mp4）的实现过程，实现播放视频、暂停/继续播放视频和停止播放视频文件的功能。本例仍然读取了 data 目录下的视频文件，所以需要事先将 video.mp4 文件放到工程的 rawfile 目录下，然后使用 9.1.1 节中实现的 Tools.extractFile 方法将 video.mp4 文件复制到 data 目录下。

因为播放视频需要 UI，所以先看一下本例中 UI 的实现，下面是布局代码。

代码位置：src/main/resources/base/layout/video_ability.xml

```
<?xml version="1.0" encoding="utf-8"?>
<DirectionalLayout
   xmlns:ohos="http://schemas.huawei.com/res/ohos"
   ohos:height="match_parent"
   ohos:width="match_parent"
   ohos:alignment="center"
   ohos:background_element="#CCCCCC"
   ohos:orientation="vertical">
   <DirectionalLayout ohos:id="$+id:playLayout"
                  ohos:width="match_parent"
                  ohos:background_element="#E1E1E1"
                  ohos:orientation="vertical"
                  ohos:height="200vp">
   </DirectionalLayout>
   <Button
      ohos:id="$+id:button1"
      ohos:background_element="$graphic:shape_text_bac"
      ohos:height="40vp"
      ohos:width="match_parent"
      ohos:text_color="#9b9b9b"
      ohos:text_size="17fp"
      ohos:text_alignment="center"
      ohos:text="播放视频"
      ohos:top_margin="10vp"
      ohos:right_margin="60vp"
      ohos:left_margin="60vp"/>
   <Button
      ohos:id="$+id:button2"
      ohos:background_element="$graphic:shape_text_bac"
      ohos:height="40vp"
      ohos:width="match_parent"
      ohos:text_color="#9b9b9b"
      ohos:text_size="17fp"
      ohos:text_alignment="center"
      ohos:text="暂停播放"
      ohos:right_margin="60vp"
      ohos:left_margin="60vp"
      ohos:top_margin="10vp"/>
   <Button
      ohos:id="$+id:button3"
```

```
        ohos:background_element="$graphic:shape_text_bac"
        ohos:height="40vp"
        ohos:width="match_parent"
        ohos:text_color="#9b9b9b"
        ohos:text_size="17fp"
        ohos:text_alignment="center"
        ohos:text="停止播放"
        ohos:right_margin="60vp"
        ohos:left_margin="60vp"
        ohos:top_margin="10vp"/>
</DirectionalLayout>
```

图 9-3 视频播放主界面

在这段布局代码中，最上方放置了一个 DirectionalLayout 组件，用于显示视频的帧画面，在代码中会动态添加 SurfaceProvider 组件，下面放置了 3 个控制按钮，效果如图 9-3 所示。

最后看一下完整的实现代码。

代码位置：src/main/java/com/unitymarvel/demo/VideoAbility.java

```
package com.unitymarvel.demo;

import ohos.aafwk.ability.Ability;
import ohos.aafwk.content.Intent;
import ohos.agp.components.Button;
import ohos.agp.components.Component;
import ohos.agp.components.DependentLayout;
import ohos.agp.components.DirectionalLayout;
import ohos.agp.components.surfaceprovider.SurfaceProvider;
import ohos.agp.graphics.SurfaceOps;
import ohos.app.Context;
import ohos.media.common.Source;
import ohos.media.player.Player;
import java.io.*;
import java.util.concurrent.atomic.AtomicBoolean;

public class VideoAbility extends Ability {
   private AtomicBoolean isPlaying = new AtomicBoolean(false);
   private static final Object LOCK = new Object();
   private Player player;
   // 用于显示视频帧画面的组件
   private SurfaceProvider surfaceView;
   // 包含显示视频帧画面组件的容器
   private DirectionalLayout mPlayerLayout;
   // 用于创建、添加 SurfaceProvider 组件，并将该组件与 Player 对象绑定
   private void addSurfaceView() {
      DirectionalLayout.LayoutConfig surfaceViewLayoutConfig = new
      DirectionalLayout.LayoutConfig(DependentLayout.LayoutConfig.MATCH_PARENT,
      DependentLayout.LayoutConfig.MATCH_PARENT);
      // 创建显示视频帧画面的组件
      surfaceView = new SurfaceProvider(this);
      // 将组件的 ZOrder 调整到最顶端
      surfaceView.pinToZTop(true);
      // 开始与 Player 对象绑定
```

```
        surfaceView.getSurfaceOps().get().addCallback(new SurfaceOps.Callback() {
            @Override
            public void surfaceCreated(SurfaceOps surfaceOps) {
                // 与 Player 对象绑定
                player.setSurfaceOps(surfaceOps);
                // 开始播放视频
                player.play();
            }
            @Override
            public void surfaceChanged(SurfaceOps surfaceOps, int i, int i1, int i2) {
            }
            @Override
            public void surfaceDestroyed(SurfaceOps surfaceOps) {

            }
        });
        surfaceView.setLayoutConfig(surfaceViewLayoutConfig);
        // 将 SurfaceProvider 组件添加到父布局中
        mPlayerLayout.addComponent(surfaceView);
        mPlayerLayout.setVisibility(Component.VISIBLE);

    }
    // 用于监听视频状态的回调对象
    private Player.IPlayerCallback mplayerCallback = new Player.IPlayerCallback() {
        // 调用 Player.prepared 方法时触发
        @Override
        public void onPrepared() {
        }
        @Override
        public void onMessage(int i, int i1)
        {
            Tools.print("video player:" + String.valueOf(i) + ":" + String.valueOf(i1));
        }

        @Override
        public void onError(int i, int i1) {
        }
        // 视频分辨率发生变化时触发
        @Override
        public void onResolutionChanged(int i, int i1) {
        }
        // 视频播放完后触发，通常在该事件中销毁视频播放窗口
        @Override
        public void onPlayBackComplete() {
            if (player != null) {
                player.stop();
                getUITaskDispatcher().asyncDispatch(new Runnable() {
                    @Override
                    public void run() {
                        surfaceView.removeFromWindow();
                    }
                });

                player = null;
```

```
            }
        }
        @Override
        public void onRewindToComplete() {
        }
        @Override
        public void onBufferingChange(int i) {
        }
        @Override
        public void onNewTimedMetaData(Player.MediaTimedMetaData mediaTimedMetaData) {
        }
        @Override
        public void onMediaTimeIncontinuity(Player.MediaTimeInfo mediaTimeInfo) {
        }
    };
    // 播放视频
    private void playVideo(Context context) {
        try {
            player = new Player(context);
            // 打开 data 目录下的视频文件
            FileInputStream fis = new FileInputStream(getDataDir() + "/video.mp4");
            // 获取文件描述符
            FileDescriptor fd = fis.getFD();
            // 创建视频源对象
            Source source = new Source(fd);
            // 为 Player 对象指定视频源
            player.setSource(source);
            // 装载播放视频需要的资源
            player.prepare();
            // Player 与 SurfaceProvider 绑定
            addSurfaceView();
            // 设置用于监听播放状态的回调对象
            player.setPlayerCallback(mplayerCallback);
            // 开始播放视频
            player.play();
        } catch (Exception e) {
            e.printStackTrace();
        }
    }
    @Override
    public void onStart(Intent intent) {

        super.onStart(intent);
        super.setUIContent(ResourceTable.Layout_video_ability);
        mPlayerLayout = (DirectionalLayout)
                  findComponentById(ResourceTable.Id_playLayout);
        // 处理“播放视频”按钮
        Button button1 = (Button)findComponentById(ResourceTable.Id_button1);
        if(button1 != null) {
            button1.setClickedListener(new Component.ClickedListener() {
                // 播放本地视频文件
                @Override
                public void onClick(Component component) {
```

```
                // 将 rawfile/video.mp4 文件复制到 data 目录下
                Tools.extractFile(VideoAbility.this, "video.mp4");
                // 播放视频
                playVideo(VideoAbility.this);
            }
        });
    }
    // 处理“暂停播放/继续播放”按钮
    Button button2 = (Button)findComponentById(ResourceTable.Id_button2);
    if(button2 != null) {
        button2.setClickedListener(new Component.ClickedListener() {
            // 暂停/继续播放
            @Override
            public void onClick(Component component) {
                if(player != null) {
                    if(button2.getText().equals("暂停播放")) {
                        player.pause();
                        button2.setText("继续播放");
                    } else {
                        player.play();
                        button2.setText("暂停播放");
                    }
                }
            }
        });
    }
    // 处理“停止播放”按钮
    Button button3 = (Button)findComponentById(ResourceTable.Id_button3);
    if(button3 != null) {
        button3.setClickedListener(new Component.ClickedListener() {
            // 停止播放
            @Override
            public void onClick(Component component) {
                if(player != null) {
                    // 停止播放视频
                    player.stop();
                    // 移除显示视频帧画面的窗口
                    surfaceView.removeFromWindow();
                    player = null;
                    Tools.showTip(VideoAbility.this,"已经停止播放");
                }
            }
        });
    }
  }
}
```

运行程序，单击“播放视频”按钮，会在按钮上方区域播放视频，效果如图 9-4 所示。单击“暂停播放”按钮，视频会暂停播放，并且按钮文本会变成“继续播放”。单击“停止播放”按钮，视频会停止播放。

图 9-4　播放视频的效果

9.3 相机

本节将介绍如何使用 HarmonyOS 的相关 API 操作相机进行拍照，并给出一个完整的拍照案例。

9.3.1 拍照 API 的使用方式

HarmonyOS 可以控制手机中的相机进行拍照，使用相机的方式与播放视频的方式类似，同样需要 SurfaceProvider 组件，利用该组件显示相机实时采集到的帧画面。

实现拍照功能需要使用 Camera 对象，下面是操作该对象的核心步骤。

（1）创建用于显示拍照帧画面的 SurfaceProvider 对象。通过相机采集的视频流与播放的视频流一样，都由帧画面组成。所以显示这些帧画面同样需要 SurfaceProvider 组件。创建 SurfaceProvider 对象的代码如下：

```
SurfaceProvider surfaceProvider = new SurfaceProvider(this);
```

（2）实现 SurfaceOps.Callback 接口。SurfaceOps.Callback 接口是用于监听 SurfaceProvider 状态的，先来实现该接口，代码如下：

```
class MySurfaceCallback implements SurfaceOps.Callback{
   @Override
   public void surfaceCreated(SurfaceOps surfaceOps) {
   }
   @Override
   public void surfaceChanged(SurfaceOps surfaceOps, int i, int i1, int i2) {
   }
   @Override
   public void surfaceDestroyed(SurfaceOps surfaceOps)
   {
   }
}
```

（3）为 SurfaceProvider 组件指定状态监听对象。在这一步我们需要将上一步编写的用于监听 SurfaceProvider 状态的 MySurfaceCallback 类与 SurfaceProvider 组件绑定，代码如下：

```
surfaceProvider.getSurfaceOps().get().addCallback(new MySurfaceCallback());
```

（4）获取相机 ID。HarmonyOS 设备上可能有多部相机（例如，手机就有前置相机和后置相机），所以在使用相机拍照之前，首先要让用户选择使用哪一部相机，代码如下：

```
// 获取 CameraKit 对象
CameraKit camerakit = CameraKit.getInstance(this);
// 如果 CameraKit 对象为 null，表示当前 HarmonyOS 没有相机模块
if (camerakit == null) {
   return;
}
// 获取 HarmonyOS 设备中相机 ID 的列表
String[] cameraIdList = camerakit.getCameraIds();
if (cameraIdList.length <= 0) {
```

```
        return;
    }
    // 取 HarmonyOS 设备中相机列表的第一部相机
    String cameraId = cameraIdList[0];
```

（5）创建相机状态监听对象。CameraStateCallback 对象用于监听相机的状态，所以需要创建该对象，代码如下：

```
CameraStateCallback cameraStateCallback = new CameraStateCallback() {
    @Override
    public void onCreated(Camera camera) {
        super.onCreated(camera);
        // 获取 Camera 对象
        mCamera = camera;
        CameraConfig.Builder cameraConfigBuilder = camera.getCameraConfigBuilder();
        if (cameraConfigBuilder == null) {
            return;
        }
        cameraConfigBuilder.addSurface(previewSurface);
        camera.configure(cameraConfigBuilder.build());
    }
    ...
}
```

从这段代码可以看出，在 onCreated 方法中可以获取 Camera 对象，也就是说，前面提到的 Camera 对象并不是使用 new 创建的，而是通过 onCreated 方法返回的。

（6）打开相机。用户在拍照之前，要先打开相机。可以将第（4）步和第（5）步的代码封装在用于打开相机的 openCamera 方法中。openCamera 方法的原型如下：

```
private void openCamera(Surface previewSurface);
```

openCamera 方法有一个 Surface 类型的参数，该参数值需要从 SurfaceProvider 组件中获取，所以通常需要在第（2）步实现的 surfaceCreated 方法中使用 openCamera 方法打开相机，代码如下：

```
public void surfaceCreated(SurfaceOps surfaceOps) {
    // 设置帧画面的尺寸
    surfaceProvider.getSurfaceOps().get().setFixedSize(144, 176);
    // 打开相机
    openCamera(surfaceOps.getSurface());
}
```

（7）实现监听拍照的 ImageReceiver.IImageArrivalListener 接口。如果要将拍照后的照片保存到本地，我们还需要先实现用于监听拍照动作的 ImageReceiver. IImageArrivalListener 接口，代码如下：

```
ImageReceiver.IImageArrivalListener imageArrivalListener = new
                                        ImageReceiver.IImageArrivalListener() {
    @Override
    public void onImageArrival(ImageReceiver imageReceiver) {
        ...
        // 在这里将照片保存到本地
    }
};
```

（8）为图片接收器绑定监听器。因为 ImageReceiver 对象用于接收拍照后的数据，所以要想在拍照后对照片进行处理，需要 ImageReceiver 对象，并且为该对象指定上一步创建的监听对象，核心实现代码如下：

```
ImageReceiver imageReceiver = ImageReceiver.create(Math.max(mPictureSize.width,
   mPictureSize.height), Math.min(mPictureSize.width, mPictureSize.height), ImageFormat.
   JPEG, 5);
mImageReceiver.setImageArrivalListener(imageArrivalListener);
```

ImageReceiver 除了用于接收拍照数据，还用于指定照片的尺寸、格式等属性。

（9）启动拍照。其实前面 8 个步骤都是准备工作，这一步我们启动拍照，并将所有的代码都封装在 capture 方法中，代码如下：

```
private void capture() {
   // 获取拍照配置模板
   FrameConfig.Builder pictureFrameConfigBuilder =
       mCamera.getFrameConfigBuilder(Camera.FrameConfigType.FRAME_CONFIG_PICTURE);
   // 配置拍照 surface
   pictureFrameConfigBuilder.addSurface(imageReceiver.getRecevingSurface());
   // 配置拍照其他参数
   pictureFrameConfigBuilder.setImageRotation(90);
   try {
       // 启动拍照
       mCamera.triggerSingleCapture(pictureFrameConfigBuilder.build());
   }catch (Exception e){
       Tools.showTip(CameraAbility.this, e.getMessage());
       e.printStackTrace();
   }
}
```

9.3.2 使用相机需要申请的权限

在用相机拍照并保存照片时，需要在 config.json 文件中申请如下权限：

```
"reqPermissions": [
   {
     "name": "ohos.permission.CAMERA"
   }
   {
     "name": "ohos.permission.READ_USER_STORAGE"
   },
   {
     "name": "ohos.permission.WRITE_USER_STORAGE"
   }
]
```

同时需要使用下面的 Java 代码申请相机和数据存储权限：

```
private void requestPermission() {
      String[] permission = {"ohos.permission.READ_USER_STORAGE",
             "ohos.permission.WRITE_USER_STORAGE",
             "ohos.permission.CAMERA",
};
```

```
        List<String> applyPermissions = new ArrayList<>();
        for (String element : permission) {
            if (verifySelfPermission(element) != 0) {
                if (canRequestPermission(element)) {
                    applyPermissions.add(element);
                } else {
                }
            } else {
            }
        }
        requestPermissionsFromUser(applyPermissions.toArray(new String[0]), 0);
    }
```

通常会在 onStart 方法中执行 requestPermission 方法来申请权限。

9.3.3 完整的拍照案例

9.3.1 节讲解了实现拍照功能的核心步骤，不过只有这些步骤并不能编写出完整的拍照程序，所以本节会给出一个完整的案例，实现启动 HarmonyOS 设备中的相机，完成拍照，最后将照片保存在本机的图像目录下。运行本案例的程序，会显示图 9-5 所示的拍照主界面。

单击屏幕正上方的“拍照”按钮，就会将下方相机中显示的当前画面保存在本机的图像目录下。

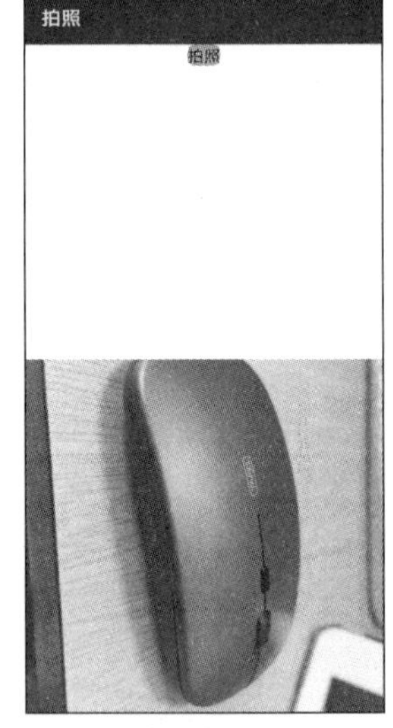

图 9-5 拍照主界面

【例 9.3】 实现一个拍照 App，在运行 App 后，屏幕下方 50%的区域会显示相机实时画面，上方有一个“拍照”按钮，单击该按钮，就会将拍摄的照片保存在本机的图像目录下。

代码位置：src/main/resources/base/layout/camera_ability.xml

```
package com.unitymarvel.demo;
...
public class CameraAbility extends Ability {

    private static final int COLOR_WHITE = 0xFFFFFFFF;
    private static final String TAG = CameraAbility.class.getSimpleName();
    private DependentLayout myLayout = new DependentLayout(this);
    private String mCameraId;  // 用于保存当前使用的相机 ID
    private Camera mCamera;    // 定义用于拍照的 Camera 对象
    private ohos.media.camera.device.CameraAbility mCameraAbility;
    private SurfaceProvider surfaceProvider;  // 用于显示相机实时画面
    private ImageReceiver mImageReceiver;
    private EventHandler mEventHandler;
    private Size mPictureSize;

    // 拍照回调对象
    private final ImageReceiver.IImageArrivalListener mImageArrivalListener = new
    ImageReceiver.IImageArrivalListener() {
        // 在这里将拍摄的照片保存到本地
```

```
        @Override
        public void onImageArrival(ImageReceiver imageReceiver) {
            StringBuffer fileName = new StringBuffer("picture_");
            // 生成本地图像文件的文件名
            fileName.append(UUID.randomUUID()).append(".jpg");
            // 定义图像文件存储的路径（本机的图像目录）
            File dirFile =
            CameraAbility.this.getExternalFilesDir(Environment.DIRECTORY_PICTURES);
            if (!dirFile.exists()) {
                dirFile.mkdir();
            }
            // 组合文件存储路径和文件名形成完整的图像文件路径
            File file = new File(dirFile, fileName.toString());

            // 创建一个读写线程任务用于保存图片
            ImageSaver imageSaver = new ImageSaver(imageReceiver.readNextImage(), file,
            fileName.toString());
            mEventHandler.postTask(imageSaver);
        }
    };


    @Override
    protected void onStart(Intent intent) {
        super.onStart(intent);

        ComponentContainer.LayoutConfig config = new ComponentContainer.LayoutConfig
        (ComponentContainer.LayoutConfig.MATCH_PARENT, ComponentContainer.LayoutConfig.
        MATCH_PARENT);
        myLayout.setLayoutConfig(config);

        // 设置布局背景
        ShapeElement shapeElement = new ShapeElement();
        shapeElement.setShape(ShapeElement.RECTANGLE);
        shapeElement.setRgbColor(new RgbColor(COLOR_WHITE));
        myLayout.setBackground(shapeElement);

        // 添加 Textview
        Text text = new Text(this);
        text.setText("Hello Camera");
        text.setTextColor(Color.BLACK);
        DependentLayout.LayoutConfig textLayoutConfig = new
        DependentLayout.LayoutConfig(DependentLayout.LayoutConfig.MATCH_CONTENT,
        DependentLayout.LayoutConfig.MATCH_CONTENT);
        textLayoutConfig.addRule(DependentLayout.LayoutConfig.CENTER_IN_PARENT);
        text.setLayoutConfig(textLayoutConfig);
        myLayout.addComponent(text);

        DependentLayout.LayoutConfig surfaceProviderLayoutConfig = new DependentLayout.
        LayoutConfig(1440, 1080);
        surfaceProviderLayoutConfig.addRule(DependentLayout.LayoutConfig.HORIZONTAL_
        CENTER);
        surfaceProviderLayoutConfig.addRule(DependentLayout.LayoutConfig.ALIGN_PARENT_
        BOTTOM);
        surfaceProvider = new SurfaceProvider(this);
```

```
        surfaceProvider.pinToZTop(true);


        surfaceProvider.getSurfaceOps().get().addCallback(new MySurfaceCallback());

        surfaceProvider.setLayoutConfig(surfaceProviderLayoutConfig);
        myLayout.addComponent(surfaceProvider);

        // 添加拍照按钮
        Button button = new Button(this);
        button.setText("拍照");
        button.setTextSize(50);
        ShapeElement buttonBackgroundShapeElement = new ShapeElement();
        buttonBackgroundShapeElement.setRgbColor(new RgbColor(0xFF51A8DD));
        buttonBackgroundShapeElement.setCornerRadius(25);
        button.setBackground(buttonBackgroundShapeElement);
        DependentLayout.LayoutConfig buttonLayoutConfig = new
        DependentLayout.LayoutConfig(DependentLayout.LayoutConfig.MATCH_CONTENT,
        DependentLayout.LayoutConfig.MATCH_CONTENT);
        buttonLayoutConfig.addRule(DependentLayout.LayoutConfig.ALIGN_PARENT_TOP);
        buttonLayoutConfig.addRule(DependentLayout.LayoutConfig.HORIZONTAL_CENTER);
        button.setLayoutConfig(buttonLayoutConfig);
        button.setClickedListener(new Component.ClickedListener() {
            @Override
            public void onClick(Component component) {
                try {
                    // 拍照
                    capture();
                }catch (Exception e) {
                    Tools.showTip(CameraAbility.this, e.getMessage());
                }
            }
        });
        myLayout.addComponent(button);
        super.setUIContent(myLayout);
    }
    // 用于监听 SurfaceProvider 状态的监听器
    class MySurfaceCallback implements SurfaceOps.Callback{
        @Override
        public void surfaceCreated(SurfaceOps surfaceOps) {
            surfaceProvider.getSurfaceOps().get().setFixedSize(144, 176);
            // 打开相机
            openCamera(surfaceOps.getSurface());
        }
        @Override
        public void surfaceChanged(SurfaceOps surfaceOps, int i, int i1, int i2) {
        }
        @Override
        public void surfaceDestroyed(SurfaceOps surfaceOps)
        {
        }
    }
    @Override
    protected void onActive() {
```

```
        super.onActive();
    }

    @Override
    protected void onForeground(Intent intent) {
        super.onForeground(intent);
    }
    @Override
    protected void onStop() {
        super.onStop();
        // 停止拍照后，移除用于显示相机实时画面的窗口
        surfaceProvider.removeFromWindow();
    }
    // 用于拍照的方法
    private void capture() {
        if(mCamera == null) {
            Tools.showTip(this, "mcamera null");
            return;
        }
        // 获取拍照配置模板
        FrameConfig.Builder pictureFrameConfigBuilder =
        mCamera.getFrameConfigBuilder(Camera.FrameConfigType.FRAME_CONFIG_PICTURE);
        // 配置拍照 surface
        pictureFrameConfigBuilder.addSurface(mImageReceiver.getRecevingSurface());
        // 配置拍照其他参数
        pictureFrameConfigBuilder.setImageRotation(90);
        try {
            // 启动拍照
            mCamera.triggerSingleCapture(pictureFrameConfigBuilder.build());
        }catch (Exception e){
            Tools.showTip(CameraAbility.this, e.getMessage());
            e.printStackTrace();
        }
    }
    private Size getPictureSizes(List<Size> pictureSizes) {
        return new Size(1440, 1080);
    }
    // 拍照初始化
    private void takePictureInit() {
        // 获取拍照支持的分辨率列表
        List<Size> pictureSizes = mCameraAbility.getSupportedSizes(ImageFormat.JPEG);
        // 根据拍照要求选择合适的分辨率
        mPictureSize = getPictureSizes(pictureSizes);
        // 创建 ImageReceiver 对象，注意 create 函数中宽度要大于高度
        // 5 为最多支持的图像数，请根据实际设置
        mImageReceiver = ImageReceiver.create(Math.max(mPictureSize.width,
        mPictureSize.height), Math.min(mPictureSize.width, mPictureSize.height),
        ImageFormat.JPEG, 5);
        // 设置拍照回调
        mImageReceiver.setImageArrivalListener(mImageArrivalListener);
    }
    // 打开相机
    private void openCamera(Surface previewSurface) {
      CameraKit camerakit = CameraKit.getInstance(this);
```

```
if (camerakit == null) {
    //LogUtil.info(TAG, "cameraKit is null.");
    return;
}
// 获取 HarmonyOS 设备所有相机的 ID
String[] cameraIdList = camerakit.getCameraIds();
if (cameraIdList.length <= 0) {
    return;
}
// 取第一个相机的 ID
mCameraId = cameraIdList[0];
// 定义相机状态监听对象，在该对象中获取 Camera 对象
CameraStateCallback cameraStateCallback = new CameraStateCallback() {
    @Override
    public void onCreated(Camera camera) {
        super.onCreated(camera);
        // 获取 Camera 对象
        mCamera = camera;
        CameraConfig.Builder cameraConfigBuilder =
                                            camera.getCameraConfigBuilder();
        if (cameraConfigBuilder == null) {
            return;
        }
        // 绑定预览实时画面的窗口
        cameraConfigBuilder.addSurface(previewSurface);
    }

    @Override
    public void onCreateFailed(String cameraId, int errorCode) {
        super.onCreateFailed(cameraId, errorCode);
    }
    @Override
    public void onConfigured(Camera camera) {
        super.onConfigured(camera);
        FrameConfig.Builder frameConfigBuilder = camera.getFrameConfigBuilder
        (Camera.FrameConfigType.FRAME_CONFIG_PREVIEW);
        frameConfigBuilder.addSurface(previewSurface);
        FrameConfig previewFrameConfig = frameConfigBuilder.build();
        int triggerId = camera.triggerLoopingCapture(previewFrameConfig);
    }
    @Override
    public void onConfigureFailed(Camera camera, int errorCode) {
        super.onConfigureFailed(camera, errorCode);
    }
    @Override
    public void onReleased(Camera camera) {
        super.onReleased(camera);
    }
    @Override
    public void onFatalError(Camera camera, int errorCode) {
        super.onFatalError(camera, errorCode);
    }
};
```

```
        mEventHandler = new EventHandler(EventRunner.current());
        mCameraAbility = camerakit.getCameraAbility(mCameraId);

        List<Size> previewSizes = mCameraAbility.getSupportedSizes(SurfaceOps.class);
        // 拍照初始化
        takePictureInit();
        camerakit.createCamera(mCameraId, cameraStateCallback, mEventHandler);
    }
    // 用于保存拍摄图像的类
    class ImageSaver implements Runnable {
        private final Image myImage;
        private final File myFile;
        private String myFileName;
        ImageSaver(Image image, File file,String fileName) {
            myImage = image;
            myFile = file;
            myFileName = fileName;
        }
        @Override
        public void run() {
            Image.Component component =
                 myImage.getComponent(ImageFormat.ComponentType.JPEG);
            byte[] bytes = new byte[component.remaining()];
            // 获取拍摄图像的字节流
            component.read(bytes);
            // 下面的代码将图像保存到本机的图像目录下
            // 插入 picture 记录，并且获取 FD
            DataAbilityHelper helper = DataAbilityHelper.creator(CameraAbility.this);
            FileDescriptor fileDescriptor = null;
            try {
                ValuesBucket value = new ValuesBucket();
                value.putString(AVStorage.Images.Media.DISPLAY_NAME,myFileName);
                int result = helper.insert(AVStorage.Images.Media.EXTERNAL_DATA_ABILITY_URI,
                value);

                if (result != 1) {
                    return
                }
                // 查询记录
                ResultSet set = null;
                Uri uri = null;
                DataAbilityPredicates predicates = new DataAbilityPredicates
                ("_display_name = '"+myFileName+"'");
                set = helper.query(AVStorage.Images.Media.EXTERNAL_DATA_ABILITY_URI, new
                String[]{AVStorage.Images.Media.ID}, predicates);
                if (set != null && set.goToFirstRow()) {
                    int columnId = set.getColumnIndexForName(AVStorage.Images.Media.ID);
                    int mediaId = set.getInt(columnId);
                    uri = Uri.appendEncodedPathToUri(AVStorage.Images.Media.
                    EXTERNAL_DATA_ABILITY_URI, "" + mediaId);

                    // 获取待写入图像文件的 FileDescriptor 对象
                    fileDescriptor = helper.openFile(uri, "rw");
                }
```

```
        }catch (Exception e){
            e.printStackTrace();
        }
        FileOutputStream output = null;
        FileOutputStream outputPublic = null;
        try {
            output = new FileOutputStream(myFile);
            output.write(bytes);
        // 开始写入图像数据
            outputPublic = new FileOutputStream(fileDescriptor);
            outputPublic.write(bytes);
        }catch (Exception e){
            e.printStackTrace();
        }finally {
            myImage.release();
            try {
                output.close();
                outputPublic.close();
            }catch (Exception e){
                e.printStackTrace();
            }
        }

    }
  }
}
```

运行程序，单击“拍照”按钮，就会在本机的图像目录下看到保存的图像文件。

9.4 总结与回顾

本章先介绍了音频、视频处理的基本功能，包括播放本地音频、在线音频和本地视频。视频使用的是 mp4 文件格式。目前 HarmonyOS 只支持一些常用的音视频格式，如 wav、mp3、mp4 等，如果用户要播放 HarmonyOS 不支持的音视频格式，就只能借助第三方的库了。

对于相机，基本的操作包括拍摄照片和保存照片，本章也详细地介绍了这些操作的实现过程，并提供了完整的案例供读者学习。

第 10 章　其他高级技术

HarmonyOS 还有很多实用的技术，在本章我们会挑出一些比较常用的技术与大家分享，这些技术涉及人工智能（AI）、传感器、定位、蓝牙等方面。

本章的源代码在配套资源中 others 目录下。

通过阅读本章，读者可以掌握：

- 如何使用 AI 接口；
- 如何使用传感器；
- 如何在各种场景中进行定位；
- 如何控制蓝牙；
- 如何发现蓝牙设备；
- 如何实现蓝牙设备配对；
- 如何通过系统拨号盘拨打电话。

10.1 AI 接口

HarmonyOS 提供了一系列 AI 接口，通过这些接口，我们可以完成很多有趣的功能，例如，分词、提取关键字、词性标注等。HarmonyOS 本身已经提供了这些 AI 接口使用的模型，所以 HarmonyOS 终端设备只需要直接使用这些 AI 模型完成对应的工作即可。

10.1.1 初始化 AI 引擎

AI 引擎的核心类是 ohos.ai.nlu.NluClient，在使用之前，需要通过 NluClient.init 方法初始化 AI 引擎，该方法的原型如下：

```
public void init(Context context, OnResultListener<Integer> listener,
                                                    boolean isLoadModel);
```

init 方法有 3 个参数，下面是这些参数的含义。

- context：应用上下文信息，应为 ohos.aafwk.ability.Ability 或 ohos.aafwk.ability.AbilitySlice 的实例或子类的实例。
- listener：初始化结果的回调对象，可以为 null。

- isLoadModel：是否加载 AI 模型。如果参数值是 true，则在初始化时加载 AI 模型；如果参数值是 false，则在初始化时不加载 AI 模型。

初始化 AI 引擎的代码如下：

```
NluClient.getInstance().init(context, new OnResultListener<Integer>(){
        @Override
        public void onResult(Integer result){
           // 初始化成功回调，在服务初始化成功调用该函数
        }
}, true);
```

AI 引擎只能初始化一次，所以通常会将初始化 AI 引擎的代码放到 onStart 方法中。

10.1.2 分词

分词是将一个句子拆分成若干独立的单词，与编译器的工作过程一样，首先需要将源代码拆解成独立的词元（称为 Token），这叫作词法分析。然后利用这些 Token，再进行语法和语义分析。因为编程语言的代码都是有一定规则的，所以非常容易拆解。但自然语言就不同了，像英语一样的自然语言比较简单，因为在书写时，每个单词都是用空格分开的，所以分析英语一般不涉及分词的问题。但中文就复杂得多，现代中文中标点符号最多只能分句，不能分词，但我们需要将一个没有标点符号的句子拆解成一个个独立的词组，这样才能进行下面的语义分析，这个过程就叫作分词。

分词有很多种方法，最简单的方法是直接利用传统算法进行分析，但这种分析方式准确率较低，没有太大的使用价值，目前最流行的分词方式是使用深度学习技术进行分词，这种分词技术需要依赖大量的语料（一般是已经分好词的文章）以及特定的 AI 模型。利用语料来训练 AI 模型，最终让 AI 模型的参数达到最佳状态。然后将待分词的句子输入 AI 模型，就会得到准确率较高的结果。也就是说，使用深度学习技术进行分词是利用了已经有结果的数据训练 AI 模型，在训练的过程中调整 AI 模型的参数，然后利用训练好的 AI 模型去处理结果未知的数据。

深度学习的基本原理很简单，但是利用海量数据训练 AI 模型需要一样东西，那就是钱。因为需要大量的算力才能在短时间内利用海量的数据训练完复杂的 AI 模型，这些算力都需要购买大量的服务器。不过在 HarmonyOS 中使用 AI 模型就不需要我们花钱购买算力了，因为华为已经为我们训练好了 AI 模型，我们直接用就可以了。使用 AI 模型是不需要消耗多少算力的，就像我们使用的数学公式，当年在推导这些数学公式时，可能是某个顶级数学家花费 10 年时间推导出来的，而我们只需要 10 秒就可以学会这个公式。

现在回到分词上，其实在 HarmonyOS 中分词，只需要两行代码就可以实现（最后一行用于显示分词结果），使用简单易用的 AI 接口。其中核心方法是 getWordSegment 方法，实现代码如下：

```
// 指定待分词的字符串
String requestData = "{\"text\":\"今天天气太热了，穿多了，走了一天，累死了！\\n\",\"type\":0}";
// 分词
ResponseResult responseResult = NluClient.getInstance().getWordSegment(requestData,
NluRequestType.REQUEST_TYPE_LOCAL);
// 显示分词结果
Tools.showTip(MainAbility.this, responseResult.getResponseResult());
```

其中，getWordSegment 方法的第 2 个参数值目前只能是 NluRequestType.REQUEST_TYPE_LOCAL，表示利用本地的 AI 模型进行分词。getWordSegment 方法的返回值是 ResponseResult 对象，该对象有一个 getResponseResult 方法，用于以 JSON 格式返回处理结果。

在指定待分词字符串时，也需要使用 JSON 格式，其中 text 属性表示待分词的字符串，type 属性表示分词的粒度，type 属性可取的值及其描述如表 10-1 所示。

表 10-1 type 属性可取的值及其描述

type 属性值	是否为默认值	描述
0	是	基本词，粒度较小。例如“我要看美女与野兽”，分成“我/要/看/美女/与/野兽”
1	否	在基本词的基础上，做实体合并。例如“我要去沈阳恒大中央广场看美女与野兽”，分成“我/要/去/沈阳恒大中央广场/看/美女/与/野兽”。对于没有可合成实体的文本信息，其分词效果与 type 为 0 的分词效果相同。例如：“明天下午 5 点一起看电影”，分成“明天/下午/5 点/一起/看/电影”
9223372036854775807（2 的 63 次方减 1）	否	在 type 为 1 的基础上，把实体时间、地点等整体结构合并，出现符号隔开不合并，并把一些常用短语合并，例如“形容词+的”“单字动词+单字名词”等，简化句子成分。例如“明天下午五点到七点我在沈阳中街光陆影城看电影”，分成“明天下午五点/到/七点/我/在/沈阳中街光陆影城/看/电影”

现在运行程序，单击“分词”按钮，会显示图 10-1 所示的结果。

图 10-1 显示分词结果

我们可以看到，返回的结果是 JSON 格式，其中 words 属性是一个 JSON array 格式，每一个数组元素就是一个单词，例如“今天”“太”“累死”等。

注意，AI 与人类一样，并不能保证 100%准确率。评价 AI 模型的效果，并不是用对错，而是用准确率。所以读者不要期望 AI 模型会返回一个 100%准确的结果。而且目前 HarmonyOS 中的 AI 分词有一定的局限性，只能完成特定领域的分词，例如，电影、动漫、球队、快递单号等领域。如果句子中的词组超出了 AI 模型理解的领域范畴，可能就无法准确分词了，AI 分词支持的完整领域列表请参考 https://developer.harmonyos.com/cn/docs/documentation/doc-guides/ai-word-segmentation-guidelines-0000001050732478。

10.1.3 词性标注

词性就是句子中每一个词的语法特征，如名词、动词、形容词等，词性标注也是对自然语言进行语义分析的重要一步。

词性标注需要使用 NluClient.getWordPos 方法，实现代码如下：

```
String requestData = "{\"text\":\"我要看美女与野兽\",\"type\":0}";
ResponseResult responseResult = NluClient.getInstance().getWordPos(requestData,
                                            NluRequestType.REQUEST_TYPE_LOCAL);
Tools.showTip(MainAbility.this, responseResult.getResponseResult());
```

getWordPos 方法的参数与上一节介绍的 getWordSegment 方法的参数及其含义完全相同。requestData 中 type 属性的含义与表 10-1 描述的含义完全相同。

运行程序，单击“词性标注”按钮，会显示图 10-2 所示的结果。

从词性标注的结果可以看出，pos 属性表示具体的词性标注结果。pos 属性是 JSON array 格式，每一个数组元素是一个 JSON object 类型，其中 word 表示分词的结果，tag 属性表示具体的词性，例如，rr 表示人称代词，v 表示动词，n 表示名词，cc 表示并列连词。完整的词性描述请查看 https://developer.harmonyos.com/cn/docs/documentation/doc-guides/ai-pos-tagging-guidelines-0000001050732512 中的“表 1 词性说明”。

图 10-2 显示词性标注结果

10.1.4 意图分析

意图分析已经是语义分析的范畴了。意图分析是根据文本的上下文分析出文本的含义，例如，前面有“账单”这个词，后面是一个数字，那么 AI 引擎就会认为后面的数字表示钱，并用 moneyInfo 属性表示钱的相关信息。如果句子中包含“银行”这个词，AI 引擎就会认为在“银行”前面的一个名词是银行的名字，并用 bank 属性表示银行。而且 AI 引擎还可以根据文本分析出具体的意图，并用 intentions 属性表示。例如，分析出意图是 repayNotice，就表明这是一条大多数人都不希望看到的信息——信用卡还款提醒。再结合 bank 和 moneyInfo 的值，就是 XXXX 银行提醒用户该还款了，而且明确了还款的金额。

对一段文本进行意图分析，需要使用 NluClient.getChatIntention 方法，实现代码如下：

```
String requestJson = "{text:'您个人信用卡 12 月账单￥32198.00,还款日 12 月 15 日【XXXX 银行】'}";
// 对文本进行意图分析
ResponseResult responseResult = NluClient.getInstance().getChatIntention(requestJson,
NluRequestType.REQUEST_TYPE_LOCAL);
Tools.showTip(MainAbility.this, responseResult.getResponseResult());
```

运行程序，单击“聊天意图”按钮，会显示图 10-3 所示的结果。

从分析结果可以看出，意图是 repayNotice，具体日期是 2021 年 12 月 15 日。其中，因为文本中并未指定具体的年份，所以 AI 引擎就将当前年份当作日期的年份了。分析结果中只有一条关于钱的信息，也就是 moneyInfo 属性的元素值。其中，amount 属性表示具体的金额，值是 32198.00。bank 属性表示具体的银行，值是 XXXX 银行。综合意图和关键词，可以分析出一个结果——XXXX 银行要你在 2021 年 12 月 15 日前还款 32198 元。

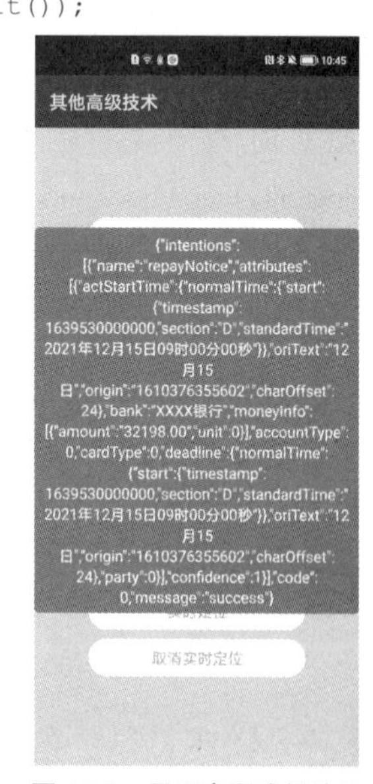

图 10-3 显示意图分析结果

10.1.5 关键词提取

通过 NluClient.getKeywords 方法可以提取一段话中的关键词，实现代码如下：

```
String requestData= "{number:2,body:'今天是我生日，有人给我红包吗？
坐等'}";
```

```
    ResponseResult responseResult = NluClient.getInstance().getKeywords(requestData,
NluRequestType.REQUEST_TYPE_LOCAL);
    Tools.showTip(MainAbility.this, responseResult.getResponseResult());
```

运行程序，单击“关键字提取”按钮，会显示图 10-4 所示的结果。

从分析结果可以看出，“红包”和“坐等”是 body 属性指定的文本中的关键词。

图 10-4　显示关键词提取的结果

10.1.6　实体识别

实体识别也属于分词技术，只是会从分出来的多个单词中挑出一些实体，忽略掉其他非实体部分。这里的实体是指具有特定意义的词，例如，电影（movie）、名字（name）、时间（time）、团队（team）、电子邮件（email）等。完整的实体种类参见 https://developer.harmonyos.com/cn/docs/documentation/doc-guides/ai-entity-recognition-guidelines-0000001050814419。

通过 NluClient.getEntity 方法可以获取文本中代表实体的词组，实现代码如下：

```
    String requestData= "{text:'本来这个星期六我要去看电影星际迷航，结果同事王军病了，我要替他的班！',
module:'movie'}";
    ResponseResult responseResult = NluClient.getInstance().getEntity(requestData,
NluRequestType.REQUEST_TYPE_LOCAL);
    Tools.showTip(MainAbility.this, responseResult.getResponseResult());
```

requestData 变量指定了待分析的文本以及相关属性，text 属性表示待分析的文本，是必须指定值的属性，而 module 表示要分析的实体的种类，是可选指定值的，如果不指定 module 属性值，则分析 AI 引擎支持的所有实体，本例指定了 movie，所以只分析与电影相关的实体。

现在运行程序，单击“实体识别”按钮，会显示图 10-5 所示的结果。

从显示结果可以看出，只分析出了一个电影实体——“星际迷航”。这是因为 module 属性值是 movie，如果将 module 属性去掉，再次运行程序，单击“实体识别”按钮，就会得到图 10-6 所示的结果。

从分析结果可以看出，这段文本中包含了 movie 实体、name 实体和 time 实体。

图 10-5　显示实体识别结果

图 10-6　显示所有的实体

10.2 传感器

传感器是手机的重要组成部分，有了传感器，我们就可以让手机变得多功能，例如，可以利用重力传感器实现支持感应的游戏，利用环境光传感器实现背光自动调节。下面是 HarmonyOS 支持的传感器。

- 方向传感器数据：可以感知设备当前的朝向，从而达到为用户指明方向的目的。
- 重力传感器：可以感知设备倾斜的程度。
- 陀螺仪传感器：可以感知设备的旋转量，通过重力传感器和陀螺仪传感器，可以提升用户在游戏场景中的体验。
- 接近光传感器：可以感知设备与遮挡物之间的距离，使设备能够自动亮屏和灭屏，达到防误触的目的。
- 气压计传感器：可以准确地判断设备当前所处的海拔。
- 环境光传感器：让设备能够实现背光自动调节。
- 霍尔传感器：可以用于实现智能皮套（磁皮套）。也就是说，扣上皮套后屏幕就会在皮套上留出的小窗口中出现一个界面，用来接听来电或阅读短信。或者更简单一点，合上皮套，手机自动休眠，打开皮套，手机自动唤醒。

10.2.1 获取当前设备支持的传感器

尽管 HarmonyOS 支持很多传感器，但并不是所有的 HarmonyOS 设备都内置了这些传感器，如果硬件设备不支持，那么 HarmonyOS 支持也没用，所以在使用特定传感器之前，最好检测当前设备支持哪些传感器。

要想获取当前设备支持的传感器列表，首先要创建 CategoryOrientationAgent 对象，然后调用 CategoryOrientationAgent.getAllSensors 方法获取传感器列表，实现代码如下：

```
CategoryOrientationAgent categoryOrientationAgent = new CategoryOrientationAgent();
// 获取传感器列表
List<CategoryOrientation> sensorList = categoryOrientationAgent.getAllSensors();
String sensors = "";
// 将列表中所有传感器的名称组合成一个字符串
for (CategoryOrientation sensor : sensorList) {
   sensors += sensor.getName() + "\r\n";
}
// 显示当前设备支持的所有传感器的名称
Tools.showTip(MainAbility.this, sensors.trim());
```

运行程序，单击“获取支持的传感器”按钮，就会显示类似图 10-7 的结果。注意，在读者的 HarmonyOS 设备上可能会得到不同的结果。

从这个列表可以看出，当前设备并没有支持 HarmonyOS 支持的全部传感器。

图 10-7 显示当前设备支持的所有传感器的名称

10.2.2 订阅方向传感器

在 HarmonyOS 中使用订阅的方式来接收传感器的数据，本节会通过最常用的传感器——方向传感器来演示这一过程。方向传感器是几乎所有智能手机都会支持的传感器，所以非常方便做实验。

订阅方向传感器需要如下 3 步。

（1）创建用于接收传感器数据的回调对象。

（2）获取方向传感器对象。

（3）为方向传感器指定接收数据的回调对象。

为了实现第（1）步，需要实现 ICategoryOrientationDataCallback 接口，该接口中的 onSensorDataModified 方法用于接收传感器的数据，实现代码如下：

```
int matrix_length = 3;
int rotationVectorLength = 3;
ICategoryOrientationDataCallback orientationDataCallback =
                           new ICategoryOrientationDataCallback() {
   @Override
   public void onSensorDataModified(CategoryOrientationData categoryOrientationData) {
      // 解析和使用接收的 categoryOrientationData 传感器数据对象
      // 获取传感器的维度信息
      int dim = categoryOrientationData.getSensorDataDim();
      // 获取方向类传感器的第一维数据
      float degree = categoryOrientationData.getValues()[0];
      float[] rotationMatrix = new float[matrix_length];
      // 根据旋转矢量传感器的数据获得旋转矩阵
      CategoryOrientationData.getDeviceRotationMatrix(rotationMatrix,
      categoryOrientationData.values);
      float[] rotationAngle = new float[rotationVectorLength];
      // 根据计算出来的旋转矩阵获取设备的方向
      rotationAngle = CategoryOrientationData.getDeviceOrientation(rotationMatrix,
      rotationAngle);
      // rotationAngle 数组的 3 个元素值分别是方位角、俯仰角和旋转角
      Tools.print("orientation sensor:" + rotationAngle[0] +
               ":" + rotationAngle[1] +
               ":" + rotationAngle[2]);
   }

   @Override
   public void onAccuracyDataModified(CategoryOrientation categoryOrientation, int i) {
      // 使用变化的精度
   }

   @Override
   public void onCommandCompleted(CategoryOrientation categoryOrientation) {
```

```
        // 传感器执行命令回调
    }
};
```

在 onSensorDataModified 方法中进行了一系列计算，最后得出下面 3 个角度。

- rotationAngle[0]：方位角。
- rotationAngle[1]：俯仰角。
- rotationAngle[2]：旋转角。

```
// 获取传感器数据的时间间隔
long interval = 100000000;
// 获取方向传感器
CategoryOrientation orientationSensor = categoryOrientationAgent.getSingleSensor(
        CategoryOrientation.SENSOR_TYPE_ORIENTATION);
if (orientationSensor != null) {
    // 设置用于监听方向传感器数据的回调对象
    categoryOrientationAgent.setSensorDataCallback(
            orientationDataCallback, orientationSensor, interval);
}
Tools.showTip(MainAbility.this, "成功订阅传感器");
```

在这段代码中，setSensorDataCallback 方法用于设置监听方向传感器数据的回调对象，其中第 3 个参数是获取传感器数据的时间间隔，单位是纳秒，1 秒相当于 10 亿纳秒。而本例指定的值是 1 亿纳秒，也就是 100 毫秒。

现在运行程序，单击“订阅方向传感器”按钮，晃动手机，在 HiLog 视图中就会输出不同方向的角度。

如果要取消订阅，执行下面的代码即可：

```
categoryOrientationAgent.releaseSensorDataCallback(
                        orientationDataCallback, orientationSensor);
```

10.3 定位

定位的作用是获取设备当前所在位置的坐标，也就是经度（longitude）和纬度（latitude）。为了获取经度和纬度，HarmonyOS 设备会利用不同的定位技术，例如，全球导航卫星系统（Global Navigation Satellite System，GNSS）就是一种被普遍认可、广泛接受的追踪定位技术。除了 GNSS，还可以采用基站定位、WLAN 定位等技术，也可能采用多种定位技术的结合体。

HarmonyOS 提供了一套 API 来使用定位功能。定位的核心类是 Locator 类，所以需要先创建 Locator 类的实例。然后通过 Locator.startLocating 方法开启定位追踪，这是一个持续的过程，也就是说，并不是定位一次就完成了，而是需要持续进行跟踪。如果读者使用过高德地图、Google 地图等应用，对定位就不会陌生。当我们移动时，在地图上指示当前位置的小箭头也会沿着地图移动，这就是持续跟踪定位的结果。

持续跟踪的基本原理肯定不是天上的卫星主动将位置信息发给地面上的设备的，而是地面上的设备主动获取的。那么这就需要确定获取位置信息的触发条件。通常，只要满足下面两个条件之一就可以主动获取位置信息。

- 超过一段时间后。
- 超过一段距离后。

也就是说，Locator 对象会根据时间间隔和距离变化两个条件来决定什么时候获取位置信息。例如，时间间隔是 5 秒，位置变化了 5 米。如果距上次获取位置信息的时间已经超过了 5 秒，那么就会再次获取位置信息，即使没有超过 5 秒，但在 5 秒内，距离上次定位的位置已经变化了 5 米，那么也会再次获取位置信息。也就是说，只要这两个条件（时间间隔超过 5 秒或位置变化超过 5 米）满足一个，就会再次获取位置信息。

为了实现这个功能，需要通过 RequestParam 类的构造方法来指定获取位置信息的条件。为了方便，RequestParam 类提供了一些常量来描述特定的场景，例如，RequestParam.SCENE_NAVIGATION 主要适用于导航场景。

除了需要确定获取位置信息的条件，在获取位置信息后，还需要接收这些信息，所以需要实现 LocatorCallback 接口，该接口的 onLocationReport 方法用来接收位置信息。完整的定位以及输出经纬度的代码如下：

```
// 实现 LocatorCallback 接口，用来接收位置信息
class MyLocatorCallback implements LocatorCallback {
    // 这个方法只要满足特定条件就会调用，所以会不断在 HiLog 视图中输出经纬度
    @Override
    public void onLocationReport(Location location) {
        Tools.print("MyLocatorCallback",
        "经度（Longitude）: " + String.valueOf(location.getLongitude()) +
        "纬度（Latitude）:" + String.valueOf(location.getLatitude()));
    }
    // 状态变化时触发
    @Override
    public void onStatusChanged(int type) {
        Tools.print("MyLocatorCallback", String.valueOf(type));
    }

    @Override
    public void onErrorReport(int type) {
    }
}
// 创建 MyLcoatorCallback 对象
MyLocatorCallback locatorCallback = new MyLocatorCallback();
// 创建用于定位的 Locator 对象
Locator locator = new Locator(this);
// 设置与定位相关的参数，RequestParam.SCENE_NAVIGATION 适用于导航场景
RequestParam requestParam = new RequestParam(RequestParam.SCENE_NAVIGATION);
// 开始定位（会根据是否超出时间和距离来决定是否获取新的位置信息）
locator.startLocating(requestParam, locatorCallback);
Tools.showTip(MainAbility.this, "开始实时定位");
```

运行程序，单击“实时定位”按钮，会发现每隔一段时间，就会在 HiLog 视图中输出一行位置信息（经纬度），如图 10-8 所示。

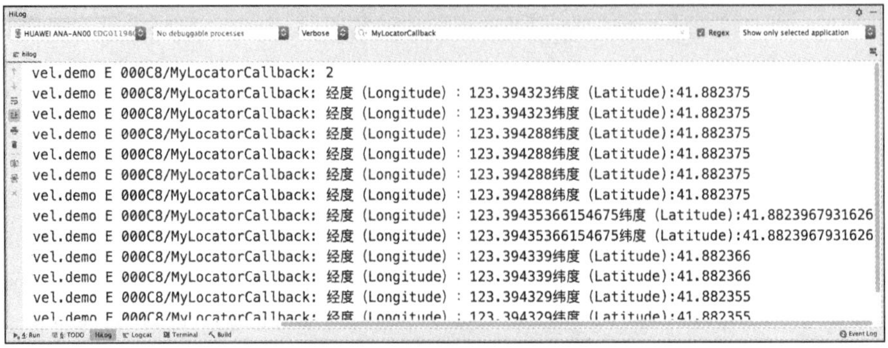

```
vel.demo E 000C8/MyLocatorCallback: 2
vel.demo E 000C8/MyLocatorCallback: 经度 (Longitude) : 123.394323纬度 (Latitude):41.882375
vel.demo E 000C8/MyLocatorCallback: 经度 (Longitude) : 123.394323纬度 (Latitude):41.882375
vel.demo E 000C8/MyLocatorCallback: 经度 (Longitude) : 123.394288纬度 (Latitude):41.882375
vel.demo E 000C8/MyLocatorCallback: 经度 (Longitude) : 123.394288纬度 (Latitude):41.882375
vel.demo E 000C8/MyLocatorCallback: 经度 (Longitude) : 123.394288纬度 (Latitude):41.882375
vel.demo E 000C8/MyLocatorCallback: 经度 (Longitude) : 123.394288纬度 (Latitude):41.882375
vel.demo E 000C8/MyLocatorCallback: 经度 (Longitude) : 123.39435366154675纬度 (Latitude):41.8823967931626
vel.demo E 000C8/MyLocatorCallback: 经度 (Longitude) : 123.39435366154675纬度 (Latitude):41.8823967931626
vel.demo E 000C8/MyLocatorCallback: 经度 (Longitude) : 123.394339纬度 (Latitude):41.882366
vel.demo E 000C8/MyLocatorCallback: 经度 (Longitude) : 123.394339纬度 (Latitude):41.882366
vel.demo E 000C8/MyLocatorCallback: 经度 (Longitude) : 123.394329纬度 (Latitude):41.882355
vel.demo E 000C8/MyLocatorCallback: 经度 (Longitude) : 123.394329纬度 (Latitude):41.882355
```

图 10-8　输出当前位置的经纬度

如果想关闭实时定位，可以使用下面的代码：

```
// 停止实时定位
locator.stopLocating(locatorCallback);
```

使用定位服务必须在 config.json 文件中申请如下权限：

```
"reqPermissions": [
    {
        "name": "ohos.permission.LOCATION"
    }
]
```

同时还需要使用 Java 代码申请 ohos.permission.LOCATION 权限。

RequestParam 类的构造方法有两种重载形式，它们的原型如下：

```
public RequestParam(int scenario);
public RequestParam(int priority, int timeInterval, int distanceInterval);
```

第 1 种重载形式只有一个 int 类型的参数。本例设置的 RequestParam.SCENE_NAVIGATION 就使用了第 1 种重载形式，表示特定的场景。RequestParam.SCENE_NAVIGATION 适用于在户外实时定位的场景，如车载、步行导航。在此场景下，为保证系统提供最优的位置结果精度，主要使用 GNSS 定位技术提供定位服务。结合场景特点，在导航启动之初，用户很可能处于室内、车库等遮蔽环境中，GNSS 技术很难提供位置服务。为解决此问题，HarmonyOS 会在 GNSS 提供稳定位置结果之前，使用系统网络定位技术，向 App 提供位置服务，在导航初始阶段提升用户体验。

如果使用第 1 种重载形式，那么系统会以最短时间（1 秒）间隔不断获取位置信息，虽然有可能出于各种原因，系统响应会超过 1 秒，但总体获取位置信息的时间间隔是非常短的。

第 2 种重载形式有 3 个参数，下面是这些参数的含义。

- priority：如果第 1 种重载形式不能满足需求，可以使用精度优先策略，例如，PRIORITY_ACCURACY 表示精度优先。精度优先策略主要以 GNSS 定位为主，在开阔场景下可以提供米级的定位精度，具体性能指标依赖用户设备的定位硬件能力，但在室内等强遮蔽定位场景下，无法提供准确的位置服务；PRIORITY_FAST_FIRST_FIX 表示快速定位优先策略，会同时使用 GNSS 定位、基站定位、WLAN 定位、蓝牙定位等定位技术，实现快速定位。即使在室内，也可以实现定位。但定位精度并不高。

- timeInterval：获取位置信息的时间间隔，单位是秒，每隔 timeInterval 秒，会获取一次位置信息。
- distanceInterval：获取位置信息的距离间隔，单位是米，如果当前位置离上次定位的位置超过了 distanceInterval 米，则获取一次位置信息。

下面的代码使用了第 2 种重载形式实现了实时定位的功能。

```
// 创建 MyLcoatorCallback 对象
MyLocatorCallback locatorCallback = new MyLocatorCallback();
// 创建用于定位的 Locator 对象
Locator locator = new Locator(this);
// 设置与定位相关的参数，每 2 秒或超过上次定位位置 5 米时获取一次位置信息
RequestParam requestParam = new RequestParam(RequestParam.PRIORITY_ACCURACY,2, 5);
// 开始定位（会根据是否超出时间或距离来决定是否获取新的位置信息）
locator.startLocating(requestParam, locatorCallback);
Tools.showTip(MainAbility.this, "开始实时定位");
```

更详细的定位策略参见 https://developer.harmonyos.com/cn/docs/documentation/doc-guides/device-location-info-0000000000031900。

10.4 蓝牙

蓝牙是一种通信协议，用于传输数据。蓝牙的数据传输速度要大于 NFC。有效连接距离在没有障碍物的情况下，通常在 10 米以内。传输速度根据版本，从每秒几兆字节到每秒几十兆字节不等。例如，蓝牙 4.0 的传输速度是 24MB/s。蓝牙的应用很广，通常用于连接无线外设。例如，蓝牙鼠标、蓝牙键盘、蓝牙音箱，以及手机之间传输数据等。本节将介绍 HarmonyOS 中使用蓝牙的基本方式。

10.4.1 打开和关闭蓝牙

要想使用蓝牙，首先要打开手机中的蓝牙功能。用户可以通过选择“设置”→“蓝牙”打开或关闭蓝牙功能，如图 10-9 所示。

图 10-9 打开或关闭蓝牙功能

HarmonyOS 也提供了 API，可以在程序中通过 Java 代码在不显示设置界面的前提下打开或关闭蓝牙功能，代码如下：

```
// 创建 BluetoothHost 对象
BluetoothHost bluetoothHost = BluetoothHost.getDefaultHost(this);
Button button1 = (Button)findComponentById(ResourceTable.Id_button1);
if(button1 != null) {
   button1.setClickedListener(new Component.ClickedListener() {
      @Override
      public void onClick(Component component) {
         if(button1.getText().equals("打开蓝牙")) {
            // 打开蓝牙
            bluetoothHost.enableBt();
```

```
                button1.setText("关闭蓝牙");
                Tools.showTip(BluetoothAbility.this, "成功打开蓝牙");
            } else if(button1.getText().equals("关闭蓝牙")) {
                // 关闭蓝牙
                bluetoothHost.disableBt();
                button1.setText("打开蓝牙");
                Tools.showTip(BluetoothAbility.this, "成功关闭蓝牙");
            }
        }
    });
    // 蓝牙处于关闭状态
    if(bluetoothHost.getBtState() == 0) {
        button1.setText("打开蓝牙");
    } else if(bluetoothHost.getBtState() == 2) {
        button1.setText("关闭蓝牙");
    }
}
```

阅读这段代码，需要了解以下内容。

- 如需使用设备中的蓝牙功能，要创建 BluetoothHost 对象，BluetoothHost 类提供了一个 BluetoothHost 静态方法，用于以 singleton 模式获取 BluetoothHost 对象，保证 BluetoothHost 对象在整个 App 中只有一个。
- BluetoothHost.enableBt 方法用于打开蓝牙，BluetoothHost.disableBt 方法用于关闭蓝牙。
- BluetoothHost.getBtState 方法用于获取蓝牙的当前状态，返回值为 0 表示蓝牙处于关闭状态，返回值为 2 表示蓝牙处于打开状态。在这段代码中，会检测蓝牙的当前状态，并为按钮设置相应的文本。例如，如果蓝牙处于关闭状态，按钮文本为“打开蓝牙”，否则为“关闭蓝牙”。

在 HarmonyOS 中使用蓝牙，需要在 config.json 文件中申请下面的权限：

```
"reqPermissions": [
  {
    "name": "ohos.permission.USE_BLUETOOTH"
  },
  {
    "name": "ohos.permission.DISCOVER_BLUETOOTH"
  },
  {
    "name": "ohos.permission.MANAGE_BLUETOOTH"
  }
]
```

现在运行程序，然后单击“关闭蓝牙”或“打开蓝牙”按钮，会在窗口下方显示图 10-10 所示的询问对话框，单击“允许”按钮，就会关闭或打开蓝牙。

图 10-10　询问是否要关闭蓝牙

10.4.2　发现和连接蓝牙设备

在连接蓝牙设备之前，首先要发现蓝牙设备。调用 BluetoothHost.startBtDiscovery 方法可以搜索周围的蓝牙设备。如果周围设备中的蓝牙功能已经打开，并且允许被搜索到，那么当前设备就会发现这些蓝牙设备。

HarmonyOS 采用了订阅的方式来接收已经搜索到的蓝牙设备。在订阅的接收方会实现一个继承自 CommonEventSubscriber 的类。该类有一个 onReceiveEvent 方法，每发现一个蓝牙设备，就会调用该方法一次。onReceiveEvent 方法的原型如下：

```
public void onReceiveEvent(CommonEventData var);
```

通过 var 参数可以获取蓝牙设备的信息，获取信息的具体方式请看后面完整的源代码。最后需要使用 CommonEventManager.subscribeCommonEvent 方法来订阅蓝牙设备发现事件。完成订阅后，运行程序，单击“扫描蓝牙设备”按钮，如果周围有可被发现的蓝牙设备，就会在按钮下方的 ListContainer 组件中显示蓝牙设备的名称和设备地址，如图 10-11 所示。注意，HarmonyOS 不仅能发现其他支持蓝牙的 HarmonyOS 设备，还可以搜索到很多其他类型的蓝牙设备（如 Android 手机、iPhone 等），例如，图 10-11 中我们搜索到了 HarmonyOS 手机、平板电脑，还有几部 Android 手机和一部 iPhone。

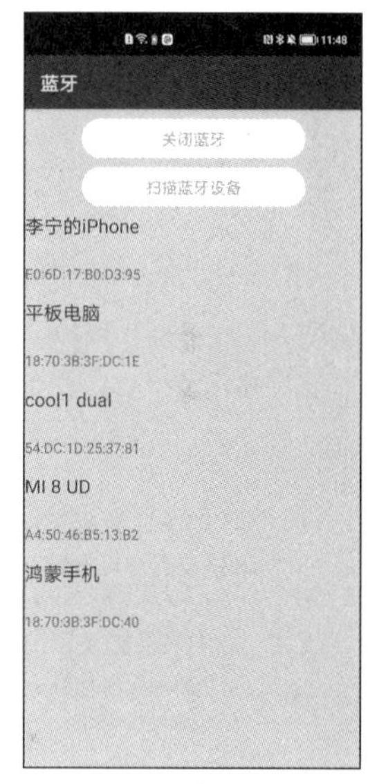

图 10-11 已经发现的蓝牙设备

发现蓝牙设备的完整代码如下：

```
package com.unitymarvel.demo;

import ohos.aafwk.ability.Ability;
import ohos.aafwk.content.Intent;
import ohos.aafwk.content.IntentParams;
import ohos.agp.components.*;
import ohos.bluetooth.BluetoothHost;
import ohos.bluetooth.BluetoothRemoteDevice;
import ohos.event.commonevent.*;
import java.util.ArrayList;
import java.util.List;

public class BluetoothAbility extends Ability {
   // 控制蓝牙的 BluetoothHost 对象
   private BluetoothHost bluetoothHost;
   // 用于显示被发现的蓝牙设备信息的列表组件
   private ListContainer blueDevices;
   // 用于保存被发现的蓝牙设备对象
   private List<BluetoothRemoteDevice> bluetoothDeviceList;
   // 列表组件的数据源
   private RecycleItemProvider provider;
   // 定义用于监听发现蓝牙设备动作的类（订阅蓝牙发现动作的类）
   public class BluetoothEventSubscriber extends CommonEventSubscriber {
      public BluetoothEventSubscriber(CommonEventSubscribeInfo subscribeInfo) {
         super(subscribeInfo);
      }
      // 每发现一个蓝牙设备，就会调用一次该方法
      @Override
      public void onReceiveEvent(CommonEventData var){
         try {
```

```
            // 蓝牙设备信息是通过 Intent 对象返回的
            Intent info = var.getIntent();
            if (info == null) return;
            // 每发现一个蓝牙设备，系统会发送一个广播，所以要先获取这个广播的 Action
            String action = info.getAction();
            // 判断触发该事件的是否是发现蓝牙的广播，因为可能多个广播共用该方法
            if (action == BluetoothRemoteDevice.EVENT_DEVICE_DISCOVERED) {
               IntentParams myParam = info.getParams();
               // 获取被发现的蓝牙设备的 BluetoothRemoteDevice 对象
               BluetoothRemoteDevice device = (BluetoothRemoteDevice)
            myParam.getParam(BluetoothRemoteDevice.REMOTE_DEVICE_PARAM_DEVICE);
               // 并不是所有的蓝牙设备都有名字，所以这里要忽略没有名字的蓝牙设备
               if(device.getDeviceName().isPresent()) {
                  // 保存当前被发现的蓝牙设备的 BluetoothRemoteDevice 对象
                  bluetoothDeviceList.add(device);
                  // 通知列表组件，数据源发生了变化，要重新显示列表中的内容
                  provider.notifyDataChanged();
               }
            }
         } catch (Exception e) {
            Tools.print("bluetooth device error:" + e.getMessage());
         }
      }
   }
   // 用于订阅蓝牙设备发现事件
   private void subscribeEvent() {
      MatchingSkills matchingSkills = new MatchingSkills();
      // 添加蓝牙设备发现事件
      matchingSkills.addEvent(BluetoothRemoteDevice.EVENT_DEVICE_DISCOVERED);
      CommonEventSubscribeInfo subscribeInfo = new
                                 CommonEventSubscribeInfo(matchingSkills);
      BluetoothEventSubscriber subscriber = new
                                 BluetoothEventSubscriber(subscribeInfo);
      try {
         // 订阅蓝牙设备发现事件
         CommonEventManager.subscribeCommonEvent(subscriber);
      } catch (Exception e) {
         Tools.print("bluetooth device error:" + e.getMessage());
      }
   }
   @Override
   public void onStart(Intent intent) {
      super.onStart(intent);
      super.setUIContent(ResourceTable.Layout_bluetooth_ability);
      // 获取 BluetoothHost 对象
      bluetoothHost = BluetoothHost.getDefaultHost(this);
      // 订阅蓝牙设备发现事件
      subscribeEvent();
      // 处理“关闭蓝牙”或“打开蓝牙”按钮
      Button button1 = (Button)findComponentById(ResourceTable.Id_button1);
      if(button1 != null) {
         button1.setClickedListener(new Component.ClickedListener() {
            @Override
            public void onClick(Component component) {
```

```
            if(button1.getText().equals("打开蓝牙")) {
               bluetoothHost.enableBt();
               button1.setText("关闭蓝牙");
               Tools.showTip(BluetoothAbility.this, "成功打开蓝牙");
            } else if(button1.getText().equals("关闭蓝牙")) {
               bluetoothHost.disableBt();
               button1.setText("打开蓝牙");
               Tools.showTip(BluetoothAbility.this, "成功关闭蓝牙");
            }
         }
      });
      // 蓝牙处于关闭状态
      if(bluetoothHost.getBtState() == 0) {
         button1.setText("打开蓝牙");
      } else if(bluetoothHost.getBtState() == 2) {   // 蓝牙处于打开状态
         button1.setText("关闭蓝牙");
      }
   }

   // 处理“扫描蓝牙设备”按钮
   Button button2 = (Button)findComponentById(ResourceTable.Id_button2);
   if(button2 != null) {
      button2.setClickedListener(new Component.ClickedListener() {
         @Override
         public void onClick(Component component) {
            try {
               // 清除列表中的数据
               bluetoothDeviceList.clear();
               // 通知列表组件数据已经变化了，需要重新加载数据
               provider.notifyDataChanged();
               // 开始扫描蓝牙设备
               bluetoothHost.startBtDiscovery();
               Tools.showTip(BluetoothAbility.this,"开始扫描");

            } catch (Exception e) {
               Tools.showTip(BluetoothAbility.this, e.getMessage());
            }
         }
      });
   }
   // 创建用于保存 BluetoothRemoteDevice 对象的列表
   bluetoothDeviceList = new ArrayList<>();
   // 下面的代码处理显示蓝牙设备的列表组件
   blueDevices = (ListContainer)findComponentById(ResourceTable.Id_bluedevices);
   if(blueDevices != null) {
      provider = new RecycleItemProvider() {
         @Override
         public int getCount() {
            return bluetoothDeviceList.size();
         }

         @Override
         public Object getItem(int i) {
            return bluetoothDeviceList.get(i);
```

```
            }

            @Override
            public long getItemId(int i) {
                return i;
            }

            @Override
            public Component getComponent(int i, Component component,
                    ComponentContainer componentContainer) {
              DirectionalLayout view = (DirectionalLayout)
              LayoutScatter.getInstance(getContext()).parse(ResourceTable.Layout_
              list_item, null, false);

              Text textDeviceName = (Text) view.findComponentById(ResourceTable.
              Id_text_device_name);
              Text textDeviceAddress = (Text) view.findComponentById(ResourceTable.
              Id_text_device_address);

            // 显示蓝牙设备名
            textDeviceName.setText(bluetoothDeviceList.get(i).getDeviceName().get());
            // 显示蓝牙设备的地址
            textDeviceAddress.setText(bluetoothDeviceList.get(i).getDeviceAddr());
                return view;
            }
        };
         blueDevices.setItemProvider(provider);
    }
  }
}
```

如果想连接蓝牙设备，需要先配对。在本例中，bluetoothDeviceList 保存了所有被发现的蓝牙设备的 BluetoothRemoteDevice 对象，要想与某个蓝牙设备配对，只需要获取该设备对应的 BluetoothRemoteDevice 对象，然后获取蓝牙设备的地址，并使用下面的代码将该设备与另一部蓝牙设备配对。

```
BluetoothRemoteDevice device = bluetoothHost.getRemoteDev(bluetoothAddress);
// 开始配对
device.startPair();
```

其中，bluetoothAddress 变量表示蓝牙设备的地址，即在 getComponent 方法中通过 Bluetooth RemoteDevice.getDeviceAddr 方法获取的值。

10.5 拨打电话

HarmonyOS 可以通过调用系统拨号盘的方式拨打电话，要实现这个功能，需要使用系统的 Action——ohos.intent.action.dial。实现代码如下：

```
Intent intent = new Intent();
// 指定调用系统拨号盘的 Action
```

```
intent.setAction("ohos.intent.action.dial");
// 指定要拨打的电话
intent.setUri(Uri.parse("tel:12345678"));
// 显示系统拨号盘
startAbility(intent, 0);
```

执行这段代码，会显示图 10-12 所示的系统拨号盘，然后就可以在这个界面拨打电话了。

图 10-12 系统拨号盘

10.6 总结与回顾

本章讲解的内容很多，包括 AI 接口、传感器、定位、蓝牙、拨打电话等。这里要着重提一下 HarmonyOS 中的 AI 接口。在其他系统（如 Android、iOS）中，这些功能都要通过第三方组件来实现，而 HarmonyOS 本身就提供了大量丰富的 AI 接口，开发人员只用几行代码，就可以实现复杂的功能，如分词、关键字提取、实体识别等，这极大地方便了 App 的开发。这些 AI 接口使用的模型都是本地的，不过在使用时通常需要连网来获取一些必要的信息。

HarmonyOS 设备（手机、平板电脑等）中通常会有大量的硬件模块，其中传感器模块、定位模块和蓝牙模块是比较常用的模块。本章讲解了各个模块的应用技术，最后还介绍了如何实现拨打电话的功能。

第 11 章　应用类 App 项目：跨设备在线电子词典

在本章中我们将实现一个完整的 HarmonyOS 项目，这个项目是一款电子词典 App，可以用离线和在线两种方式查询英文单词，并显示查询结果。本章将讲解这款电子词典 App 的完整实现过程。

本章的源代码在配套资源的 edict 目录下。

通过阅读本章，读者可以掌握：

- 跨设备的原理和方法；
- 如何部署和使用 SQLite 数据库；
- 如何访问网络；
- 如何分析网站的 HTML 代码；
- 如何用网络爬虫抓取数据；
- HarmonyOS 如何引用第三方库。

11.1 功能需求分析

在编码之前，必须要了解这款电子词典 App 的主要功能，然后才能按功能说明书来实现这款 App。从章名我们可以推断这款 App 的主要功能，并将章名分成如下 3 个部分：

- 跨设备；
- 电子词典；
- 在线。

现在分别解释一下这 3 个部分。

1. 跨设备

HarmonyOS 支持多种类型的设备，包括华为智慧屏、智能手表、智能手机、车载电脑、平板电脑等。在 DevEco Studio 的 Virtual Device Manager 中可以看到目前 HarmonyOS 支持的所有设备的列表，如图 11-1 所示。

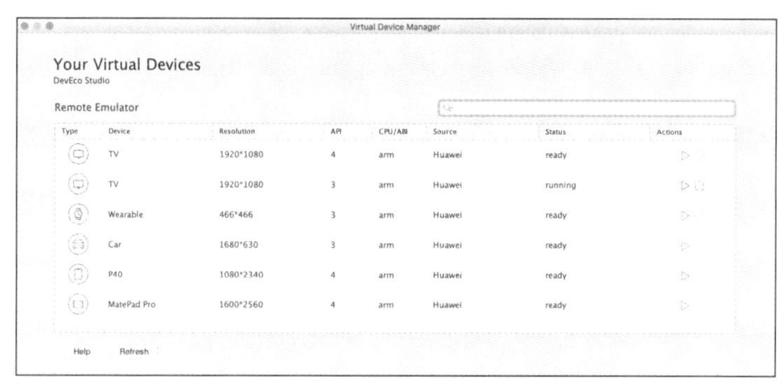

图 11-1 HarmonyOS 支持的设备列表

对一款 App 来说，不仅要考虑功能实现，还要考虑 App 的覆盖率，这个覆盖率既包括用户的覆盖率，也包括设备的覆盖率。这里只讨论设备的覆盖率。

一款 App 要想让尽可能多的人使用，那么覆盖尽可能多的终端设备是非常必要的。所以本章实现的电子词典 App 就需要尽可能覆盖 HarmonyOS 支持的设备。为了方便讲解，我们选择了 3 种非常典型的设备。

- ❑ 华为智慧屏（TV）：这是目前 HarmonyOS 支持的屏幕最大的设备，而且是横屏的。
- ❑ 智能手机（Phone）：手机是最常用的 HarmonyOS 设备，竖屏和横屏都支持。
- ❑ 智能手表（Wearable）：智能手表是目前 HarmonyOS 支持的最小屏幕的设备（除了部分 IoT 设备外）。

这 3 类设备覆盖了大屏、横屏、竖屏、小屏这 4 种最常见的屏幕类型。而对电子词典 App 来说，在这些设备上使用的 API 是相同的，只是屏幕尺寸不同。所以实现跨设备 App，只要适应不同屏幕尺寸即可。通常一款 App 只要适应这 4 种类型的屏幕，就可以很容易适应各种类型的设备。

下面我们先看一下电子词典 App 在这 3 类设备上运行的效果。图 11-2 所示是 App 在 TV 上的运行效果。

图 11-3 所示是 App 在智能手机上的运行效果。

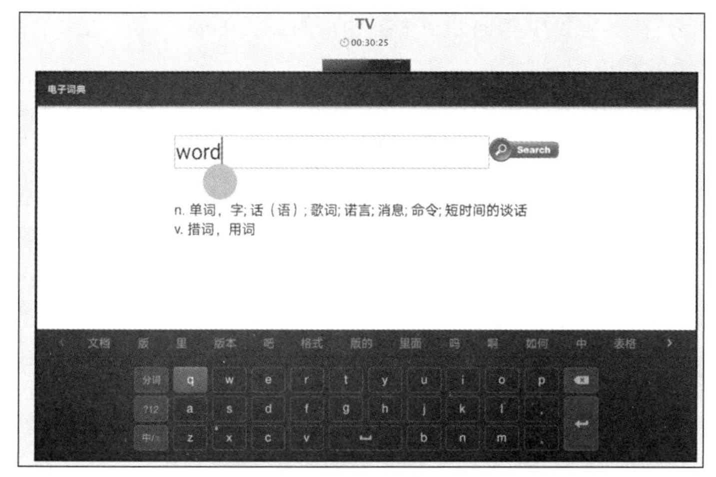

图 11-2 App 在 TV 上的运行效果

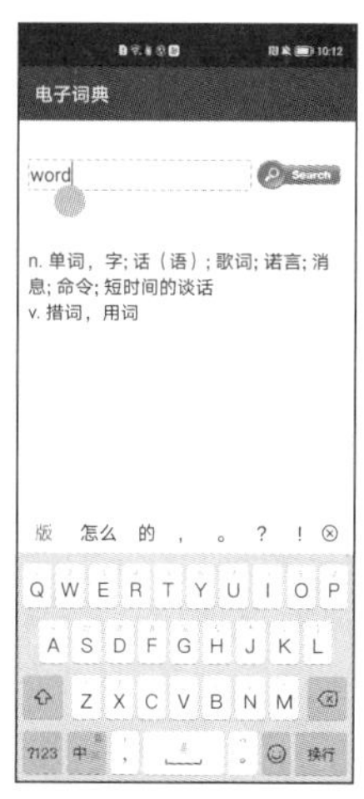

图 11-3 App 在智能手机上的运行效果

图 11-4 和图 11-5 所示是 App 在智能手表上的运行效果。

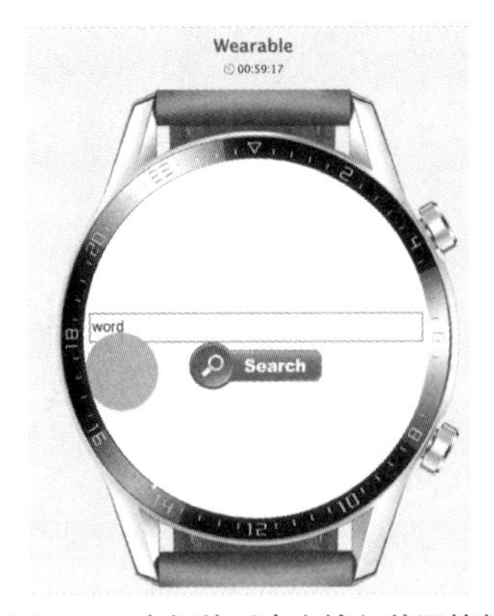

图 11-4 App 在智能手表上输入单词的效果

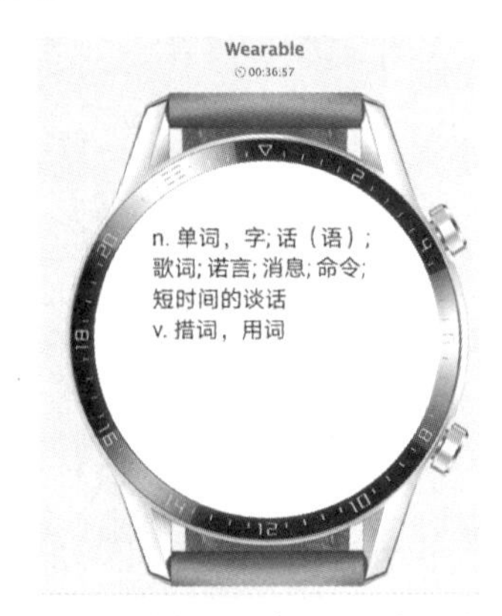

图 11-5 App 在智能手表上显示查询结果的效果

在 TV 和智能手机上查询单词时都会直接在查询输入文本框下方显示查询结果，而智能手表用另一个窗口来显示查询结果，这是因为智能手表的屏幕尺寸太小，在同一个窗口中查询结果无法显示完整，所以会将查询结果显示在另外一个窗口中。

实现跨设备的 App 有多种方式，最直接的方式就是为每一类设备单独做一个 App，不过这样会造成很多代码冗余。例如，对于本章实现的电子词典 App 来说，跨 3 种设备，只是布局有一定的差异，99%的 Java 代码是相同的，所以使用这种方式并不划算。

另外一种方式是利用 HarmonyOS 布局文件的自适应功能，让组件自动适应不同的屏幕尺寸。使用这种方式处理 TV 和智能手机是可以的，不过要同时适应智能手表有些困难。因为手表的屏幕尺寸太小了，无论是分辨率和物理尺寸，都与 TV 和智能手机不是一个数量级的。

因此，电子词典 App 采用了另外一种跨设备的方式，就是单独为每一种设备准备一套布局。在 App 启动时会检测当前运行的设备，然后会根据设备类型装载不同的布局文件。

2. 电子词典

查词是电子词典 App 的核心功能，在本项目中查词是指根据输入的英文单词，从数据库中查询出对应的中文解释，并显示在窗口中的相应位置。

本项目使用 SQLite 数据库，在数据库中已经预先保存了一些单词（本项目提供的数据库大概有两千多个英文单词）。现在有一个问题，就是如何让 SQLite 数据库与 App 一起发布呢？当然，可以在第一次运行时从网络上下载数据库文件，但是这种方式不适合离线的情况。如果当时没有网络，那该怎么办呢？所以最好的方式就是将 SQLite 数据库嵌入到 hap 文件中，随着 App 一同发布。这样在查询数据库中英文单词的时候就不需要连接网络了。但还有一个问题，就是 HarmonyOS API 是不能直接打开嵌入到 hap 文件中的 SQLite 数据库的，那该怎么办呢？这也是本章要解决的问题之一，要知道详情，就继续往后看吧！

3. 在线

任何离线的词库都不可能包含全部的英文单词，那么遇到 SQLite 数据库中没有的英文单词该怎么办呢？为了让电子词典 App 可以查询所有正确的英文单词，我们在电子词典中加入了在线查词的功能。也就是说，如果本地词库不存在某个单词，就会从网络词库中查询，如果从网络词库中查询到该单词，就会将查询结果添加到本地词库，并显示单词对应的中文含义。这样在第 2 次查询该单词时，就不需要再从网络词库中查询了。

在这段描述中涉及多种技术，如 JSoup 库、数据库读写技术、网络爬虫技术等。这里要着重提一下网络爬虫技术。

尽管现在有很多在线词典都提供了 API，但要么功能有限，要么是收费的。当然，如果读者已经购买了在线词典的收费服务，那么可以直接在 HarmonyOS 中使用这些服务。对于无法使用这些查词服务的读者来说，唯一的希望就是网络爬虫了。因为网络爬虫可以直接通过网页版在线词典抓取英文单词的相关信息，而网页版在线词典是没有任何限制的。

很多读者可能会发现一个问题，前面提到 SQLite 数据库中已经预存了两千多个英文单词，那么这两千多个英文单词是如何放到数据库中的呢？当然，你可以直接从网上下载现成的词库，不过大多数情况下这种方式并不合适。所以最好的方式是创建一个空的 SQLite 数据库和相应的表，然后从网络上抓取词库，插入到 SQLite 数据库中，其实这也用到了网络爬虫。所以电子词典 App 有两个地方用到了网络爬虫：生成本地词库和在线查词。在本章后面的内容中我们会详细介绍如何使用网络爬虫完成这两项工作。

现在来总结一下，电子词典 App 的特点是跨设备、本地查词和在线查词，其涉及的技术主要有以下几项：

- 跨设备；
- SQLite 数据库；
- 文件读写；
- 读写资源目录；
- JSoup 库；
- HTML 代码分析；
- 网络爬虫。

本章后面的内容会结合项目，详细介绍这些技术的实现。

11.2 搭建项目框架

在开发电子词典项目之前，要先创建项目工程，并做相应的配置。

11.2.1 创建项目工程

尽管电子词典项目可以跨设备运行，但在创建 HarmonyOS 工程时，只能先选择一个设备。选择 File→New→New Project 菜单项，会显示图 11-6 所示的 Create HarmonyOS Project 对话框。

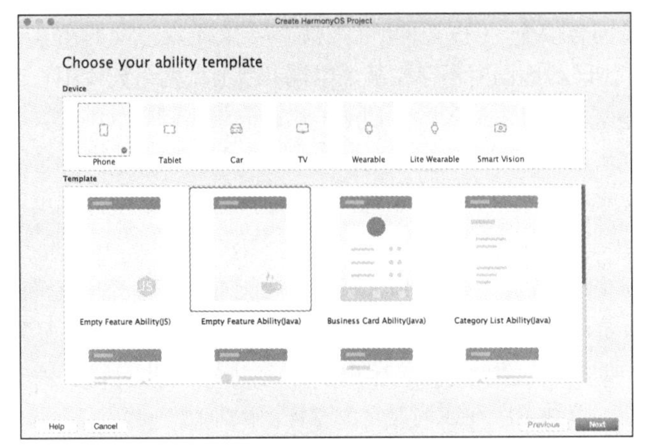

图 11-6 创建电子词典项目工程

电子词典项目可以跨 TV、智能手机和智能手表运行，在创建工程时，只需选择一类设备即可。在本例中我们选择了 Phone，在模板列表中选择 Empty Feature Ability（Java）。然后单击 Next 按钮进入图 11-7 所示的设置工程信息页面。在这个设置页面中输入合法的工程名、包名和工程保存位置，API 版本可以选择版本 3 或版本 4。

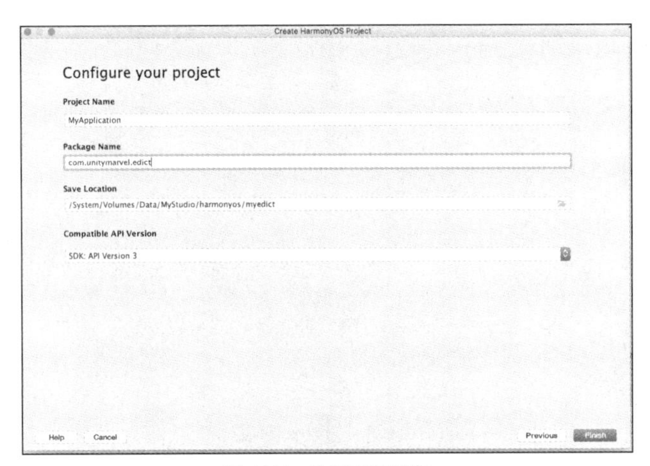

图 11-7 设置工程信息

11.2.2 让工程可以在多个设备上运行

目前工程还不能在多个设备上运行，如果读者选择的设备是智能手机，那么目前只能在智能手机上运行。为了让工程可以在多个设备上运行，打开 config.json 文件，找到 deviceType 属性，添加 wearable 和 tv，代码如下：

```
"deviceType": [
   "tv",
   "wearable",
   "phone"
]
```

读者可以启动不同类型设备的虚拟机，然后在虚拟机上运行刚创建的工程。如果要在真机上运行，还需要选择 File→Project Structure 菜单项，打开 Project Structure 对话框。在对话框左侧选择 Modules 列表项，在对话框右侧选择上方的 Signing Configs 标签，在页面中的 debug 部分填写签名信息。如果要发布 App，需要在 release 部分填写签名信息。配置签名信息如图 11-8 所示。

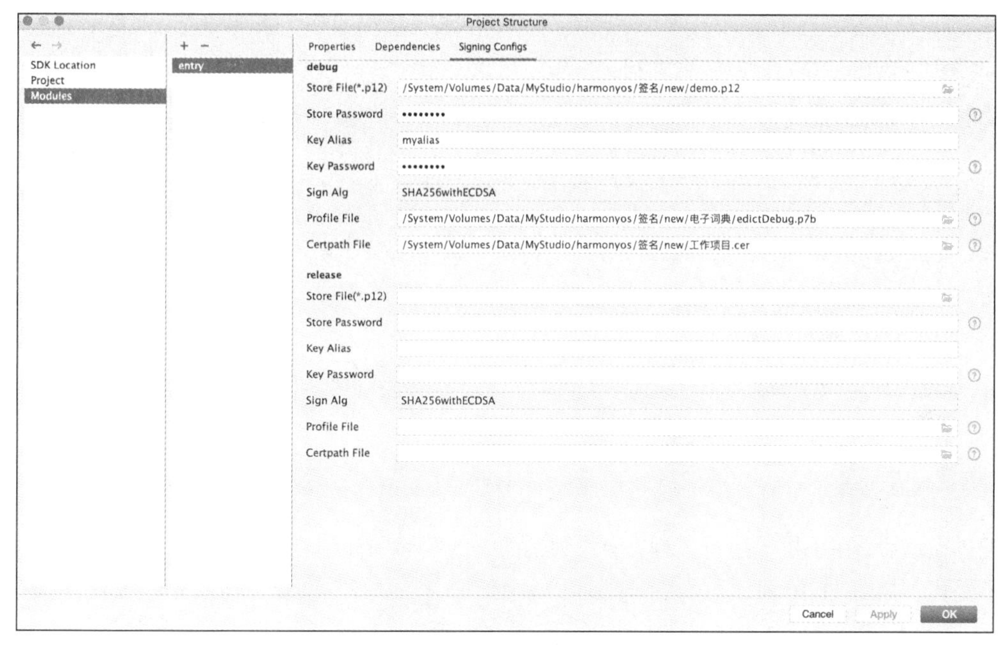

图 11-8 配置签名信息

对 App 签名需要 3 个文件，关于签名的详细步骤，读者可以参考 1.7 节的内容。

最后运行程序，在设备选择列表中选择要部署的设备，如图 11-9 所示。

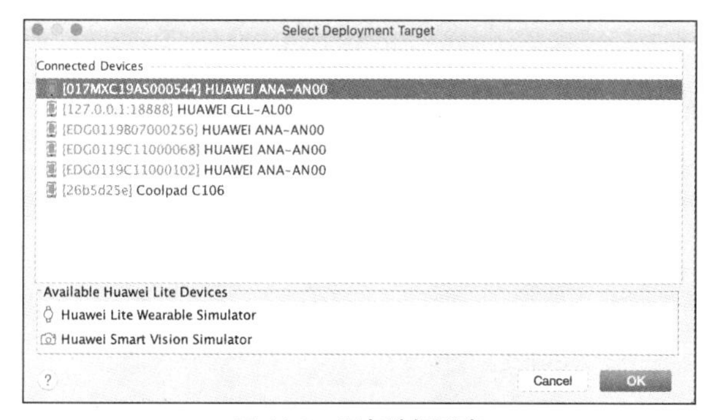

图 11-9 设备选择列表

11.2.3 配置 App 图标和名称

为了让电子词典 App 看起来像正式发布的应用，最好改一下 App 的图标和名称（图标下方的文字），首先准备一个图像文件（在本例中文件名为 edict.png），然后打开 config.json 文件，找到

abilities 部分。在新创建的工程中，只有一个 Page Ability，现在修改这个 Page Ability 的 icon 属性和 label 属性，修改后的结果如下：

```
"abilities": [
  {
    "skills": [
      {
        "entities": [
          "entity.system.home"
        ],
        "actions": [
          "action.system.home"
        ]
      }
    ],
    "orientation": "unspecified",
    "formEnabled": false,
    "name": "com.unitymarvel.edict.MainAbility",
    "icon": "$media:edict",
    "description": "$string:mainability_description",
    "label": "电子词典",
    "type": "page",
    "launchType": "standard"
  }
]
```

配置完成后，在手机上运行工程，会看到桌面上的图标和标题发生了变化，如图 11-10 所示。

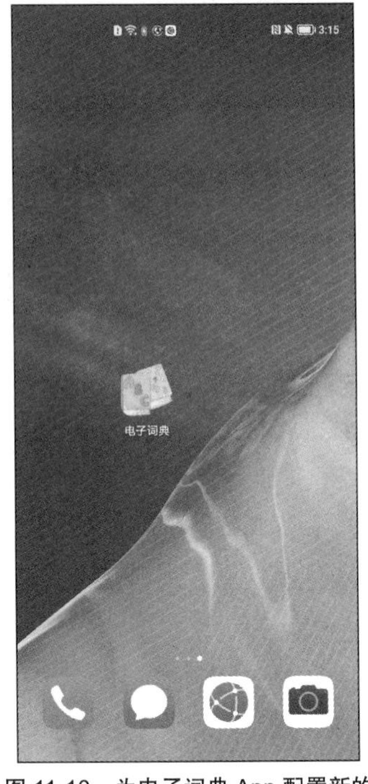

图 11-10 为电子词典 App 配置新的图标和标题

11.2.4 添加权限

因为电子词典 App 要访问网络需要在 config.json 文件的 module 属性中申请网络访问权限，所以要添加一个 reqPermissions 属性，代码如下：

```
"reqPermissions": [
  {
    "name": "ohos.permission.INTERNET",
    "usedScene": {
      "ability": [
        "com.unitymarvel.edict.MainAbility"
      ],
      "when": "always"
    }
  }
],
```

读者需要将 ability 属性中的 Ability 类全名改成自己创建的 Ability 的全名。

11.3 利用网络爬虫生成本地词库

在本节中我们将学习本地词库的生成，本地词库的生成需要使用网络爬虫从 Web 版词库中抓取单词信息，并插入 SQLite 数据库中。

11.3.1 分析 Web 版词库的 HTML 代码

要想抓取英文单词信息，首先要找一个在线英文词库，本例中我们选择了英语 4 级的部分单词，页面地址为 https://www.eol.cn/html/en/cetwords/cet4.shtml。进到这个页面，会看到图 11-11 所示的内容。

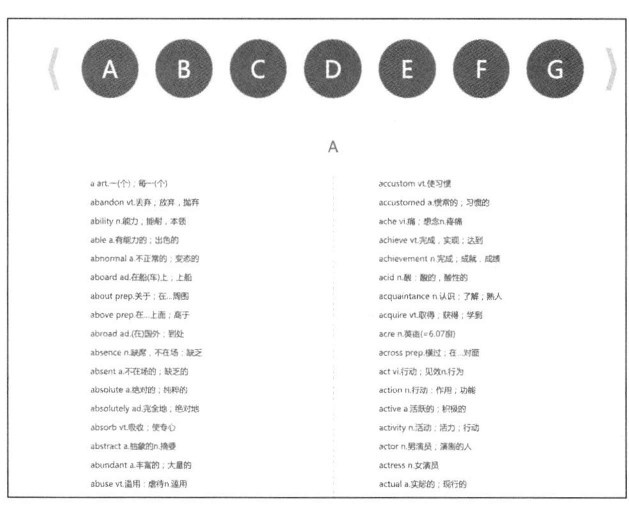

图 11-11 英语 4 级单词页面

在页面中单词分成两列显示，我们单击鼠标右键，选择“检查”菜单项，如图 11-12 所示，可以启动浏览器的调试功能显示页面的 HTML 代码和控制台。注意，本例中使用的是 Chrome 浏览器，如果读者使用其他浏览器，其他浏览器也会有类似的功能。

接下来的任务是找到抓取页面中单词信息的办法。进入浏览器的调试界面后，单击界面左上角小箭头启动代码跟踪功能，如图 11-13 所示。

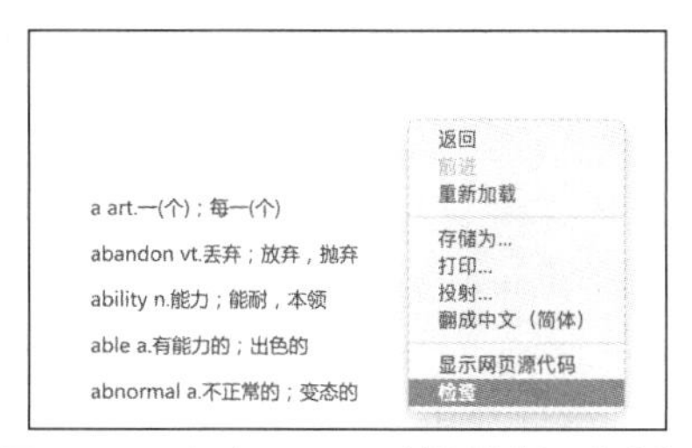

图 11-12 启动 Chrome 浏览器的调试功能

图 11-13 启动代码跟踪功能

然后将鼠标指针放到某个单词上，在 Elements 页面中就会自动定位到该单词对应的 HTML 代码，如图 11-14 所示。

要抓取页面中每一个单词的信息，就要清楚显示这些单词的 HTML 代码是什么，因为网络爬虫直接访问的是 HTML 代码。

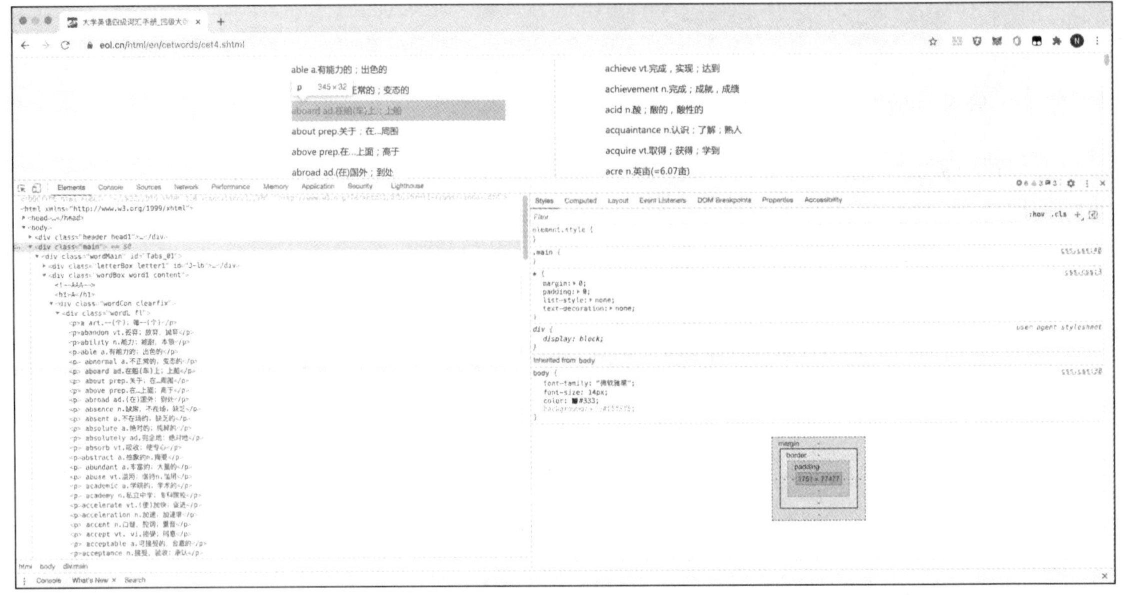

图 11-14　自动定位与单词对应的 HTML 代码

经过分析，可以得出结论，页面中的英文单词是以 26 个字母组织的。也就是说，所有单词首字母为 a 的放在一起，首字母为 b 的放在一起，以此类推。每一组所有的单词都在一个<div>标签中，如图 11-15 所示。

然后展开任意一组单词，如首字母为 a 的英文单词，会发现每一个<div>标签中都包含另外一个<div>标签，而这个<div>标签中包含了所有首字母为 a 的英文单词，每一个英文单词以及对应的中文解释都包含在一个<p>标签中，如图 11-16 所示。

```
▶<div class="wordBox word1 content">…</div>
▶<div class="wordBox word1 content">…</div>
▶<div class="wordBox word1 content">…</div>
▶<div class="wordBox word1 content">…</div>
▶<div class="wordBox word1 content">…</div>
▶<div class="wordBox word1 content">…</div>
▶<div class="wordBox word1 content">…</div>
▶<div class="wordBox word1 content">…</div>
▶<div class="wordBox word1 content">…</div>
▶<div class="wordBox word1 content">…</div>
▶<div class="wordBox word1 content">…</div>
▶<div class="wordBox word1 content">…</div>
▶<div class="wordBox word1 content">…</div>
▶<div class="wordBox word1 content">…</div>
▶<div class="wordBox word1 content">…</div>
▶<div class="wordBox word1 content">…</div>
▶<div class="wordBox word1 content">…</div>
▶<div class="wordBox word1 content">…</div>
▶<div class="wordBox word1 content">…</div>
▶<div class="wordBox word1 content">…</div>
▶<div class="wordBox word1 content">…</div>
▶<div class="wordBox word1 content">…</div>
▶<div class="wordBox word1 content">…</div>
▶<div class="wordBox word1 content">…</div>
▶<div class="wordBox word1 content">…</div>
▶<div class="wordBox word1 content">…</div>
```

图 11-15　26 组英文单词

图 11-16　每一组单词的 HTML 代码

现在我们对页面的 HTML 代码结构已经非常清楚了，只要获取这些包含英文单词信息的<p>标签，就可以获取页面中所有的英文单词信息。

11.3.2 利用网络爬虫生成本地词库

理论上，任何编程语言都可以实现网络爬虫，本例中选择目前非常流行的 Python 语言来实现网络爬虫。如果读者对 Python 语言不熟悉，可以看我的《Python 从菜鸟到高手》一书，或通过我的微信公众号“极客起源”学习 Python 编程。本节只对实现网络爬虫的关键技术点进行讲解，并给出完整的源代码。

现在进一步分析上一节得到的 HTML 代码。分析这些 HTML 代码的目的是获取这些<p>标签，肯定不能获取 HTML 代码中的所有<p>标签，因为并不是每一个<p>标签都包含单词信息。

经过分析发现，包含这些<p>标签的<div>标签都有一个特性，就是都使用了 wordL 和 fl 这两个样式，所以我们的方法是先过滤出所有包含 wordL 样式和 fl 样式的<div>标签，然后分别获取每一个<div>标签中所有的<p>标签，这样就可以准确地获取所有包含单词信息的<p>标签了。

在抓取每一个英文单词的 HTML 代码后，仍然需要提取出更细粒度的信息，例如，英文单词、词性、中文解释等。下面以一个典型的英文单词 adult 为例，它对应的 HTML 代码如下：

```
<p> abstract a.抽象的 n.摘要</p>
```

这行 HTML 代码除了<p>标签外，可以分为如下 3 个部分。

- 英文单词：abstract。
- 词性：a.、n.。
- 中文解释：抽象的、摘要。

现在需要分别提取出这 3 个部分。这里采用了正则表达式。我们发现，英文单词和中文解释之间实际上是通过词性分隔的，也就是“a.”和“n.”可以作为分隔符，所以可以使用下面的正则表达式表示词性：

```
[a-z]+\.
```

注意，每一个词性后面有一个点（.）。

如果使用上面的正则表达式作为分隔符，为“abstract a.抽象的 n.摘要”生成一个 String 类型的数组，那么数组的第 1 个元素就是英文单词本身，后面的所有元素值都是该单词的中文解释。然后再利用上面的正则表达式提取出所有的词性。

下面看一下 Python 网络爬虫的具体实现。

本节实现的 Python 网络爬虫使用 requests 网络库，这是目前 Python 库中易使用且强大的网络库之一。爬虫的实现分为两个部分，先从页面的 HTML 代码中提取出英文单词的相关信息，然后再将这些信息插入 SQLite 数据库的 t_words 表中。Python 工程会自动创建 SQLite 数据库。完整的代码如下：

```
# 引用 request 库
import requests
# 引用正则表达式库
import re
# 定义一个字典，用于保存抓取的单词信息
words = {}
# 分析单词信息（<p>标签里面的内容），并将分析结果添加到 words 字典中
```

```
def add_word(word_line):
    # 指定 words 为全局变量
    global words
    # 用词性作为分隔符分隔字符串
    result1 = re.split('[a-z]+\.', word_line)
    # 定义用于提取词性的模式
    pattern = re.compile('[a-z]+\.')
    # 获取所有词性
    result2 = pattern.findall(word_line)
    # 获取英文单词
    word = result1[0].strip()
    # 保存单词的解释，以词性作为 key，中文解释作为 value
    meanings = {}
    for i in range(0, len(result2)):
        key = result2[i].strip()                # 获取当前词性
        value = result1[i + 1].strip()          # 获取对应的中文解释
        meanings[key] = value                   # 保存词性和中文解释
    words[word] = meanings                      # 保存英文单词的全部词性和中文解释
# 通过 HTTP GET 方式访问单词页面
r = requests.get('https://www.eol.cn/html/en/cetwords/cet4.shtml')
# 获取单词页面中的所有 HTML 代码
html=r.content
# 将 HTML 代码转换为 utf-8 格式
html_doc=str(html,'utf-8')
# 使用 BeautifulSoup 库对 HTML 代码进行分析
from bs4 import BeautifulSoup
soup = BeautifulSoup(html_doc,'lxml')
# 查找所有包含 wordL 样式和 fl 样式的标签，本例中只有<div>标签
tags = soup.find_all(attrs = {'class':'wordL fl'})
# 扫描所有满足条件的标签，并获取这些标签中所有的<p>标签，然后调用 add_word 函数分析单词信息
for tag in tags:
    p_list = tag.select('p')
    for p in p_list:
        add_word(p.text)
print(words)
print('单词抓取完毕')
############## 下面的代码生成本地词库（SQLite 数据库）##############
import sqlite3
import os
# 定义数据库名
db_path = 'dict.sqlite'
# 如果数据库文件已存在，则删除数据库文件
if os.path.exists(db_path):
    os.remove(db_path)
# 连接 SQLite 数据库，如果数据库文件不存在，会创建一个新的数据库文件
conn = sqlite3.connect(db_path)
# 获取用于执行 SQL 语句的 Cursor 对象
c = conn.cursor()
# 通过 SQL 语句创建 t_words 表
c.execute('''create table t_words
       (id INTEGER primary key   autoincrement  not null,
       word         varchar(30)    not null,
       type         varchar(10)   not null,
       meanings     text  not null);''')
```

```
# 为 t_words 表在 word 字段上创建一个索引，该字段用于存储英文单词本身
c.execute('create index word_index on t_words(word)')
# 提交修改，必须调用 commit 方法才能将执行结果存回数据库
conn.commit()
# 关闭 SQLite 数据库
conn.close()
print('创建数据库成功')
# 再次连接数据库，这时数据库文件已经存在，所以直接打开 SQLite 数据库
conn = sqlite3.connect(db_path)
c = conn.cursor()
# 扫描 words 中的单词信息，然后将信息逐条插入 t_words 表
for word in words:
    value = words[word]
    for type in value:
        meanings = value[type]
        # 定义将单词插入 t_words 表的 SQL 语句
        sql = f'insert into t_words(word,type,meanings) \
             values("{word}","{type}","{meanings}")';
        # 执行 insert 语句插入单词信息
        c.execute(sql)
    # 提交修改
    conn.commit()
# 关闭 SQLite 数据库
conn.close()
print('电子词典数据库生成完毕')
```

这段代码中我们使用了 BeautifulSoup 库来分析 HTML 代码，因为这个库提供了丰富的 API，可以通过多种方式对 HTML 代码中我们感兴趣的数据进行分析和提取。当然，读者也可以自己手动通过算法进行分析，不过非常费时费力。所以推荐读者使用类似的库来完成复杂的分析工作。

现在运行这段代码，如果在 Console 视图输出图 11-17 所示的内容，表示已经成功生成了本地词库。生成本地词库后，会发现在当前目录下多了一个名为 dict.sqlite 的文件，这就是本地数据库文件。

```
Debug: word
Debugger  Console
{'a': {'art.': '一(个); 每一(个)'}, 'abandon':
单词抓取完毕
创建数据库成功
电子词典数据库生成完毕
```

图 11-17　成功生成本地词库

11.3.3 管理本地词库

在上一节我们生成了本地词库，在这一节我们将打开这个 SQLite 数据库看一看成果。管理 SQLite 数据库的工具很多，这里选择一款跨平台的 SQLite 管理工具——DB Browser for SQLite。读者可以到该工具的官方页面下载 DB Browser for SQLite 的最新版。

现在使用 DB Browser for SQLite 打开 dict.sqlite 文件，会看到 t_words 表，结果如图 11-18 所示。

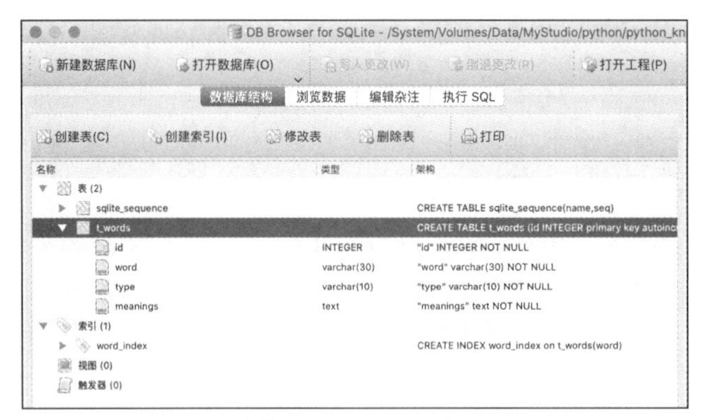

图 11-18 t_words 表

这个表有 4 个字段，下面是字段的含义。

- ❑ id：记录的 ID，唯一标识每条记录。
- ❑ word：英文单词，在该字段上有索引，但并不是唯一索引。这是因为有一些单词，有两个或两个以上的词性，在 t_words 表中，一个单词的每一个词性都是一条记录。例如，abstract 有两个词性就是两条记录。
- ❑ type：单词的词性。
- ❑ meanings：单词的中文解释。

现在切换到“浏览数据”页面，并选择 t_words 表，会看到 t_words 表中的记录，如图 11-19 所示。因为单词可能有多个词性，所以一个单词有可能占用了两条或两条以上的记录，因此，t_words 表中的记录数要多于实际的单词数。

读者可以用下面的 SQL 语句统计一下单词的个数，统计结果是 2226 个单词：

	id	word	type	meanings
1	1	a	art.	一(个)；每一(个)
2	2	abandon	vt.	丢弃；放弃，抛弃
3	3	ability	n.	能力；能耐，本领
4	4	able	a.	有能力的；出色的
5	5	abnormal	a.	不正常的；变态的
6	6	aboard	ad.	在船(车)上；上船
7	7	about	prep.	关于；在...周围
8	8	above	prep.	在...上面；高于
9	9	abroad	ad.	(在)国外；到处
10	10	absence	n.	缺席，不在场；缺乏
11	11	absent	a.	不在场的；缺乏的
12	12	absolute	a.	绝对的；纯粹的
13	13	absolutely	ad.	完全地；绝对地
14	14	absorb	vt.	吸收；使专心
15	15	abstract	a.	抽象的
16	16	abstract	n.	摘要

1 - 17 / 2616

图 11-19 本地词库中的英文单词

```
select count(*) as c from t_words group by word
```

11.4 在本地词库中查询

本地词库创建完成后，需要将本地词库嵌入到 hap 文件中。在程序运行后，打开这个本地词库，根据用户输入的英文单词进行查询。本节将以智能手机作为实验设备，后面再考虑跨设备。

11.4.1 主界面布局

电子词典 App 的主界面如图 11-20 所示。

电子词典 App 的主界面非常简单，只显示了 3 个组件：1 个 TextField 组件和 2 个 Image 组件。TextField 组件用于输入要查询的单词，右侧的 Search 按钮是 1 个 Image 组件，下方的金牌也是 1 个 Image 组件。实际的组件要多于 3 个，只是没有显示而已。当查询英文单词时，不管是否查询到，都会隐藏下方的 Image 组件，并显示一个 Text 组件，展示查询结果。

图 11-20 电子词典 App 的主界面

现在看一下手机版主界面的布局实现。

代码位置：src/main/resources/base/layout/edict_phone.xml

```
<?xml version="1.0" encoding="utf-8"?>
<DirectionalLayout
   xmlns:ohos="http://schemas.huawei.com/res/ohos"
   ohos:height="match_parent"
   ohos:width="match_parent"
   ohos:background_element="#E1E1E1"
   ohos:orientation="vertical">
   <DirectionalLayout
      ohos:height="match_parent"
      ohos:width="match_parent"
      ohos:left_margin="10vp"
      ohos:orientation="vertical"
      ohos:right_margin="10vp"
      ohos:top_margin="40vp">
      <DirectionalLayout
         ohos:height="50vp"
         ohos:width="match_parent"
         ohos:orientation="horizontal">
         <TextField
            ohos:id="$+id:textfield_word"
            ohos:height="match_content"
            ohos:width="match_parent"
            ohos:background_element="$graphic:border"
            ohos:hint="请输入要查询的单词"
            ohos:padding="2vp"
            ohos:text_size="20fp"
            ohos:weight="1"/>
         <Image
            ohos:id="$+id:image_search"
            ohos:height="30vp"
            ohos:width="100vp"
            ohos:image_src="$media:search_word"
            ohos:scale_mode="zoom_center"/>
      </DirectionalLayout>
      <Text
         ohos:id="$+id:text_search_result"
         ohos:height="match_parent"
         ohos:width="match_parent"
         ohos:multiple_lines="true"
```

```
            ohos:visibility="hide"
            ohos:top_margin="40vp"
            ohos:text_size="20vp"
            ohos:text_alignment="left|top"/>
        <Image
            ohos:id="$+id:image"
            ohos:height="match_parent"
            ohos:width="match_parent"
            ohos:padding="60vp"
            ohos:image_src="$media:gold_medal"
            ohos:scale_mode="zoom_center"/>
    </DirectionalLayout>
</DirectionalLayout>
```

从布局文件可以看出，除了前面描述的 1 个 TextField 组件和 2 个 Image 组件外，还有 1 个 Text 组件用于显示查询结果，不过默认这个 Text 组件的 ohos:visibility 属性值为 hide，说明这个组件默认不可见，直到单击查询输入文本框右侧的 Search 按钮（1 个 Image 组件），才会隐藏下方的 Image 组件，并显示这个 Text 组件。

11.4.2 如何让本地词库与 App 一同发布

我们期望将 dict.sqlite 文件（即本地词库）与 hap 文件一同发布，所以需要将 dict.sqlite 文件放到工程的 resources/rawfile 目录下，如图 11-21 所示。

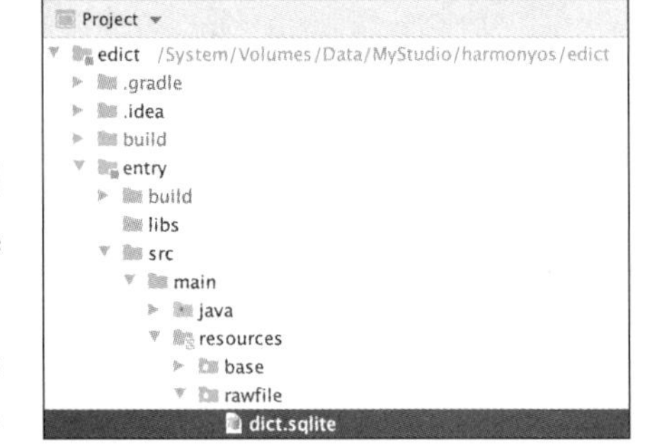

图 11-21 dict.sqlite 文件的位置

如果将 dict.sqlite 文件放在 rawfile 目录下，那么将工程打包成 hap 文件时，dict.sqlite 文件就会成为 hap 文件的一部分。

11.4.3 打开 rawfile 目录下的 SQLite 数据库文件

HarmonyOS API 是无法直接打开 rawfile 目录下的 SQLite 数据库文件的，所以在 App 第一次运行时，需要将 dict.sqlite 文件从 rawfile 目录提取出来，并放到 App 的 data 目录或 sdcard 目录下。本例中我们选择将 dict.sqlite 文件放在 App 的 data 目录下，因为操作这个目录不需要任何权限。

通过 ResourceManager.getRawFileEntry 方法可以获得 rawfile 目录下任何文件的实体，然后通过输入输出流，可以将 rawfile 目录下的任何文件保存到手机的任何目录（需要有写入权限）下。

将 dict.sqlite 文件放到 data 目录下，剩下的工作就变得简单了，使用 DatabaseHelper 对象的相关 API 就可以对 SQLite 数据库进行操作。对 DatabaseHelper 不熟悉的读者可以查看本书 6.2 节的内容，这里只给出完整的实现代码。

代码位置：src/main/java/com/unitymarvel/edict/common/MyDict.java

```
// 完成与字典相关的操作
public class MyDict {
```

```
private AbilityContext context;
// dict.sqlite 文件所在目录的绝对路径
private File dictPath;
// dict.sqlite 文件在 data 目录的绝对路径
private File dbPath;
private RdbStore store;
// 用于打开 data 目录下 SQLite 数据库的路径
private StoreConfig config = StoreConfig.newDefaultConfig("dict.sqlite");
// 打开 SQLite 数据库时的回调对象
private static final RdbOpenCallback callback = new RdbOpenCallback() {
   @Override
   public void onCreate(RdbStore store) {
   }
   @Override
   public void onUpgrade(RdbStore store, int oldVersion, int newVersion) {
   }
};
public MyDict(AbilityContext context) {
   this.context = context;
   // 获得 data 目录下数据库的专属目录
   dictPath =  new File(context.getDataDir().toString() +
                                    "/MainAbility/databases/db");
   // 如果目录不存在，则创建该目录
   if(!dictPath.exists()) {
      dictPath.mkdirs();
   }
   // 初始化 dict.sqlite 文件的绝对路径
   dbPath = new File(Paths.get(dictPath.toString(),"dict.sqlite").toString());

}
// 将 dict.sqlite 文件从 rawfile 目录下复制到 data 目录的相应子目录下
private void extractDB() throws  IOException {
   // 获取 dict.sqlite 文件的 Resource 对象
   Resource resource = context.getResourceManager().
         getRawFileEntry("resources/rawfile/dict.sqlite").
         openRawFile();
   // 如果文件存在，先删除该文件
   if(dbPath.exists()) {
      dbPath.delete();
   }
   // 创建一个输出流，用于将 dict.sqlite 文件复制到 data 目录的相应子目录下
   FileOutputStream fos = new FileOutputStream(dbPath);
   // 每次复制 4KB 数据
   byte[] buffer = new byte[4096];
   int count = 0;
   // 用循环以每次复制 4KB 的速度将 rawfile 目录下的 dict.sqlite 文件复制到 data 目录下
   // 相应子目录
   while((count = resource.read(buffer)) >= 0) {
      fos.write(buffer,0, count);
   }
   // 关闭输入流
   resource.close();
   // 关闭输出流
   fos.close();
}
```

```
    // 用于初始化本地词库
    public void init() throws  IOException {
        // 将 dict.sqlite 文件从 rawfile 目录下复制到 App 的 dict 目录下
        extractDB();
        // 打开已经复制到 data 目录下的 dict.sqlite 数据库文件
        DatabaseHelper helper = new DatabaseHelper(context);
        store = helper.getRdbStore(config,
                1,
                callback,
                null);
    }
}
```

在 MyDict 类中的 init 方法是外部需要调用的方法，在主窗口类的 onStart 方法中调用 init 方法就可以直接打开 dict.sqlite 数据库。

11.4.4 在本地词库中查询

在本地词库中查询可以通过 MyDict.searchLocalDict 方法实现，需要将待查询的英文单词传入该方法，该方法将返回一个 ArrayList 类型对象，元素类型是 WordData，用来表示单词相关信息。WordData 类的代码如下：

```
package com.unitymarvel.edict.common;

public class WordData {
    public String type;              // 词性
    public String meanings;          // 词义
}
```

searchLocalDict 方法的代码如下：

```
public class MyDict {
    ...

    //在本地词库中查询
    public ArrayList<WordData> searchLocalDict(String word) {
        // 将英文单词转换为小写
        word = word.toLowerCase();
        // 将 word 放到 String 数组中，作为 SQL 语句的参数使用
        String[] args = new String[]{word};
        // 从 t_words 表中查询 word 指定的英文单词
        ResultSet resultSet = store.querySql("select * from t_words where word = ?",args);

        ArrayList<WordData> result = new ArrayList<>();
        int i = 0;
        // 如果找到，创建若干个 WordData 对象，并设置相应的信息
        while(resultSet.goToNextRow())
        {
            WordData wordData = new WordData();
            wordData.type = resultSet.getString(2);
            wordData.meanings = resultSet.getString(3);
            result.add(wordData);
        }
```

```
        resultSet.close();
        // 返回查询结果
        return result;
    }
}
```

如果英文单词有多个词性，searchLocalDict 方法会分别返回多个词性和对应的中文解释，也就是说，返回的 ArrayList 对象会有多个元素值。

11.4.5 在主界面中显示查询结果

在本地词库中查询的最后一步，就是单击文本输入框右侧的 Search 按钮，然后在单击事件中调用 MyDict. searchLocalDict 方法在本地词库中查询英文单词，核心代码如下：

```
// 创建 MyDict 对象
MyDict myDict = new MyDict(this);
Image image = (Image)findComponentById(ResourceTable.Id_image);

imageSearch = (Image)findComponentById(ResourceTable.Id_image_search);
if(imageSearch != null) {
    // 让 Image 组件允许单击
    imageSearch.setClickable(true);
    // 设置 Image 组件的单击监听器
    imageSearch.setClickedListener(new Component.ClickedListener() {
        @Override
        public void onClick(Component component) {
            // 隐藏显示金牌的 Image 组件
            image.setVisibility(Component.HIDE);
            // 显示展示查询结果的 Text 组件
            textSearchResult.setVisibility(Component.VISIBLE);
            // 调用 searchLocalDict 方法在本地词库中查询
            ArrayList<WordData> result = myDict.searchLocalDict(textfieldWord.getText());
            if(result.size() > 0) {
                // 先清空 Text 组件
                textSearchResult.setText("");
                // 输出查询结果
                for (WordData wordData : result) {
                    // 将中文解释追加到 Text 组件（每个词性一行）
                    textSearchResult.append(wordData.type + " " +
                                                      wordData.meanings + "\r\n");
                }
            } else {
                // 如果未查到英文单词，则显示提示信息
                textSearchResult.setText("本地词库没有查到此单词，正在查找网络词库...");
            }
        }
    });
}
```

现在运行程序，输入一个英文单词，然后单击右侧的 Search 按钮，如果单词存在，会显示类似图 11-22 的查询结果。

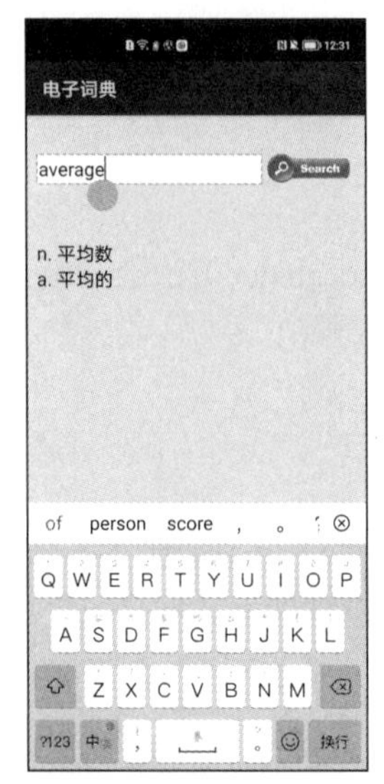

图 11-22　在手机的本地词库查词

11.5 实现跨设备运行

到现在为止，电子词典 App 只能在 HarmonyOS 手机上正常运行，如果在 TV 或智能手表上运行，界面就会变形。在这一节我们将让这款 App 可以在 TV 和智能手表上正常运行。

11.5.1 不同的设备使用不同的布局文件

因为智能手机与 TV 和智能手表在屏幕尺寸和分辨率上差异很大，所以在本节我们将为不同类型的设备单独设计布局文件。适应智能手机的布局文件代码在 11.4.1 节已经给出，在本节我们将给出适应 TV 和智能手表的布局文件代码。

TV 与智能手机在显示查询结果的形式上相同，都是在查询输入文本框下方显示查询结果，只是对字体、字间距以及组件位置有所调整。下面是适合于 TV 的布局代码。

代码位置：src/main/resources/base/layout/edict_tv.xml

```
<?xml version="1.0" encoding="utf-8"?>
<DirectionalLayout
   xmlns:ohos="http://schemas.huawei.com/res/ohos"
   ohos:height="match_parent"
   ohos:width="match_parent"
   ohos:background_element="#E1E1E1"
   ohos:orientation="vertical">

   <DirectionalLayout
      ohos:height="match_parent"
      ohos:width="match_parent"
      ohos:left_margin="200vp"
```

```
        ohos:orientation="vertical"
        ohos:right_margin="200vp"

        ohos:top_margin="40vp">
        <DirectionalLayout
            ohos:height="50vp"
            ohos:width="match_parent"
            ohos:orientation="horizontal">
            <TextField
                ohos:id="$+id:textfield_word"
                ohos:height="match_content"
                ohos:width="match_parent"
                ohos:background_element="$graphic:border"
                ohos:hint="请输入要查询的单词"
                ohos:padding="2vp"
                ohos:text_size="30fp"
                ohos:weight="1"/>
            <Image
                ohos:id="$+id:image_search"
                ohos:height="40vp"
                ohos:width="100vp"
                ohos:image_src="$media:search_word"
                ohos:scale_mode="zoom_center"/>
        </DirectionalLayout>
        <Text
            ohos:id="$+id:text_search_result"
            ohos:height="match_parent"
            ohos:width="match_parent"
            ohos:multiple_lines="true"
            ohos:visibility="hide"
            ohos:top_margin="40vp"
            ohos:text_size="20vp"
            ohos:text_alignment="left|top"/>
        <Image
            ohos:id="$+id:image"
            ohos:height="match_parent"
            ohos:width="match_parent"
            ohos:padding="60vp"
            ohos:image_src="$media:gold_medal"
            ohos:scale_mode="zoom_center"/>
    </DirectionalLayout>
</DirectionalLayout>
```

我们可以看到，edict_tv.xml 与 edict_phone.xml 这两个布局文件的代码类似，只是 edit_tv.xml 布局文件中的查询输入文本框与 Search 按钮距离左右两侧边界更远，显示的效果如图 11-23 所示。

图 11-23 在 TV 上的主界面

下面再看一下适应智能手表的布局文件（edict_wearable.xml）。

代码位置：src/main/resources/base/layout/edict_wearable.xml

```
<?xml version="1.0" encoding="utf-8"?>
<DirectionalLayout
    xmlns:ohos="http://schemas.huawei.com/res/ohos"
    ohos:height="match_parent"
    ohos:width="match_parent"
    ohos:background_element="#E1E1E1"
    ohos:orientation="vertical">
    <DirectionalLayout
        xmlns:ohos="http://schemas.huawei.com/res/ohos"
        ohos:height="match_parent"
        ohos:width="match_parent"
        ohos:left_margin="5vp"
        ohos:orientation="vertical"
        ohos:right_margin="5vp"
        ohos:top_margin="100vp">
        <DirectionalLayout
            ohos:height="50vp"
            ohos:width="match_parent"
            ohos:alignment="center"
            ohos:orientation="vertical">
            <TextField
                ohos:id="$+id:textfield_word"
                ohos:height="match_content"
                ohos:width="match_parent"
                ohos:background_element="$graphic:border"
                ohos:hint="请输入要查询的单词"
                ohos:padding="2vp"
                ohos:text_size="20"/>
            <Image
                ohos:id="$+id:image_search"
                ohos:height="30vp"
                ohos:width="100vp"
                ohos:image_src="$media:search_word"
                ohos:scale_mode="zoom_center"/>
        </DirectionalLayout>
    </DirectionalLayout>
</DirectionalLayout>
```

从 edict_wearable.xml 布局文件的代码可以看出，该窗口中并没有用于显示查询结果的 Text 组件，而且 Search 按钮在 TextField 组件的下方。其实在智能手表中，显示查询结果是由另外一个窗口完成的，本章后面的内容会详细讲解这个窗口的布局和相关代码。在智能手表中，首页的布局效果如图 11-24 所示。有的读者可能会有疑问，怎么智能手表的窗口是正方形的呢？实际上，智能手表的窗口就是正方形的，而表盘只是这个正方形的内切圆，被表盘切掉的部分就看不到了，所以视觉上是圆形的。因此，在设计智能手表布局时不能按正方形样式设计，因为在表盘中越靠近圆

图 11-24　在智能手表上的主界面

的边缘，显示的区域越有限，在两侧的组件可能看不到或显示不全。所以尽管智能手表的窗口是正方形的，在设计布局时，应该尽可能将尺寸大的组件往中间放，将比较"苗条"的组件往两边放。

11.5.2 代码选择布局文件

我们已经为智能手机、TV 和智能手表分别创建了布局文件（edict_phone.xml、edict_tv.xml 和 edict_wearable.xml），现在的问题是如何让 App 根据不同类型的设备选择与其对应的布局文件。最简单的方式就是直接通过 Java 代码控制。

DeviceInfo.getDeviceType 方法可以返回 App 所在设备的类型，也就是 config.json 文件中 deviceType 属性的某一个值，通过这个返回值，就可以很容易地确定应该使用哪一个布局文件，实现代码如下：

```
if(DeviceInfo.getDeviceType().equals("wearable")) {
    // 装载智能手表的布局文件
    super.setUIContent(ResourceTable.Layout_edict_wearable);
} else if(DeviceInfo.getDeviceType().equals("tv")){
    // 装载 TV 的布局文件
    super.setUIContent(ResourceTable.Layout_edict_tv);
} else {
    // 装载智能手机的布局文件
    super.setUIContent(ResourceTable.Layout_edict_phone);
}
```

如果在代码的其他部分，需要专门针对特定的设备进行处理，也可以使用同样的处理方式。

11.5.3 跨设备在本地词库中查询

对于 TV 和智能手机，单击 Search 按钮，查询结果就会显示在查询输入文本框下方，而对于智能手表来说，单击 Search 按钮，查询结果会在另外一个窗口显示，所以需要改进 Search 按钮的单击事件方法。改进的基本思路是判断当前设备是否为智能手表，如果不是智能手表，仍然在查询输入文本框下方显示查询结果；如果是智能手表，则运行另外的代码来显示一个新窗口，并将查询结果发送给新窗口。

代码位置：src/main/java/com/unitymarvel/edict/slice/MainAbilitySlice.java

```
imageSearch.setClickedListener(new Component.ClickedListener() {
    @Override
    public void onClick(Component component) {
        // 判断当前设备是否为智能手表
        if(!DeviceInfo.getDeviceType().equals("wearable")) {
            // 不是智能手表（TV 或智能手机），隐藏下方的 Image 组件，并显示用于显示查询结果的 Text 组件
            image.setVisibility(Component.HIDE);
            textSearchResult.setVisibility(Component.VISIBLE);
        }
        ArrayList<WordData> result = myDict.searchLocalDict(textfieldWord.getText());
        if(result.size() > 0) {
            if(DeviceInfo.getDeviceType().equals("wearable")) {
```

```
            // 在智能手表上显示查询结果，通过 showSearchResult 方法为新窗口传递数据，并显示新窗口
            showSearchResult(result);
        } else {   // TV 或智能手机
            textSearchResult.setText("");
            // 输出查询结果
            for (WordData wordData : result) {
                textSearchResult.append(wordData.type + " " +
                                        wordData.meanings + "\r\n");
            }
        }

    } else {
        if(!DeviceInfo.getDeviceType().equals("wearable")) {
            // 只有在 TV 或智能手机中运行 App，才会在查询输入文本框下方显示未查询到的提示
            textSearchResult.setText("本地词库没有查到此单词，正在查找网络词库...");
        }
    }
  }
});
```

在这段代码中使用了一个名为 showSearchResult 的方法，该方法为新窗口传递数据，并显示这个窗口，该方法的实现代码如下：

```
public void showSearchResult(List<WordData> result) {
    Intent intent = new Intent();
    // 用于保存所有的词性
    ArrayList<String> typeList = new ArrayList<>();
    // 用于保存所有的中文解释，与词性的个数和位置一一对应
    ArrayList<String> meaningList = new ArrayList<>();
    // 转换词性和中文解释的格式
    for(WordData wordData:result) {
        typeList.add(wordData.type);
        meaningList.add(wordData.meanings);
    }
    // 将词性列表保存到 Intent 对象中
    intent.setStringArrayListParam("typeList",typeList);
    // 将中文解释列表保存到 Intent 对象中
    intent.setStringArrayListParam("meaningList",meaningList);
    // 切换到用于显示查询结果的 AbilitySlice
    present(new EDictsSearchResultAbilitySlice(),intent);
}
```

阅读这段代码，需要了解如下几点：

- 为了方便接收方处理，在 showSearchResult 方法中将词性与中文解释拆开进行传递；
- 本例实现在智能手表上显示查询结果时，并未使用独立的窗口（Page Ability），而使用了嵌在窗口内的 AbilitySlice 来显示查询结果，所以要使用 present 方法切换到另一个 AbilitySlice。读者也可以使用 Page Ability 显示查询结果。

11.5.4　在智能手表上显示查询结果

在上一节我们使用了 AbilitySlice，即实现了 EDictsSearchResultAbilitySlice 类，在该类中处理

传过来的词性与中文解释数据，并将查询结果显示在 Text 组件中，显示效果如图 11-25 所示。

图 11-25 在智能手表中显示查询结果

下面看一下 EDictsSearchResultAbilitySlice 类的实现代码。

```
package com.unitymarvel.edict.slice;

import com.unitymarvel.edict.ResourceTable;
import ohos.aafwk.ability.AbilitySlice;
import ohos.aafwk.content.Intent;
import ohos.agp.components.Text;
import java.util.ArrayList;
public class EDictsSearchResultAbilitySlice extends AbilitySlice {
   private Text textSearchResult;

   @Override
   public void onStart(Intent intent) {
      super.onStart(intent);
      super.setUIContent(ResourceTable.Layout_edict_search_result_wearable);
      textSearchResult = (Text)findComponentById(ResourceTable.Id_text_search_result);
      if(textSearchResult != null) {
         textSearchResult.setText("");
         // 获取词性数据
         ArrayList<String> typeList = intent.getStringArrayListParam("typeList");
         // 获取中文解释数据
         ArrayList<String> meaningList =
                        intent.getStringArrayListParam("meaningList");
         // 在 Text 组件中显示查询结果
         for(int i = 0; i < typeList.size();i++) {
            textSearchResult.append(typeList.get(i)+ " " +
                                   meaningList.get(i) + "\r\n");
         }
         // 如果未查到单词，显示提示信息
         if(typeList.size() == 0) {
            textSearchResult.setText("单词没有查到，请确认单词是否输入错误！");
         }

      }

   }

}
```

在这段代码中使用了一个名为 edict_search_result_wearable.xml 的布局文件，代码如下：

```
<?xml version="1.0" encoding="utf-8"?>
<DirectionalLayout
   xmlns:ohos="http://schemas.huawei.com/res/ohos"
   ohos:height="match_parent"
   ohos:width="match_parent"
   ohos:orientation="vertical">
      <Text
         ohos:id="$+id:text_search_result"
         ohos:height="match_parent"
         ohos:width="match_parent"
```

```
        ohos:padding="40vp"
        ohos:multiple_lines="true"
        ohos:text_size="15vp"
        ohos:text_alignment="left|top"
        />
</DirectionalLayout>
```

在布局文件中只有一个 Text 组件，用来显示查询结果。

11.6 在网络词库中查询

到现在为止，我们已经可以使用本地词库查询单词了，不过这只完成了这款 App 的一部分功能，App 的另外一项重要功能就是可以在网络词库中查询单词。在本例我们选择的网络词库是金山词霸 Web 版本。实现在网络词库中查询的核心就是通过网络爬虫抓取查词时返回的 HTML 代码，并提取出与单词相关的信息，本节将介绍这一过程的完整实现。

11.6.1 分析网络词典的 HTML 代码

在本例我们选择金山词霸作为分析对象。读者可以在金山词霸官网的查询输入文本框中任意输入一个正确的英文单词，例如 super，就可以查询与该单词对应的中文解释和例句，如图 11-26 所示。

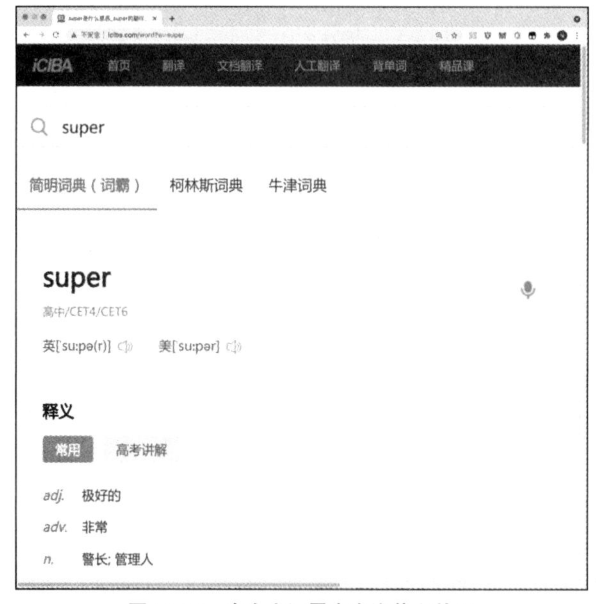

图 11-26 在金山词霸中查询英文单词

金山词霸会给我们返回很多信息，不过对于本例，我们只关注词性和中文解释，也就是图 11-26 显示的内容。我们的任务是通过网络爬虫抓取这些数据，然后作为查询结果显示。

现在按 11.3.1 节的方法定位与单词 super 的中文解释对应的 HTML 代码，如图 11-27 所示。

将对应的 HTML 代码单独提取出来，就是下面的样子：

```
<ul class="Mean_part__1RA2V">
   <li><i>adj.</i>
      <div>
         <span>极好的</span>
      </div>
   </li>
   <li>
      <i>adv.</i>
      <div>
         <span>非常</span>
      </div>
   </li>
   <li>
      <i>n.</i>
      <div>
         <span>警长; </span>
         <span>管理人</span>
      </div>
   </li>
</ul>
```

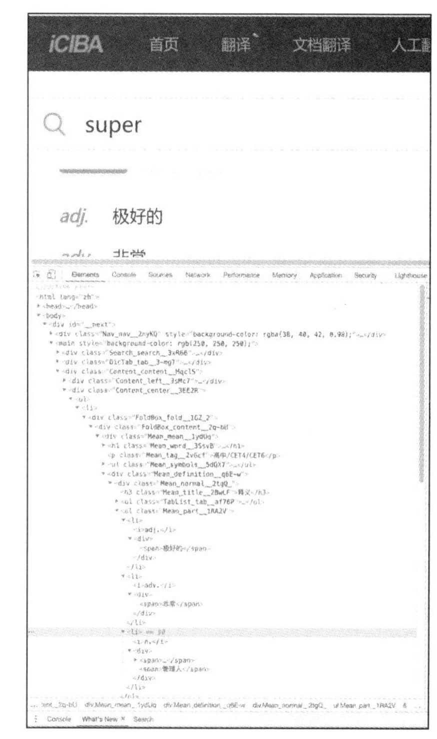

图 11-27　定位与中文解释对应的 HTML 代码

显然，中文解释都在一个<ul>标签中，而且该标签使用了一个名为 Mean_part__1RA2V 的样式。为了让分析更准确，读者可以多查询几个英文单词，会得到和本次分析同样的结果。

我们现在的任务是获取使用样式 Mean_part__1RA2V 的<ul>标签中的所有内容，然后再做进一步的分析，这些工作将在下一节完成。

最后我们看一下浏览器地址栏中 URL 的变化，最新的 URL 如下所示。很显然，如果要查询单词，只需要在参数 w 后面加要查询的单词即可。

https://www.iciba.com/word?w=super

11.6.2　在网络词库中异步查询

实现使用网络词库查询单词的关键是选择一种查询模式，通常有如下两种模式：

- 服务端查询；
- 客户端查询。

1. 服务端查询

服务端查询是指提供一个服务端程序，用来接收客户端发送过来的英文单词，然后服务端程序负责抓取和分析金山词霸返回的数据，并将结果返回给客户端。服务端可以利用任何编程语言实现，如 Python、Java、C#都可以胜任。使用这种方法最大的好处是将客户端与网络词库解耦。也就是说，客户端在查询时，只需要提供要查询的单词，并接收返回结果即可。至于服务端是在网络词库中查询，还是在本地词库中查询，客户端完全不需要操心。如果需要改变网络词库，或者增加新的网络词库，只需要改变服务端的代码即可。只要传递数据的接口不变，客户端就不需要变动。不过这需要开发人员多一项开发服务端程序的技能，如果你有一个团队，或者你是一名全栈开发人员，可以尝试一下。

但这种方式还有一个缺点，就是如果查询量非常大，可能导致服务端频繁访问网络词库，通常服务端的 IP 是不变的，所以网络词库的服务端很容易探测到异常，可能会做一些反爬虫的处理。我们可以利用缓存或定期更换 IP 来解决这一问题，但为什么不试一试另一种方式呢？

2. 客户端查询

这是本例中采用的方式，这种方式直接在客户端（HarmonyOS 设备端）抓取并分析金山词霸的查询结果，然后显示最终的分析结果。这种方式的优点是只需要使用客户端开发技术就可以完成所有工作，而且不怕服务端被封 IP，因为每部 HarmonyOS 设备都可能是不同的 IP，或设备的 IP 可能会根据不同的环境不断切换（Wi-Fi、5G 等网络的 IP 是不同的）。但缺点也很明显，如果网络词典服务端出现什么问题，就需要更换所有客户端的 App，很麻烦。不过本例只是用于演示如何利用 Java 抓取和分析 HTML 代码，所以这个缺点对本例的影响很小。

因为 HarmonyOS App 使用 Java 开发，所以需要一个 Java 版的抓取和分析 HTML 代码的库，本例中我们选择 JSoup，这个库的用法类似前面用于生成本地词库的 Python 版本的 Beautiful Soup。

读者可以直接到 JSoup 官网下载 JSoup 的最新版本。为了方便读者，在电子词典项目中已经内置了 JSoup 的源代码，如图 11-28 所示。读者只需要在工程中直接引用 JSoup 提供的 API 即可。

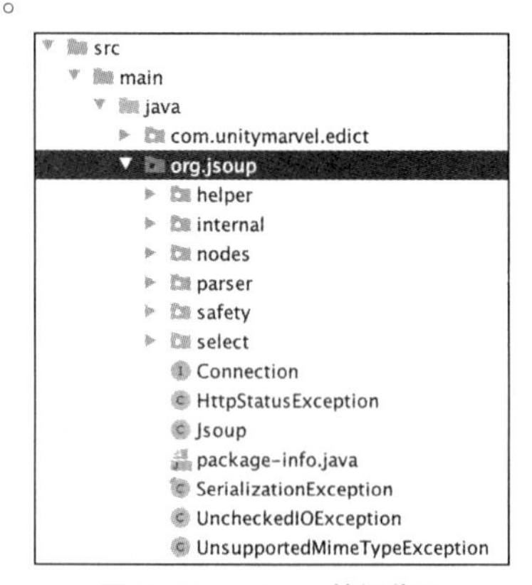

图 11-28 JSoup 的源代码

通常在网络词库中查询采用异步的方式，因为不同的网络环境，服务端的响应时间也会不同，如果采用同步的方式，可能会造成 App“假死”的现象，所以本例采用了异步的方式。

代码位置：src/main/java/com/unitymarvel/edict/common/MyDict.java

```
package com.unitymarvel.edict.common;

import com.unitymarvel.edict.SearchWordCallback;
import ohos.app.AbilityContext;
import ohos.data.DatabaseHelper;
import ohos.data.rdb.RdbOpenCallback;
import ohos.data.rdb.RdbStore;
import ohos.data.rdb.StoreConfig;
import ohos.data.resultset.ResultSet;
import ohos.global.resource.Resource;
import org.jsoup.Jsoup;
import org.jsoup.nodes.Document;
import org.jsoup.nodes.Element;
import org.jsoup.select.Elements;
```

```
import java.io.*;
import java.nio.file.Paths;
import java.util.ArrayList;
import java.util.List;
// 用于查询和分析网络词库的线程类
class AsyncSearchWord extends Thread {
   private String word;
   private RdbStore store;
   // 接收查询结果的回调对象
   private SearchWordCallback callback;
   // 线程类的构造方法，用来接收要查询的英文单词，RdbStore 对象和接收查询结果的回调对象
   public AsyncSearchWord(String word, RdbStore store, SearchWordCallback callback) {
      this.word = word;
      this.store = store;
      this.callback = callback;
   }

   // 在 run 方法中抓取网络词库查询结果的 HTML 代码，并提取出与单词相关的信息
   @Override
   public void run() {
      try {
         // 使用 JSoup 访问金山词霸用于查词的 URL
         Document doc = Jsoup.connect("https://www.iciba.com/word?w=" +  word).get();
         // 获取使用样式 Mean_part__1RA2V 的所有标签（只有一个<ul>标签）
         Elements ulElements = doc.getElementsByClass("Mean_part__1RA2V");
         // 定义用于将查询结果插入本地词库的 insert 语句
         String insertSQL = "insert into t_words(word,type,meanings) values(?,?,?);";
         // 保存查询结果的列表
         List<WordData> wordDataList = new ArrayList<>();
         for(Element ulElement: ulElements) {
            // 获取<ul>标签中的所有<li>标签，一个<li>标签是一个词性以及对应的中文解释
            Elements liElements = ulElement.getElementsByTag("li");
            // 对每一个<li>标签进行迭代
            for(Element liElement:liElements) {
               WordData wordData = new WordData();
               // 获取<li>标签中的所有<i>标签（只有一个<i>标签，表示词性）
               Elements iElements = liElement.getElementsByTag("i");
               for(Element iElement:iElements) {
                  // 获取当前词性
                  wordData.type = iElement.text();
                  break;
               }
               // 获取<li>标签中的所有<div>标签（只有一个<div>标签，包含中文解释）
               Elements divElements = liElement.getElementsByTag("div");
               for(Element divElement:divElements) {
                  // 获取中文解释
                  wordData.meanings = divElements.text();
                  break;
               }
               // 插入每一个词性和中文解释
               wordDataList.add(wordData);
               /* 将从网络词库中查询到的单词的相关信息插入到本地词库，下一次就可以直接在本地词库中
               查询单词了*/
```

```
                    store.executeSql(insertSQL,new String[]{word,
                                                  wordData.type,wordData.meanings});

                }

                break;
            }
            if(callback != null) {
                // 如果指定了回调对象，则通过 onResult 方法返回查询结果
                callback.onResult(wordDataList);
            }
        }
        catch (Exception e)
        {

        }
    }
}
public class MyDict {
    ...
    // 异步查询
    public void searchWebDict(String word, SearchWordCallback callback) {
        word = word.toLowerCase();
        new AsyncSearchWord(word, store, callback).start();
    }
}
```

在网络词库中异步查询的方法是 searchWebDict，该方法也在 MyDict 类中。该类并不直接返回查询结果，而是通过回调对象中的 onResult 方法返回查询结果。

因为在网络词库中查询需要访问网络，所以需要在 config.json 文件中添加如下的权限。

```
"reqPermissions": [
 {
   "name": "ohos.permission.INTERNET",
   "usedScene": {
     "ability": [
       "com.unitymarvel.edict.MainAbility"
     ],
     "when": "always"
   }
 }
]
```

11.6.3 同时在本地词典和网络词典中查词

最理想的状态是首先在本地词库中查询英文单词，如果本地词库中不存在这个单词，则在网络词库中查询英文单词，如果还不存在，则显示未查到英文单词的提示信息。所以需要根据前面实现的代码来改进 Search 按钮的单击事件方法，完整的代码如下：

```
imageSearch.setClickedListener(new Component.ClickedListener() {
    @Override
    public void onClick(Component component) {
```

```
        // 如果不是智能手表，隐藏下方的 Image 组件，显示 Text 组件用于显示查询结果
        if(!DeviceInfo.getDeviceType().equals("wearable")) {
            image.setVisibility(Component.HIDE);
            textSearchResult.setVisibility(Component.VISIBLE);
        }
        ArrayList<WordData> result = myDict.searchLocalDict(textfieldWord.getText());
        if(result.size() > 0) {
            // 如果是智能手表，用 AbilitySlice 显示查询结果
            if(DeviceInfo.getDeviceType().equals("wearable")) {
                showSearchResult(result);
            } else {  // 如果是 TV 或智能手机，在查询输入文本框下方显示查询结果
                textSearchResult.setText("");
                // 输出查询结果
                for (WordData wordData : result) {
                    textSearchResult.append(wordData.type + " " +
                                                   wordData.meanings + "\r\n");
                }
            }

        } else {
            // 如果是 TV 或智能手机，而且在本地词库没查到单词，先显示未查到提示
            if(!DeviceInfo.getDeviceType().equals("wearable")) {
                textSearchResult.setText("本地词库没有查到此单词，正在查找网络词库...");
            }
            // 在网络词库中异步查询
            myDict.searchWebDict(textfieldWord.getText(),new SearchWordCallbackImpl());
        }
    }
});
```

在这段代码中涉及 SearchWordCallbackImpl 类，这是传入 searchWebDict 方法的回调对象类，该类的实现代码如下：

```
private class SearchWordCallbackImpl implements SearchWordCallback {
    @Override
    public void onResult(List<WordData> result) {
        // 如果是智能手表，用 AbilitySlice 显示查询结果
        if(DeviceInfo.getDeviceType().equals("wearable")) {
            // 切换到 AbilitySlice，并显示查询结果
            showSearchResult(result);
        } else {
            // 通过 Handler 将执行权限交给 UI 线程，以便在主界面中显示查询结果
            EventRunner runnerA = EventRunner.getMainEventRunner();
            MyEventHandler handler = new MyEventHandler(runnerA, result);
            /* 将执行权限交给主线程（UI 线程），SEARCH_RESULT 是一个常量，用于鉴别是谁将执行任务交给了
            UI 线程，根据交还者不同，执行的代码也会不同*/
            handler.sendEvent(SEARCH_RESULT);
            runnerA = null;
        }

    }
}
```

如果是 TV 或智能手机，在 onResult 方法中会使用一个 EventRunner 对象将执行权限交给 UI 线程。这是由 HarmonyOS 的机制决定的。在 HarmonyOS 中，不能在非 UI 线程中更新 UI 线程的组件。也就是说，在一个线程中要更新组件，那么这些组件也必须是在这个线程中创建的，否则会抛出异常。很明显，在网络词库中异步查询的线程肯定不是创建 UI 组件的线程，所以需要使用 EventRunner 对象在 UI 线程（即主线程）中更新组件，也就是显示查询结果。这里涉及 MyEventHandler 类，实现代码如下：

```
public class MyEventHandler extends EventHandler {
   private List<WordData> dataWordList;
   // 构造方法，传入查询结果
   private MyEventHandler(EventRunner runner, List<WordData> dataWordList)
   {
      super(runner);
      this.dataWordList = dataWordList;
   }

   // processEvent 方法会在 UI 线程（主线程）中执行
   @Override
   public void processEvent(InnerEvent event) {
      super.processEvent(event);
      if (event == null) {
         return;
      }
      // 这个 ID 是在 onResult 方法中通过 sendEvent 方法执行的值
      // sendEvent 方法实际上是将执行权限交给 UI 线程，也就是执行 processEvent 方法
      int eventId = event.eventId;

      switch (eventId) {
         case SEARCH_RESULT: {    // 这就是 sendEvent 方法指定的常量
            // 在网络词库中也没有查到该单词
            if(dataWordList.size() == 0) {
               textSearchResult.setText("单词没有查到，请确认单词是否输入错误！");
            } else {
               // 在网络词库中查到了单词，并显示查询结果
               textSearchResult.setText("");
               for (WordData wordData : dataWordList) {
                  textSearchResult.append(wordData.type + " " +
                                          wordData.meanings + "\r\n");
               }
            }
            break;
         }

      }
   }
}
```

在网络词库中异步查询时使用了 Handler 更新 UI 线程中的组件，其实还有更方便的方式，就是在非 UI 线程中直接获取 UI 线程的 TaskDispatcher 对象，然后通过 TaskDispatcher.asyncDispatch

方法执行访问 UI 组件的代码。具体实现如下：

```
context.getUITaskDispatcher().asyncDispatch(new Runnable() {
   @Override
   public void run() {
      // 在这里更新 UI 线程中创建的组件
   }
});
```

这里的 context 是 Context 对象，也可以是 Ability、AbilitySlice 等对象，这些类都是 Context 类的子类。

到现在为止，电子词典 App 的基本功能都已经实现了，我们选择一个设备运行程序，就会看到本章一开始显示的运行效果，好好享受自己的成果吧！

11.7 总结与回顾

通过本章的学习，读者已经可以实现一个完整的 HarmonyOS App 项目。

本章讲述的开发项目是一个电子词典 App，可以在线查英文单词，也可以跨设备运行。这个项目使用了多种技术，是对前面 10 章知识的一次综合演练。

电子词典项目主要涉及了如下技术点：

- Page Ability；
- 权限设置；
- 跨设备布局；
- JSoup 库；
- 网络爬虫；
- 使用第三方组件；
- SQLite 数据库；
- 与 HAP 一起发布数据库；
- 异步技术。

为了生成电子词典 App 使用的本地词库（本章案例中使用 SQLite 数据库），本章还使用了 Python 来抓取英文单词数据，并将这些英文单词数据插入到 SQLite 数据库中。

尽管电子词典项目已经拥有了电子词典 App 的基本功能，但这个项目还远远没到完善的地步，我们还可以对电子词典项目进行如下改进。

- 增加英文单词朗读功能，可以利用语音自动合成技术，如讯飞语音，也可以利用真人发音，以 mp3 文件提供。这些音频文件可以是本地的，也可以是在线的。
- 为每个英文单词增加例句，支持语音朗读例句。
- 增加背单词功能，如将某一特定类型的单词放到一组，如英语四级、六级单词，循环显示这些单词，支持只显示英文，或只显示中文，或加入更复杂的背单词功能。
- 增加生词本功能，如果未掌握某个英文单词，可以将这个单词加入生词本，供以后复习。

- 增加文章查词功能，例如，打开一篇英文文章，如果不清楚文章中的某个英文单词是什么意思，可以单独定位这个英文单词，单击查词（可以单独弹出窗口显示单词的中文解释）。
- 向其他设备提供跨设备查词的 API（可以利用跨设备访问 Service Ability 实现）。

当然，还有更多的功能可以添加，各位读者可以充分发挥自己的想象力，让这款电子词典 App 更加完善。这些功能利用本书学习的知识完全可以实现，感兴趣的读者可以尝试对这款 App 进行改进。

第12章　游戏类App项目：5分钟搞定俄罗斯方块

很多读者可能会问："在5分钟内能实现一个游戏吗？"尤其是像俄罗斯方块这样的游戏。我的答案是：当然不能。就算照抄代码，在5分钟内也完不成！那本章的章名为什么指出"5分钟搞定俄罗斯方块"呢？如果从头开始做游戏当然不行，但在本章中我们已将游戏组件化，也就是说，俄罗斯方块本身已经成了一个组件（Tetris组件）。实际上，只要将这个组件放在窗口中，就已经具备了游戏的基本功能，然后再稍微加工，就可以实现一个相对复杂的俄罗斯方块游戏。想知道怎么做吗？继续看本章的内容吧！

本章的源代码在配套资源中tetris目录下。

通过阅读本章，读者可以掌握：

- 如何绘制图形；
- 如何在非UI线程更新组件；
- 自定义组件的设计和实现；
- 俄罗斯方块游戏的核心算法实现。

12.1 功能需求分析

相信大家对俄罗斯方块这款游戏应该非常熟悉，那么大家对俄罗斯方块游戏的算法熟悉吗？本节我们先从宏观上描述一下俄罗斯方块游戏的功能需求，从下一节开始，我们将会一步一步实现这款基于HarmonyOS的俄罗斯方块游戏。

一款功能完善的俄罗斯方块游戏通常会由下面几个部分组成：

- 游戏背景；
- 随机产生的方块（block）；
- 控制方块旋转和平移的按钮（或用其他方式控制）；
- 显示下一个方块；
- 游戏积分机制；

- 让方块快速下落。

如果将这几个部分组合成一款游戏，那么样式如图 12-1 所示。我们可以看到，窗口的左上角是游戏主界面和随机产生的方块，右上角显示的是积分和下一个方块，下面是控制按钮。

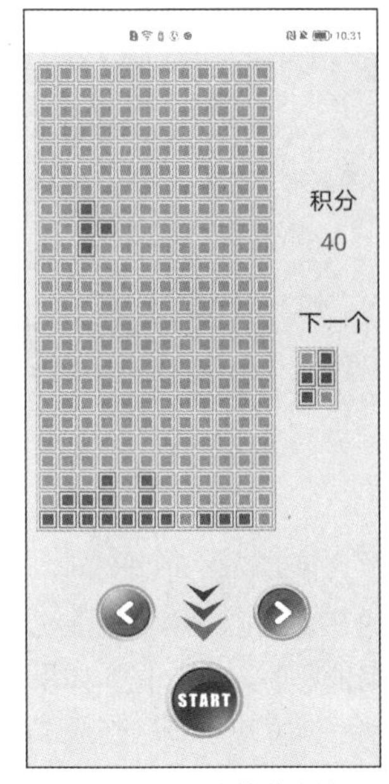

图 12-1 俄罗斯方块游戏的界面

下面分别来描述这几个部分。

（1）游戏背景。俄罗斯方块游戏除随机产生的方块是小方格以外，背景一般也是不同颜色的小方格（cell）。这主要是为了在移动方块时可以对齐下落的位置。

（2）随机产生的方块。方块是单个小方格的不同组合，方块背景颜色与游戏背景颜色的差别很大，如本例中分别为红色和深绿色（读者在实际操作时可以看到）。不同类型的俄罗斯方块有不同样式的方块。通过旋转和水平移动调整方块位置，让方块下落到合适的位置，然后游戏会消除颜色与方块相同的行（本例中是红色），被消除的行上面的所有小方格都整体下落。

（3）控制方块旋转和平移的按钮。方块的动作主要是旋转和水平移动，控制动作的方式有多种，在本例中，旋转动作通过触摸俄罗斯方块实现，触摸一下，方块顺时针旋转 90°。而左右水平移动动作通过窗口下方的左右箭头按钮控制。

（4）显示下一个方块。为了让玩家可以更好地摆放方块，通常在产生当前方块的时候，还会产生下一个方块，这样有利于玩家更合理地安排当前方块和下一个方块的位置。

（5）游戏积分机制。为了增加游戏的趣味性和用户粘度，通常会在游戏中设立积分机制（或其他更富有挑战性的机制）。本例中也为俄罗斯方块设立了积分，例如，消除的行越多，就会得到越高的积分。这会对玩家更有吸引力。

（6）让方块快速下落。如果让方块自然下落，可能需要很长时间，所以通常在对准方块后，让方块快速下落到合适的位置，这样非常节省时间。在图 12-1 所示的界面中向下的箭头用于控制方块快速下落，按下后，方块就会加快下落的速度，抬起后，方块会恢复默认的下落速度。

12.2 类的继承关系

上一节已经简要介绍了俄罗斯方块游戏 App 的核心功能。其实一款游戏就是一个可以与用户交互的绘图程序。不同游戏的区别只是绘图的复杂程度不同而已。

尽管可以将所有的功能都放到一个文件和一个类中，但这样做会让类文件太大，而且将所有的功能都混在一起，不容易维护和扩展。所以本例中将功能分散在 7 个不同的类文件中，下面是这 7 个类文件中的类及其实现的功能。

- TetrisListener：因为本章实现的案例会将俄罗斯方块游戏 App 的功能封装在组件中，而组件的一个重要特性就是事件，程序可以通过事件与外界交互。所以在这个类中定义了与事件相关的属性和方法。
- TetrisBase：俄罗斯方块游戏 App 的核心类，在这个类中实现了大部分业务逻辑，包括初始化俄罗斯方块、绘制方块、触发事件等。
- TetrisBorder：绘制游戏界面的边框。
- TetrisEvent：处理游戏的事件，如用来控制方块顺时针旋转的触摸事件。
- TetrisBackgroundCells：绘制游戏界面的背景小方格。
- TetrisProperties：定义游戏的一些属性和方法，如设置背景颜色等。
- Tetris：俄罗斯方块游戏 App 的主类，也是组件类。用户直接使用的就是 Tetris 类。这个类主要用于绘制方块。

有的读者可能会有这样的疑问，这 7 个类各司其职，但如何将它们连在一起呢？为了让这 7 个类连在一起，本例中采用了继承的方式，也就是说，这 7 个类实际上是按顺序继承的。TetrisListener 类是基类，Tetris 类是继承树中的叶节点，这 7 个类的继承关系如图 12-2 所示。

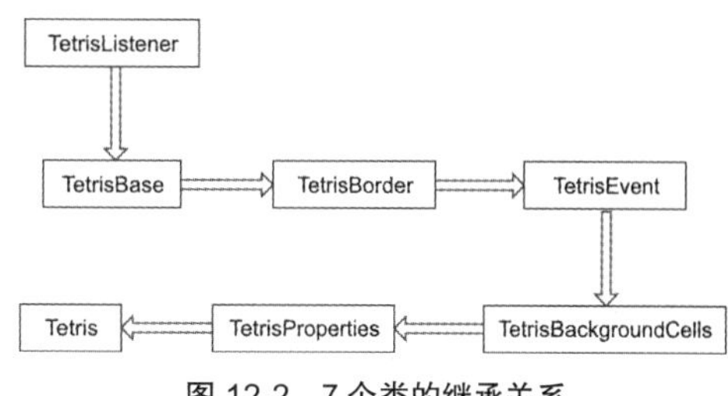

图 12-2 7 个类的继承关系

12.3 使用 Tetris 组件

Tetris 组件还没制作，为什么要先讲解如何使用 Tetris 组件呢？其实这是制作组件的一般流程。在制作组件之前，首先要搭建一个组件的基本架构，然后需要编写调用组件的代码，如果没有这些调用代码，就无法测试组件了。所以在本节我们的任务是完成 Tetris 组件的架构搭建，以及编写调用 Tetris 组件的代码。

12.3.1 搭建 Tetris 组件

Tetris 组件只由一个 Tetris 类组成，从图 12-2 所示的继承关系图可以看出，Tetris 类是继承树最末端的节点。Tetris 类的功能很少，只是负责对外联络和绘制方块。

因为其他类还没有实现，所以暂不考虑 Tetris 类继承和实现具体功能的问题，先编写一个普通的 Java 类，代码如下：

```
public class Tetris {
   public Tetris(Ability ability, Component component) {
   }
}
```

我们可以看到，Tetris 类十分简单，类中只有一个构造方法外。这个构造方法必须定义，因为在调用 Tetris 组件时，需要调用这个构造方法。

Tetris 类的构造方法有两个参数，即 ability 和 component，这两个参数的类型分别是 Ability 和 Component。包含 Ability 类型的参数，说明 Tetris 类必须在 Ability 中调用，或在可以获取 Ability 对象的地方调用。而 component 参数的主要作用是绘图，绘制俄罗斯方块游戏界面中的一切东西。component 参数可以是任何组件，但通常是一个容器，如各种布局组件。Tetris 组件的基本原理就是让 Tetris 组件与一个组件绑定，然后 Tetris 组件就会自动在该组件上绘制俄罗斯方块游戏界面中的一切图形，并可以响应单击事件（用于控制方块顺时针旋转）。

现在 Tetris 组件的架构已经搭建起来了，有了这些东西，已经足够我们使用 Tetris 组件了，下一节将讲解如何调用 Tetris 组件。

12.3.2 游戏主界面的布局

因为游戏界面是一体的，所以在本节我们将完成整个游戏界面的布局。从图 12-1 所示的效果可以看出，游戏界面分成 3 个部分，左上角是游戏主界面，右上角显示的是积分和下一个方块，下面是控制按钮。左上角的游戏主界面其实是一个 DirectionalLayout 组件，然后将 Tetris 组件与 DirectionalLayout 组件绑定，并在 DirectionalLayout 组件上绘制俄罗斯方块游戏主界面。完整的布局代码如下：

```
<?xml version="1.0" encoding="utf-8"?>
<DirectionalLayout
   xmlns:ohos="http://schemas.huawei.com/res/ohos"
   ohos:height="match_parent"
   ohos:width="match_parent"
   ohos:background_element="#E1E1E1"
   ohos:orientation="vertical">
   <DirectionalLayout
      ohos:height="500vp"
      ohos:width="match_parent"
      ohos:orientation="horizontal">
      <!-- 用于绘制俄罗斯方块游戏主界面的布局 -->
      <DirectionalLayout
         ohos:id="$+id:layout_tetris"
         ohos:height="500vp"
         ohos:width="250vp"
         ohos:top_margin="10vp"
         ohos:left_margin="10vp"
         ohos:right_margin="10vp"/>
      <!-- 右侧区域，用于显示积分和下一个方块 -->
      <DirectionalLayout
         ohos:height="match_parent"
         ohos:width="match_parent"
         ohos:weight="1"
         ohos:alignment="center"
         ohos:orientation="vertical">
         <Text
            ohos:height="match_content"
```

```
            ohos:width="match_parent"
            ohos:text_alignment="center"
            ohos:top_margin="20vp"
            ohos:text_size="25fp"
            ohos:text="积分"/>
        <!-- 显示积分 -->
        <Text
            ohos:id="$+id:text_score"
            ohos:height="match_content"
            ohos:width="match_parent"
            ohos:text_alignment="center"
            ohos:top_margin="10fp"
            ohos:text_color="#FF0000"
            ohos:text_size="25fp"
            ohos:text="0"/>
        <Text
            ohos:height="match_content"
            ohos:width="match_parent"
            ohos:text_alignment="center"
            ohos:top_margin="50fp"
            ohos:text_size="25fp"
            ohos:text="下一个"/>
        <DirectionalLayout
            ohos:height="match_content"
            ohos:width="match_parent"
            ohos:alignment="center"
            ohos:top_margin="10vp">
        <!-- 绘制下一个方块 -->
        <DirectionalLayout
            ohos:id="$+id:layout_next"
            ohos:height="80vp"
            ohos:width="80vp"/>
        </DirectionalLayout>
    </DirectionalLayout>
</DirectionalLayout>
<DirectionalLayout
    ohos:height="match_parent"
    ohos:width="match_parent"
    ohos:top_margin="20vp"
    ohos:weight="1"
    ohos:alignment="center"
    ohos:orientation="vertical">
    <!-- 下面是控制按钮区域 -->
    <DirectionalLayout
        ohos:height="60vp"
        ohos:width="match_parent"
        ohos:orientation="horizontal"
        ohos:alignment="center">
        <Image
            ohos:id="$+id:image_move_left"
```

```
                ohos:image_src="$media:move_left"
                ohos:height="60vp"
                ohos:width="60vp"
                ohos:right_margin="20vp"
                ohos:scale_mode="zoom_center"/>
            <Image
                ohos:id="$+id:image_move_quickly"
                ohos:image_src="$media:quickly"
                ohos:height="60vp"
                ohos:width="60vp"
                ohos:right_margin="20vp"
                ohos:scale_mode="zoom_center"/>
            <Image
                ohos:id="$+id:image_move_right"
                ohos:image_src="$media:move_right"
                ohos:height="60vp"
                ohos:width="60vp"
                ohos:scale_mode="zoom_center"/>
        </DirectionalLayout>
        <Image
            ohos:id="$+id:image_start"
            ohos:image_src="$media:start"
            ohos:height="80vp"
            ohos:width="80vp"
            ohos:top_margin="20vp"
            ohos:scale_mode="zoom_center"/>
    </DirectionalLayout>
</DirectionalLayout>
```

如果装载这个布局文件，会显示图 12-3 所示的效果。

在这个布局中还使用了一些图像，这些图像需要事先放到 media 目录下，如图 12-4 所示。

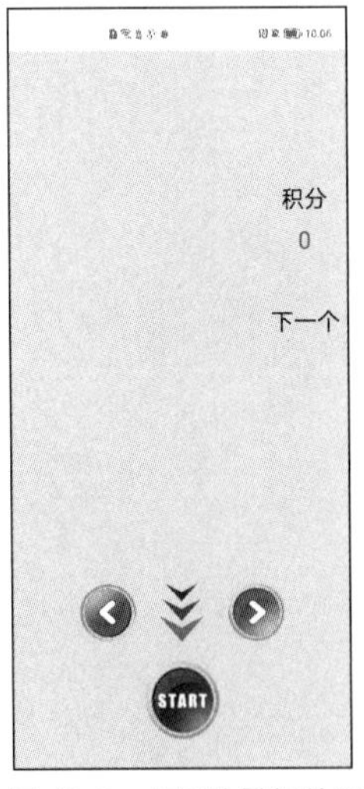

图 12-3　布局的最初效果

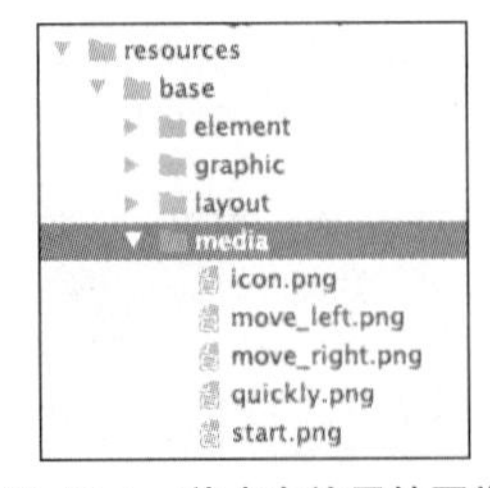

图 12-4　游戏中使用的图像

从图 12-3 所示的效果图可以看出，左上角什么都没有，其实左上角有一个 DirectionalLayout 组件，只是该组件的背景颜色与父组件的背景颜色相同，所以什么都看不出来。在游戏中，不仅游戏主界面是绘制出来的，连下一个方块也是绘制出来的，本章后面会详细介绍这一点。

12.3.3 使用 Tetris 组件

当 Tetris 组件完成后，使用 Tetris 组件就变得容易得多，只需要装载布局文件，然后创建 Tetris 对象即可，代码如下：

```
Tetris tetris = null;
DirectionalLayout layoutTetris =
          (DirectionalLayout)findComponentById(ResourceTable.Id_layout_tetris);
if(layoutTetris != null) {
  // 创建 Tetris 对象，并与 layoutTetris 绑定
  tetris = new Tetris(this,layoutTetris);
}
```

如果 Tetris 组件已经实现，运行程序，就会在 layoutTetris 组件上绘制游戏界面，不过现在运行程序，什么都不会出现，因为 Tetris 组件还是空架子，里面什么都没有。

12.4 实现 Tetris 组件

本节将详细讲解 Tetris 组件的实现过程，并给出完整的源代码。

12.4.1 定义 Tetris 组件的事件类

在 Tetris 组件中会根据不同的状态触发如下 3 个事件。

- 产生下一个方块事件：用于显示下一个方块。
- 消除行事件：用于处理积分。
- 游戏结束事件：用于提示游戏结束。

这 3 个事件的回调方法分别在如下 3 个接口中。

- TetrisNextBlockListener：产生下一个方块。
- TetrisRowRemovedListener：消除行。
- TetrisGameoverListener：游戏结束。

下面给出这 3 个接口的实现代码。

TetrisNextBlockListener.java 代码如下：

```
package com.harmonyos.tetris.listeners;
public interface TetrisNextBlockListener {
// 当产生下一个方块时触发，主要用于预告下一个方块是什么
// block 数组中的每一个值表示是否存在小方格，1 表示存在，0 表示不存在
   void onNextBlock(int[][] block);
}
```

TetrisRowRemovedListener.java 代码如下：

```
package com.harmonyos.tetris.listeners;
public interface TetrisRowRemovedListener {
   // 当 tetris 的行被消除时调用，rowCount 表示被消除的行数，可以在该事件中处理积分的操作
   void rowRemoved(int rowCount);
}
```

TetrisGameoverListener.java 代码如下：

```
package com.harmonyos.tetris.listeners;
public interface TetrisGameoverListener {
   // 当游戏结束时被触发
   void onGameover();
}
```

12.4.2 定义游戏事件属性

上一节中实现的 3 个接口用于监听 3 个事件，这些接口需要在 Tetris 组件中设置，所以需要为 Tetris 组件添加 3 个属性，用于设置这 3 个事件的监听器。

包含这些属性和相关方法的类是 TetrisListener 类，因为 TetrisListener 类是继承树的根类，所以先实现这个类。

实现游戏的核心工作就是使用画笔在画布上绘制出游戏的界面以及方块等元素。在 HarmonyOS 中，绘制图形需要实现 Component.DrawTask 接口，然后在 onDraw 方法中完成绘制工作。所以 TetrisListener 类需要实现 Component.DrawTask 接口，即使在 TetrisListener.onDraw 方法中没有任何代码，也要实现 Component.DrawTask 接口，这样后续的类如果从 TetrisListener 类继承，就可以在 onDraw 方法中绘制与该类相关的图形。

TetrisListener 类的实现代码如下：

```
package com.harmonyos.tetris;

import com.harmonyos.tetris.listeners.TetrisGameoverListener;
import com.harmonyos.tetris.listeners.TetrisNextBlockListener;
import com.harmonyos.tetris.listeners.TetrisRowRemovedListener;
import ohos.agp.components.Component;
import ohos.agp.render.Canvas;

// 处理组件的监听事件
public class TetrisListener implements Component.DrawTask {
   protected TetrisNextBlockListener tetrisNextBlockListener;
   protected TetrisRowRemovedListener tetrisRowRemovedListener;
   protected TetrisGameoverListener tetrisGameoverListener;

   @Override
   public void onDraw(Component component, Canvas canvas) {
      // 这里不需要写代码，但必须实现 onDraw 方法
   }

   public void setOnTetrisNextBlockListener(TetrisNextBlockListener listener) {
      this.tetrisNextBlockListener = listener;
   }

   public void setOnTetrisRowRemovedListener(TetrisRowRemovedListener listener) {
```

```
        this.tetrisRowRemovedListener = listener;
    }

    public void setOnTetrisGameoverListener(TetrisGameoverListener listener) {
        this.tetrisGameoverListener = listener;
    }
}
```

我们可以看到，在 TetrisListener 类中有 3 个 setter 方法，分别用来设置 3 个监听器。

12.4.3 初始化 Tetris 组件

因为 Tetris 组件中需要一些参数来绘制游戏中的各种元素，所以在本节我们先在 TetrisBase 类中完成一些初始化工作。TetrisBase 类的代码比较复杂，本章后面会讲解更多的功能。

绘制 Tetris 组件的游戏界面，需要的核心参数如下：

- 定义游戏界面中所有小方格的数据结构，可以是二维数组或列表；
- 小方格之间的距离；
- 小方格的尺寸；
- 水平方向小方格的数量（列数）；
- 垂直方向小方格的数量（行数）；
- 每一个方块的定义。

对于行数和列数，TetrisBase 类是根据与 Tetris 绑定的组件的尺寸、每一个小方格的尺寸和小方格之间的距离自动计算的，所以不需要指定，其他参数需要在程序中指定。

TetrisBase 类的实现代码如下：

```
public class TetrisBase extends TetrisListener {
    // 定义游戏界面中所有小方格的数据结构，即一个二维列表
    // 0：背景小方格    1：已经下落的方块
    protected List<List<Integer>> tetrisData = new ArrayList<>();
    protected Ability ability;
    protected Component component;
    // 小方格之间的距离
    protected int cellSpace = 10;
    // 小方格的尺寸
    protected int cellSize = 50;
    // 水平方向小方格的数量
    protected int colCount = 0;
    // 垂直方向小方格的数量
    protected int rowCount = 0;
    // 用于绘制方块的数据结构
    protected int[][] drawBlock;
    // 用于保存所有的方块的定义
    protected List<int[][]> blocks = new ArrayList<>();
    // 构造方法的第 1 种重载形式
    public TetrisBase(Ability ability, Component component, int[][] drawBlock) {
        this.drawBlock = drawBlock;
        this.ability = ability;
        // component 指定了与 Tetris 组件结合的组件，游戏界面就绘制在这个组件中
        this.component = component;
        // 设置组件的绘制任务，这样当 component 刷新时，onDraw 方法就会被调用
```

```
        component.addDrawTask(this);
        // 让组件允许单击，这样就可以通过单击组件让方块顺时针旋转了
        component.setClickable(true);
        // 获取组件的宽度
        int width = component.getWidth();
        // 获取组件的高度
        int height = component.getHeight();
        // 如果未指定 drawBlock，就是绘制整个游戏界面
        if(drawBlock == null) {
            colCount = (width - cellSpace) / (cellSize + cellSpace);
            rowCount = (height - cellSpace) / (cellSize + cellSpace);
        } else {   // 如果指定 drawBlock，就是绘制方块
            rowCount = drawBlock.length;
            colCount = drawBlock[0].length;
            currentBlock = drawBlock;
            blockCurrentRow = 0;
            blockCurrentCol = 0;
        }
        // 初始化俄罗斯方块数据，将所有的列表元素都设置为 0，表示初始状态，背景中没有任何方块
        for(int i = 0; i < rowCount;i++) {
            List<Integer> rowList = new ArrayList<>();
            for(int j = 0; j < colCount;j++) {
                rowList.add(0);
            }
            tetrisData.add(rowList);
        }

        // 初始化方块，1 表示有小方格，0 表示无小方格
        int[][] block1 = new int[][]{{1,1,0},{0,1,1}};
        blocks.add(block1);
        int[][] block2 = new int[][]{{0,1,0},{1,1,1}};
        blocks.add(block2);
        int[][] block3 = new int[][]{{1,1},{1,1}};
        blocks.add(block3);
        int[][] block4 = new int[][]{{0,0,1},{1,1,1}};
        blocks.add(block4);
        int[][] block5 = new int[][]{{1,1,1,1}};
        blocks.add(block5);
    }
    // 构造方法第 2 种重载形式
    public TetrisBase(Ability ability, Component component) {
        this(ability,component,null);
    }
}
```

在阅读这段代码时，应了解如下几点。

- TetrisBase 类需要从 TetrisListener 类继承，这是为了逐步形成继承树。
- 图 12-1 所示的游戏背景小方格其实是一个二维列表（用数组也可以），如果列表元素值是 0，表示背景小方格（深绿色）；如果是 1，表示方块小方格（红色）。如果小方格是深绿色，方块可以经过这个位置；如果小方格是红色表示该位置被占用，正在移动或旋转的方块不能经过该区域。本例中 tetrisData 变量表示这一数据结构。

- 在使用 Tetris 组件时，必须与一个 Component 组件绑定，这里的绑定并不是指将 Component 组件赋给 Tetris 组件，而是指设置 Component 组件的绘制任务（使用 addDrawTask 方法）。这样在刷新时，Component 就会调用 Tetris 组件的 onDraw 方法，从而达到绘制游戏中各个元素的目的。
- Tetris 组件在绘制游戏界面时，会根据 Component 组件的尺寸计算行数和列数。所以在布局 Component 组件时，应该使用固定的尺寸（如 500vp 等），不能使用 match_parent 和 match_content，否则无法获取 Component 组件的宽度和高度。
- TetrisBase 类构造方法有两种重载形式，其中第 1 种重载形式最后一个参数是 int 类型二维数组（drawBlock），该参数指定了一个方块。也就是说，TetrisBase 类其实兼顾两个任务，一个是绘制整个游戏界面，一个是绘制方块，本例中它用于绘制下一个方块。如果指定 drawBlock 参数，则绘制这个方块，否则就绘制整个游戏界面。
- 本例中定义了 5 个方块，这 5 个方块随机产生，它们的样式如图 12-5 所示。这 5 种样式都是最初的形态，每一种样式还有 4 种旋转形态。因为 block3 的 4 种旋转形态都相同，block5 的 4 种旋转形态中两两相同，所以一共有 15 种方块的样式。

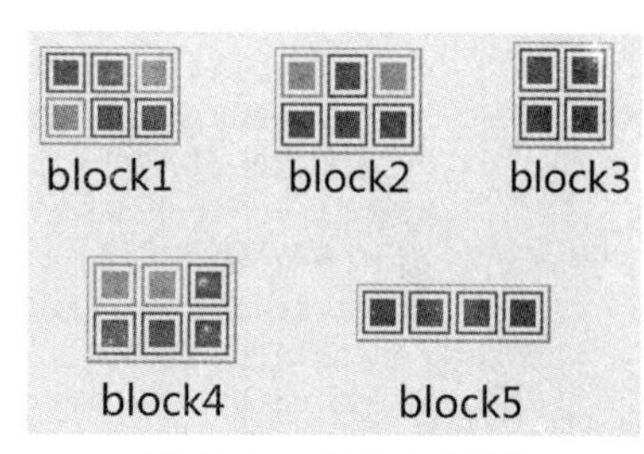

图 12-5　方块的 5 种样式

12.4.4　绘制游戏边框

绘制游戏边框是实现游戏的第一步。TetrisBorder 类负责绘制游戏界面边框。实现游戏的核心工作就是使用画笔在画布上绘制出游戏的界面以及方块等元素。在 HarmonyOS 中，绘制图形需要实现 Component.DrawTask 接口，然后在 onDraw 方法中完成绘制工作。如果 TetrisBorder 类是从 TetrisBase 继承的，也就相当于实现了 Component.DrawTask 接口。绘制是高频发生的，也就是说，操作系统每刷新一次，用户界面就会重新绘制一次，onDraw 方法就会被调用一次。所以绘制边框和绘制其他元素的过程是不断重复的，也就是说，每刷新一次，先擦除窗口中的所有东西，然后再重新绘制所有东西，这一过程是不断发生的，只是这一过程发生的频率非常高，人眼无法察觉而已。

在 TetrisBorder 类中需要指定边框的颜色和边框线的宽度。本例中边框线的颜色与背景小方格的颜色相同，都是深绿色。边框线宽度为 4 像素。

TetrisBorder 类的实现代码如下：

```
package com.harmonyos.tetris;

import ohos.aafwk.ability.Ability;
import ohos.agp.components.Component;
import ohos.agp.render.Canvas;
import ohos.agp.render.Paint;
import ohos.agp.utils.Color;
import ohos.agp.utils.RectFloat;
```

```
// 绘制俄罗斯方块的边框
public class TetrisBorder extends TetrisBase {
    // 定义边框线的颜色
    protected Color borderColor = new Color(Color.rgb(135,147,115));
    // 定义边框线的宽度（4 像素）
    protected int borderLineWidth = 4;
    // 构造方法的第 1 种重载形式
    public TetrisBorder(Ability ability, Component component,int[][] block) {
         super(ability,component,block);
    }
    // 构造方法的第 2 种重载形式
    public TetrisBorder(Ability ability, Component component) {
        super(ability, component);
    }
    // 在 onDraw 方法中绘制边框线
    public void onDraw(Component component, Canvas canvas) {
        super.onDraw(component,canvas);
        Paint paint = new Paint();
        // 设置边框线的宽度
        paint.setStrokeWidth(borderLineWidth);
        // 设置边框线为实线
        paint.setStyle(Paint.Style.STROKE_STYLE);
        // 设置边框线的颜色
        paint.setColor(borderColor);
        // 计算边框线右边的位置
        float right = colCount * cellSize + (colCount + 1) * cellSpace;
        // 计算边框线底边的位置
        float bottom = rowCount * cellSize + (rowCount + 1) * cellSpace;
        // 绘制一个矩形作为游戏界面的边框线
        canvas.drawRect(new RectFloat(0,0, right,bottom),paint);
    }
}
```

为了让调用程序可以正常绘制出边框线，需要修改 Tetris 类的代码，首先让 Tetris 类从 TetrisBorder 类继承，然后为 Tetris 类添加一个构造方法，实现代码如下：

```
package com.harmonyos.tetris;

import ohos.aafwk.ability.Ability;
import ohos.agp.components.Component;
import ohos.agp.render.Canvas;


public class Tetris extends TetrisBorder {
    public Tetris(Ability ability, Component component) {
        super(ability, component);
    }
    public Tetris(Ability ability, Component component, int[][] drawBlock)
    {
        super(ability, component,drawBlock);
    }
    public void onDraw(Component component, Canvas canvas) {
        /* 这里必须调用超类的 onDraw 方法，否则不会执行 TetrisBorder 类中的 onDraw 方法，也就不会绘
        制游戏边框了*/
        super.onDraw(component,canvas);
```

```
    }
}
```

现在运行程序，会在窗口左上角绘制一个深绿色的边框，如图 12-6 所示。绘制多大的边框，要看用于绘制的组件的尺寸是多大。

图 12-6 绘制边框的效果

12.4.5 绘制小方格

不管是绘制游戏背景，还是绘制方块，都需要绘制若干个小方格（见图 12-7），它们的区别只是背景颜色不同而已。

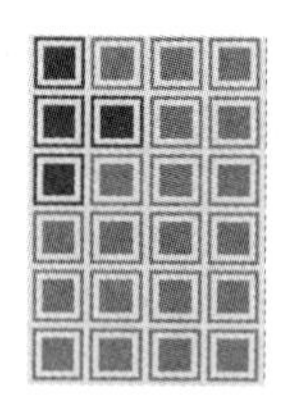

图 12-7 组成游戏的小方格

因为绘制小方格的功能是通用的，所以将这个功能封装在一个名为 drawCell 的方法中，然后封装在 TetrisBase 类中，代码如下：

```
// row：小方格的行   col：小方格的列， color：小方格的颜色
protected void drawCell(Canvas canvas, int row, int col,Color color) {
   // row 和 col 必须比游戏界面的行和列小才可以绘制
   if(row >= rowCount || col >= colCount) {
       return;
   }
   // 计算中间的实心小方格与外框的距离
   float margin = cellSize / 6;
   Paint paint = new Paint();
   // 设置线宽
   paint.setStrokeWidth(4);
   paint.setStyle(Paint.Style.STROKE_STYLE);
   paint.setColor(color);
   // 计算小方格外框的左上角横坐标
   float left = col * cellSize + cellSpace * (col + 1);
   // 计算小方格外框的左上角纵坐标
   float top = row * cellSize + cellSpace * (row + 1);
   float right = left + cellSize;
   float bottom = top + cellSize;
```

```
    // 绘制小方格外框（空心的）
    canvas.drawRect(new RectFloat(left,top,right,bottom),paint);
    paint.setStyle(Paint.Style.FILL_STYLE);
    // 绘制中心的实心小方格
    canvas.drawRect(new RectFloat(left + margin, top + margin, right - margin,bottom -
    margin),paint);
}
```

小方格由两个部分组成，一部分是中心的实心区域，另一部分是包含这一实心区域的外框。所以调用两次 drawRect 方法，分别绘制了外框和实心区域。

12.4.6 绘制游戏背景小方格

TetrisBackgroundCells 类用于绘制游戏背景小方格，也就是绘制深绿色的小方格，绘制效果如图 12-8 所示。背景有多少个小方格，由与 Tetris 组件绑定组件的尺寸决定。其实图 12-8 所示的效果外面有一个框，这个框是我们在 12.4.4 节中绘制的，为了绘制这个框，TetrisBackgroundCells 类需要从 TetrisBorder 类继承。

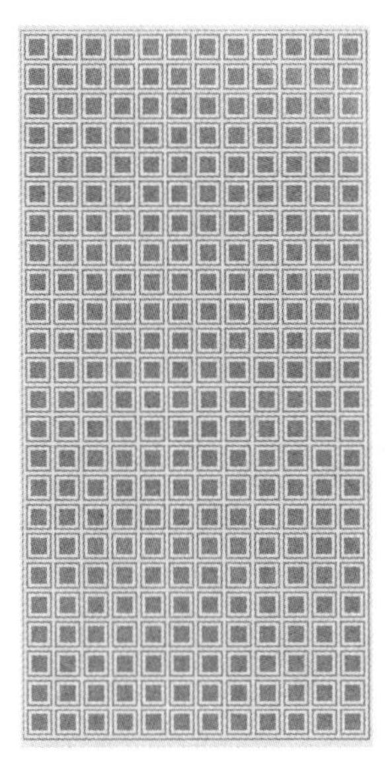

图 12-8 游戏的背景方块

绘制背景小方格的算法也非常简单，只需要用两层 for 循环对 tetrisData 的每一个元素值进行迭代，如果元素值是 0，就绘制深绿色的小方格；如果元素值是 1，就绘制红色的小方格。

TetrisBackgroundCells 类的实现代码如下：

```
package com.harmonyos.tetris;

import ohos.aafwk.ability.Ability;
import ohos.agp.components.Component;
import ohos.agp.render.Canvas;

// 绘制背景小方格
public class TetrisBackgroundCells extends TetrisBorder {

    public TetrisBackgroundCells(Ability ability, Component component) {
        super(ability, component);
    }
    public TetrisBackgroundCells(Ability ability, Component component,int[][] drawBlock)
    {
        super(ability, component,drawBlock);
    }
    // 绘制小方格,row 和 col 都是从 0 开始
    public void onDraw(Component component, Canvas canvas) {
        // 必须调用父类的 onDraw 方法，否则无法绘制边框
        super.onDraw(component,canvas);
        // 循环绘制每一个小方格
        for(int i = 0; i < rowCount;i++) {
            for(int j = 0; j < colCount;j++) {
                if(tetrisData.get(i).get(j) == 0) {
                    // 绘制深绿色的小方格组成的背景
```

```
                drawCell(canvas, i, j, backgroundCellColor);
              } else {
                // 绘制
                drawCell(canvas, i, j, blockColor);
              }
            }
          }

        }
      }
```

如果已经有很多方块落到了游戏界面的底部，那么 tetrisData 中会有很多元素的值为 1，这时绘制的游戏背景小方格就会是两个颜色。如果是 0，小方格颜色就是深绿色；如果是 1，小方格颜色就是红色，效果如图 12-9 所示。

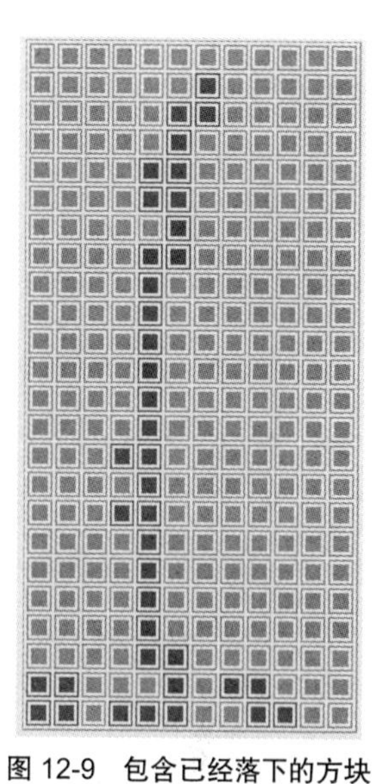

图 12-9 包含已经落下的方块的背景小方格效果

12.4.7 随机产生方块

为了增加游戏的趣味性，每一个新产生的方块都是随机的，尽管有下一个方块的提示，但之后产生的方块，玩家就不知道是什么样式了。为了实现这种随机产生方块的效果，在 TetrisBase 类中实现了 generateNextBlock 方法，用于产生随机的方块。这些随机的方块就是从前面初始化的 5 个方块中随机选择的。

generateNextBlock 函数的代码如下：

```
protected int[][] generateNextBlock() {
   Random random = new Random();
   // 随机产生方块的索引（从 0 到 4）
   int blockIndex = random.nextInt(5);
   // 随机方块的旋转角度，从 0 到 3，表示顺时针旋转 0°、90°、180° 和 270°
   int blockOrientation = random.nextInt(4);
   // 返回旋转后的方块
   return rotateBlock(blockIndex,blockOrientation);
}
// 旋转方块
public int[][] rotateBlock(int[][] block, int orientation) {
   int row = block.length;
   int col = block[0].length;
   int[][] result = null;
   switch (orientation) {
      case 0:    // 旋转 0°，也就是不旋转，直接返回
         result = block;
         break;
      case 1:    // 顺时针旋转 90°
         result = new int[col][row];
         for(int i = 0; i < row;i++) {
            for(int j = 0; j < col;j++) {
               result[j][row - i - 1] = block[i][j];
            }
         }
         break;
```

```
            case 2:      // 顺时针旋转 180°
                result = new int[row][col];
                for(int i = 0; i < row;i++) {
                    for(int j = 0; j < col;j++) {
                        result[i][j] = block[row - i - 1][col - j - 1];
                    }
                }
                break;
            case 3:      // 顺时针旋转 270°
                result = new int[col][row];
                for(int i = 0; i < row;i++) {
                    for(int j = 0; j < col;j++) {
                        result[col - j - 1][i] = block[i][j];
                    }
                }
                break;
        }
        // 重新刷新 component，也就是再次调用 onDraw 方法绘制旋转后的方块
        component.invalidate();
        return result;
    }
    // 顺时针旋转方块，blockIndex 表示 blocks 中方块的索引，orientation 表示旋转方向，从 0 到 3
    public int[][] rotateBlock(int blockIndex, int orientation) {
        int[][] block = blocks.get(blockIndex);
        return rotateBlock(block,orientation);
    }
```

generateNextBlock 方法其实做了 2 次随机处理，第 1 次是从 5 个方块中随机选择 1 个方块，第 2 次是从 4 个旋转方向中随机选择 1 个旋转方向。也就是说，产生的每一个方块，都是随机选择和随机旋转的结果。

12.4.8 消除行

因为方块一直往下落，早晚会没有地方可以落，所以游戏的一个规则是：如果某些行所有的小方格都是方块（本例中是红色），就消除这些行，而上面的所有行都会向下平移。也就是说，在图 12-10 所示的状态中，最底下一行所有的小方格都是红色，这一行就应该被消除。

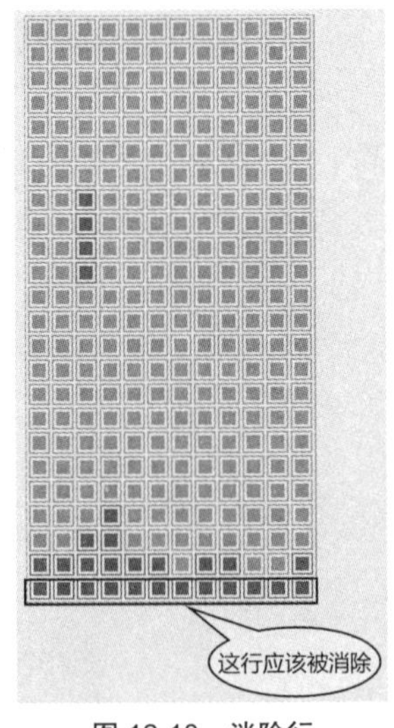

图 12-10 消除行

消除行的工作由 removeRowsFromTetris 方法完成，该方法会返回被消除的行数。消除的算法是从上到下扫描所有的行（就是扫描 tetrisData），如果发现可以消除的行，就用上一行的内容覆盖当前行的内容。例如，发现第 4 行可以消除，就用第 3 行覆盖第 4 行的内容，第 2 行覆盖第 3 行的内容，以此类推。

removeRowsFromTetris 方法的实现代码如下：

```
public int removeRowsFromTetris() {
    int removedRowCount = 0;     // 被消除的行数
    // 从第 1 行（索引为 0 的行）开始扫描
```

```
        for(int i = 0; i < rowCount;i++) {
            // 记录当前行有多少个小方格值为 1，如果个数与 colCount 相同，则消除此行
            int count = 0;
            for(int j = 0; j < colCount;j++) {
                // 一行中只要有一个小方格值为 0，就不能消除
                if(tetrisData.get(i).get(j) == 0) {
                    continue;
                }
                count++;
            }
            // 消除此行
            if(count == colCount) {
                // 消除特定的行
                removeRowFromTetris(i);
                // 消除的行数加 1
                removedRowCount++;
            }
        }
        if(removedRowCount > 0) {
            if(tetrisRowRemovedListener != null) {
                // 如果消除了至少 1 行，并且设置了消除行事件监听器，则触发消除事件
                tetrisRowRemovedListener.rowRemoved(removedRowCount);
            }
        }
        return removedRowCount;
    }
    // 消除具体的行
    public void removeRowFromTetris(int row) {
        // 行超出组件范围，不能消除
        if(row < 0 || row >= rowCount) {
            return;
        }
        // 用上一行的小方格，覆盖当前行的小方格
        if(row > 0) {
            for (int i = row; i > 0; i--) {
                for(int j = 0; j < colCount;j++) {
                    tetrisData.get(i).set(j, tetrisData.get(i - 1).get(j));
                }
            }
        }
        // 如果消除或移动第 1 行，那么就是将第 1 行小方格值全部设置为 0
        for(int j = 0; j < colCount;j++) {
            tetrisData.get(0).set(j, 0);
        }
    }
```

12.4.9 方块归位

正在移动或旋转的方块，是后期绘制到游戏界面的，而已经落下的方块，实际上是在 tetrisData 中的数据，这一部分数据都是 1。因此，在方块落下后，需要按方块落下的最后位置重新设置 tetrisData 的特定区域，可以称这一过程为方块归位。这个功能由 setBlock 方法完成，代码如下：

```
// block 归位，startRow 和 startCol 是方块在 tetrisData 中行和列的值
public void setBlock(int[][] block, int startRow, int startCol) {
    for(int i = 0; i < rowCount;i++) {
        for(int j = 0; j < colCount;j++) {
            if(i >= startRow && i < block.length + startRow) {
                if(j >= startCol && j < block[0].length + startCol) {
                    if (block[i - startRow][j - startCol] == 1) {
                        // 将特定区域设置为 1
                        tetrisData.get(i).set(j, 1);
                    }
                }
            }

        }
    }
}
```

12.4.10 判断当前位置是否可以绘制方块

在游戏界面的当前位置绘制一个方块时，需要考虑是否可以绘制。这要分如下两种情况。

- 方块超出边界。
- 方块要放置的位置已经存在其他已经落下的方块。

第 1 种情况又分为如下 4 种情况。

- 超出下边界：这种情况就是方块已经到了游戏界面底部，不能再往下移动了。
- 超出上边界：当方块一直堆砌，直到游戏界面的顶端时，例如，最上面的一个方块已经距离游戏界面最顶点只有一行了（类似于图 12-9 的效果），如果此时再随机产生一个方块，而且这个方块的行数超过了 1，就肯定无法再显示在游戏界面中了，这个新的方块就被判定为超出上边界。这种情况就认为是游戏结束。
- 超出左边界：在向左移动时，如果方块已经在最左侧一列，就不能再向左移动了，所以如果移动时超出左边界，会无法移动。这时方块的当前列不会变化。
- 超出右边界：在向右移动或顺时针旋转时，如果方块已经在最右侧一列或旋转后超出了最右侧一列，就不能再向右移动或顺时针旋转了。

判断要放置的当前位置是否存在已经落下的方块，只需要判断正在放置的方块相对于游戏背景小方格的位置是否存在值为 1 的小方格即可。

这个算法是通过 verifyBlockLocation 方法实现的，代码如下：

```
// 方块可以放到 tetris 的当前位置
protected final int TETRIS_BLOCK_CONTINUE = 0;
// 方块遇到已经落下的方块，停止移动
protected final int TETRIS_BLOCK_STOPPED = 1;
// 方块超出左边界
protected final int TETRIS_BLOCK_LEFT_OVERFLOW = 2;
// 方块超出下边界
protected final int TETRIS_BLOCK_DOWN_OVERFLOW = 3;
// 方块超出右边界
protected final int TETRIS_BLOCK_RIGHT_OVERFLOW = 4;
```

```
// 方块超出上边界
protected final int TETRIS_BLOCK_UP_OVERFLOW = 5;
// 判断当前位置是否可以放置方块，根据情况返回上面定义的 6 个常量中的一个
public int verifyBlockLocation(int[][] block, int startRow, int startCol) {
    // 获取方块的行数
    int blockRowCount = block.length;
    // 获取方块的列数
    int blockColCount = block[0].length;

    // 在双层 for 循环中同时处理了第 1 种情况和第 2 种情况

    // 扫描方块的每一个小方格，如果小方格值为 1，而且该小方格在 tetrisData 中的位置的值也是 1，
    // 那么返回 TETRIS_BLOCK_STOPPED， 如果在小方格值和在 tetrisData 中的位置有一个为 0，
    // 那么继续扫描下一个小方格
    for(int i = 0; i < blockRowCount;i++) {
        for(int j = 0; j < blockColCount;j++) {
            int rowInTetris = i + startRow;   // 计算方块在 tetris 中的行
            int colInTetris = j + startCol;   // 计算方块在 tetris 中的列
            // 超出下边界
            if(rowInTetris >= rowCount) {
                return TETRIS_BLOCK_DOWN_OVERFLOW;
            }
            // 超出上边界
            if(rowInTetris < 0) {
                return TETRIS_BLOCK_UP_OVERFLOW;
            }
            // 超出右边界
            if(colInTetris >= colCount) {
                return TETRIS_BLOCK_RIGHT_OVERFLOW;
            }
            // 超出左边界
            if(colInTetris < 0) {
                return TETRIS_BLOCK_LEFT_OVERFLOW;
            }
            if(block[i][j] == 1) {
                // 此处有已经落下的方块
                if(tetrisData.get(rowInTetris).get(colInTetris) == 1) {
                    return TETRIS_BLOCK_STOPPED;
                }

            }
        }
    }
    return TETRIS_BLOCK_CONTINUE;
}
```

verifyBlockLocation 方法会根据方块在游戏背景小方格的当前位置返回不同的常量，这些常量会在后面的代码中使用。

12.4.11 绘制方块

在 TetrisBase 类中实现了 drawBlock 方法，用于绘制方块。也就是说，在游戏中方块的移动，

其实是先绘制了背景小方格，然后在背景小方格上再绘制方块，因为每次绘制的坐标不同，所以看上去方块在移动或旋转。

drawBlock 方法的实现代码如下：

```
// 从 startRow 和 startCol 位置开始绘制方块
public void drawBlock(Canvas canvas, int[][] block, int startRow, int startCol) {
    // 如果游戏没开始，不能绘制
    if(!isPlaying && drawBlock == null) {
        return;
    }
    // 如果 drawBlock 为 null，表明绘制的是游戏背景小方格
    if(drawBlock == null) {
        // 开始绘制方块时要判断，是否可以绘制在当前位置
        // 如果可以，继续绘制，并且设置方块的属性
        int blockLocation = verifyBlockLocation(block, startRow, startCol);
        if (blockLocation == TETRIS_BLOCK_CONTINUE) {
        // 如果可以绘制方块
            // 检测方块的下一个位置（row + 1）是否可以移动，
            // 如果不可以移动，说明该位置是方块的最后一个位置
            int nextBlockLocation = verifyBlockLocation(block, startRow + 1, startCol);
            // 不能再向下移动了
            if (nextBlockLocation == TETRIS_BLOCK_DOWN_OVERFLOW ||
                        nextBlockLocation == TETRIS_BLOCK_STOPPED) {
                blockCurrentRow = 0;
                // 将产生的下一个方块作为当前的方块
                currentBlock = nextBlock;
                // 产生下一个方块
                nextBlock = generateNextBlock();
                blockCurrentCol = getBlockStartCol(block);
                if (tetrisNextBlockListener != null) {
                    // 如果设置了事件监听器，则触发下一个方块的事件
                    tetrisNextBlockListener.onNextBlock(nextBlock);
                }
                // 如果方块不能继续向下走了，就应该将正在下落的方块归位，
                // 变成游戏背景小方格的一部分
                setBlock(block, startRow, startCol);
                // 消除行
                removeRowsFromTetris();
                // 让 UI 再次刷新（重新调用 onDraw 方法），并绘制最新的游戏元素
                component.invalidate();
                return;
            }

        } else {
            // 如果不能绘制方块，要先停止游戏
            stop();
            Tools.showTip(ability, "game over");
            if (tetrisGameoverListener != null) {
                // 触发游戏结束事件
                tetrisGameoverListener.onGameover();
            }

            return;
        }
```

```
    }
    // 绘制方块，如果指定了 drawBlock，那么就只绘制方块
    // 其实这时 block 和 drawBlock 值相同，通过 block 参数也会传递 drawBlock 的值
    // 只是为了做得更通用，才通过参数传入 block
    for(int i = 0; i < block.length;i++) {
        for(int j = 0; j < block[i].length;j++) {
            if(block[i][j] == 1) {
                drawCell(canvas, startRow + i, startCol + j,blockColor);
            }
        }
    }
}
```

在 drawBlock 方法中使用了 stop 方法，该方法用于停止游戏，本章后面会详细讲解。

在实现 drawBlock 方法后，还需要改进一下 Tetris 类的 onDraw 方法。因为绘制方块是最后一个动作，所以应该最后做这个动作，因此需要将绘制方块的工作放到继承树的最末端的类中，也就是 Tetris 类，实现代码如下：

```
package com.harmonyos.tetris;

import ohos.aafwk.ability.Ability;
import ohos.agp.components.Component;
import ohos.agp.render.Canvas;

public class Tetris extends TetrisBorder {

    public Tetris(Ability ability, Component component) {
        super(ability, component);
    }
    public Tetris(Ability ability, Component component, int[][] drawBlock)
    {
        super(ability, component,drawBlock);
    }
    public void onDraw(Component component, Canvas canvas) {
        super.onDraw(component,canvas);
        if(drawBlock == null) {
            // 绘制游戏背景小方格
            drawBlock(canvas, currentBlock, blockCurrentRow, blockCurrentCol);
        } else {
            // 绘制下一个方块
            drawBlock(canvas, drawBlock, 0, 0);
        }
    }

}
```

在 onDraw 方法中，会根据是否设置了 drawBlock 决定是绘制游戏背景小方格，还是绘制下一个方块。

12.4.12 顺时针旋转方块

当单击游戏界面时，方块会顺时针旋转，如果超出边界，就不会旋转了。实现这一功能的类是

TetrisEvent 类，代码如下：

```
package com.harmonyos.tetris;

import ohos.aafwk.ability.Ability;
import ohos.agp.components.Component;

public class TetrisEvent extends TetrisBorder {
   public TetrisEvent(Ability ability, Component component,int[][] block) {
      super(ability,component,block);
   }
   public TetrisEvent(Ability ability, Component component) {
      super(ability, component);

      component.setClickedListener(new Component.ClickedListener() {
         @Override
         public void onClick(Component component) {
            if(!isPlaying) {
               return;
            }
            // 得到旋转后的方块
            int[][] block = rotateBlock(currentBlock, blockCurrentOrientation + 1);
            // 判断是否可以旋转
            int result = verifyBlockLocation(block,blockCurrentRow,blockCurrentCol);
            if(result == TETRIS_BLOCK_CONTINUE) {
            // 如果可以旋转
               // 将旋转后的方块作为当前的方块
               currentBlock = block;
               // 让 UI 刷新，也就是调用 onDraw 方法
               component.invalidate();
            }
         }
      });
   }
}
```

因为 TetrisEvent 类是 TetrisBorder 类的子类，所以需要修改 Tetris 类的继承关系，让 Tetris 类从 TetrisEvent 类继承，代码如下：

```
public class Tetris extends TetrisEvent {
   ...
}
```

12.4.13 开始和停止游戏

在默认情况下，Tetris 组件是不会自动开始和停业游戏的，需要调用 start 方法才能开始游戏，调用 stop 方法才能停止游戏。这两个方法在 TetrisBase 类中实现。

俄罗斯方块的游戏规则要求方块不断下落，这就需要一个不断执行的动作，本例中使用了定时器（Timer），将默认下落时间间隔设置为 800 毫秒，也就是说，每 800 毫秒方块会下落一格。

start 方法的代码如下：

```
// 定义一个定时器
protected Timer timer;
```

```
// 开始游戏
public void start(){
   // 每次都重新初始化游戏
   timer = new Timer();
   // 重新初始化与方块相关的变量
   blockCurrentRow = 0;
   // 产生当前方块，也就是第一个出现的方块
   currentBlock = generateNextBlock();
   // 产生下一个显示的方块，也就是第 2 个出现的方块
   nextBlock = generateNextBlock();
   if(tetrisNextBlockListener != null) {
      // 触发下一个方块的事件
      tetrisNextBlockListener.onNextBlock(nextBlock);
   }
   // 随机获取当前方块的列
   blockCurrentCol = getBlockStartCol(currentBlock);
   // 启动定时器
   timer.schedule(new TetrisTask(),0,800);
   isPlaying = true;
}
// 随机获取当前方块的列
protected int getBlockStartCol(int[][] block) {
   if(block.length > 0) {
      int blockCol = block[0].length;
      return new Random().nextInt(colCount - blockCol - 2);
   } else {
       return 0;
   }
}
```

在 start 方法中使用了 getBlockStartCol 方法，该方法会随机获取一个列，所以方块在产生时，不仅其种类和姿态（方向）是随机产生的，就连水平方向的位置也是随机产生的。

用于停止游戏的 stop 方法的代码如下：

```
public void stop() {
   // 取消定时器
   timer.cancel();
   isPlaying = false;
}
```

12.4.14 快速下落与正常下落之间的切换

将方块对准下落位置后，就可以按住快速下落按钮让方块快速下落到指定位置。如果在下落的过程中发现位置并没有对准，可以松开快速下落按钮，让方块恢复默认的下落速度，调整好位置后，再使其快速下落。

实现快速下落的原理就是加快定时器的执行时间间隔，例如，从 800 毫秒调整到 100 毫秒。恢复正常下落速度，就是恢复定时器的执行时间间隔到 800 毫秒。这两个功能分别由 moveQuickly 方法和 moveNormally 方法实现，代码如下：

```
// 快速下落
public void moveQuickly() {
```

```
        stop();
        timer = new Timer();
        // 使用执行时间间隔为 100 毫秒的定时器
        timer.schedule(new TetrisTask(),0,100);
        isPlaying = true;
    }
    // 恢复正常下落速度
    public void moveNormally() {
        stop();
        timer = new Timer();
        // 恢复使用执行时间间隔为 800 毫秒的定时器
        timer.schedule(new TetrisTask(),0,800);
        isPlaying = true;
    }
```

当开启定时器时，需要为定时器指定一个定期执行的任务，也就是 schedule 方法的第 1 个参数，本例中是 TetrisTask 类。定时器使用指定的时间频率执行 TetrisTask.run 方法。在 run 方法中会不断让方块的当前行（blockCurrentRow）加 1，这样就会让方块不断往下移动。TetrisTask 类以及相关的代码如下：

```
    public class TetrisEventHandler extends EventHandler {

        public TetrisEventHandler(EventRunner runner) {
            super(runner);
        }
        @Override
        public void processEvent(InnerEvent event) {
            super.processEvent(event);
            if (event == null) {
                return;
            }
            // 方块的当前行加 1，让方块不断往下移动
            blockCurrentRow++;
            // 刷新 UI
            component.invalidate();
        }
    };
    public class TetrisTask extends TimerTask {
        @Override
        public void run() {

            EventRunner runnerA = EventRunner.getMainEventRunner();
            TetrisEventHandler handler = new TetrisEventHandler(runnerA);
            handler.sendEvent(0);
            runnerA = null;
        }
    }
```

12.4.15　左右水平移动方块

单击窗口下方的左右箭头按钮，方块会左右水平移动，单击一下移动一个小方格。实现原理是修改方块的当前列（blockCurrentCol），让当前列的值减小（向左水平移动）或增加（向右水平移

动）。这两个方法的实现代码如下：

```
// 向左水平移动
public void moveLeft() {
   if(isPlaying) {
      int result = verifyBlockLocation(currentBlock, blockCurrentRow,
                                                  blockCurrentCol - 1);
      // 判断方块是否超出左边界
      if(result != TETRIS_BLOCK_LEFT_OVERFLOW) {
         // 未超出左边界，当前列的值减 1，方块向左水平移动
         blockCurrentCol--;
         component.invalidate();
      }
   }
}
// 向右水平移动
public void moveRight() {
   if(isPlaying) {
      int result = verifyBlockLocation(currentBlock, blockCurrentRow,
                                                   blockCurrentCol + 1);
      // 判断方块是否超出右边界
      if(result != TETRIS_BLOCK_RIGHT_OVERFLOW) {
         // 未超出右边界，当前列的值加 1，方块向右水平移动
         blockCurrentCol++;
         component.invalidate();
      }

   }
}
```

12.4.16 为 Tetris 组件增加属性

通过 TetrisProperties类可以为Tetris组件添加属性，本例中只添加了设置和获取方块背景颜色、游戏背景小方格颜色的方法（setter 和 getter），读者可以添加更多的属性。TetrisProperties 类的代码如下：

```
package com.harmonyos.tetris;

import ohos.aafwk.ability.Ability;
import ohos.agp.components.Component;
import ohos.agp.utils.Color;

public class TetrisProperties extends  TetrisBackgroundCells {
   public TetrisProperties(Ability ability, Component component) {
      super(ability, component);
   }
   public TetrisProperties(Ability ability, Component component,int[][] drawBlock) {
      super(ability, component,drawBlock);
   }
   // 获取方块背景颜色
   public Color getBlockColor() {
      return blockColor;
   }
   // 获取游戏背景小方格颜色
```

```
    public Color getBackgroundCellColor() {
        return backgroundCellColor;
    }
    // 设置方块背景颜色
    public void setBlockColor(Color blockColor) {
        this.blockColor = blockColor;
        component.invalidate();
    }
    // 设置游戏背景小方格颜色
    public void setBackgroundCellColor(Color backgroundCellColor) {
        this.backgroundCellColor = backgroundCellColor;
        component.invalidate();
    }
}
```

到现在为止，Tetris 组件中的所有类都已经实现完了，一共 7 个类，读者需要按 12.2 节讲解的继承关系重新修改这 7 个 Java 类的父类。

12.5 让游戏更完美

如果只是将 Tetris 组件放到窗口中，程序已经可以运行了。只不过功能有些单一，只支持方块向下移动和旋转（单击游戏窗口），为了让游戏更完美，本节将讲解如何利用一些组件来控制游戏。

因为已经将俄罗斯方块游戏封装在 Tetris 组件中，所以利用 Tetris 组件实现一个俄罗斯方块游戏是非常容易的，只需要几分钟时间就可以搞定。

12.5.1 开始玩游戏

Tetris 组件默认是不会开始游戏的，需要调用 start 方法才能开始游戏，代码如下：

```
Image imageStart = (Image)findComponentById(ResourceTable.Id_image_start);
if(imageStart != null) {
    imageStart.setClickedListener(new Component.ClickedListener() {
        @Override
        public void onClick(Component component) {
            if(!tetris.isPlaying) {
                Tools.showTip(MainAbility.this,"游戏开始");
                // 开始游戏
                tetris.start();
            }
        }
    });
}
```

现在运行程序，单击窗口最下方的红色 Start 按钮（见图 12-11），方块就会自动在游戏窗口的上方出现，并不断下落。

然后窗口中会弹出一个 Toast 信息框用来提示游戏已经开始，如图 12-12 所示。

图 12-11 开始按钮

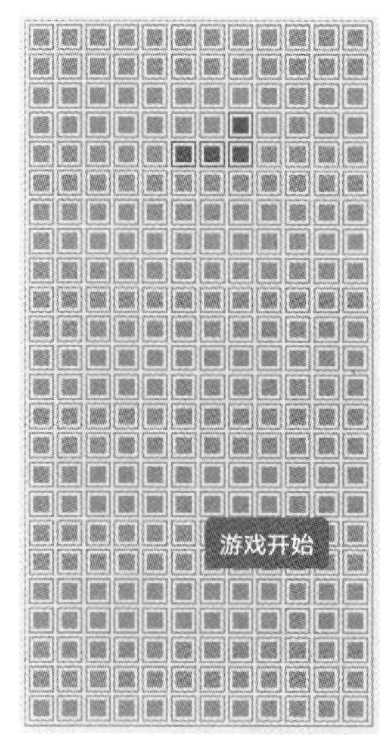

图 12-12 开始游戏

12.5.2 显示下一个方块

Tetris 组件提供了获取下一个方块的事件，只要使用该事件的 setter 方法为其指定监听器即可。显示下一个方块其实也使用了 Tetris 组件，只是为 Tetris 组件指定了一个方块，所以 Tetris 组件只绘制这个方块，实现代码如下：

```
// 为 Tetris 组件指定监听下一个方块的监听器
tetris.setOnTetrisNextBlockListener(new TetrisNextBlockListener() {
   // 每产生下一个方块，就会调用一次该方法
   @Override
   public void onNextBlock(int[][] block) {
      getUITaskDispatcher().asyncDispatch(new Runnable() {
         @Override
         public void run() {
           // 下一个方块就绘制在这个布局上
           DirectionalLayout layoutNext =
            (DirectionalLayout)findComponentById(ResourceTable.Id_layout_next);
           // 创建 Tetris 组件，用来显示下一个方块
           Tetris nextTetris = new
                       Tetris(MainAbility.this,layoutNext,tetris.nextBlock);

         }
      });
    }
});
```

我们可以看到，实际代码只有 run 方法中的两行，如果不为 Tetris 组件指定方块，就相当于在窗口中放置了另外一个俄罗斯方块游戏。

现在运行程序，开始游戏后，就会在窗口的右侧显示下一个方块，如图 12-13 所示。

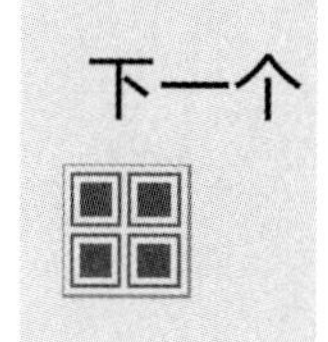

图 12-13 显示下一个方块

12.5.3 控制方块左右水平移动

单击窗口下方的左右箭头，block 会向左水平移动或向右水平移动，要完成这个功能，只要在左右箭头的单击事件中分别调用 moveLeft 方法和 moveRight 方法即可，实现代码如下：

```
Image imageMoveLeft = (Image)findComponentById(ResourceTable.Id_image_move_left);
if(imageMoveLeft != null) {
   imageMoveLeft.setClickedListener(new Component.ClickedListener() {
      @Override
      public void onClick(Component component) {
         // 向左水平移动
         tetris.moveLeft();
      }
   });
}
Image imageMoveRight = (Image)findComponentById(ResourceTable.Id_image_move_right);
if(imageMoveRight != null) {
   imageMoveRight.setClickedListener(new Component.ClickedListener() {
      @Override
      public void onClick(Component component) {
         // 向右水平移动
         tetris.moveRight();
      }
    });
}
```

运行程序，然后开始游戏，单击图 12-14 所示的左箭头按钮和右箭头按钮，方块就会向左或向右水平移动。

图 12-14 向左水平移动和向右水平移动按钮

12.5.4 控制方块快速下落

让方块快速下落需要调用 moveQuickly 方法，恢复正常下落速度需要调用 moveNormally 方法。但需要在快速下落按钮的触摸事件中对这两个方法进行调用，因为在触摸事件里可以区分手指按下和抬起的动作，实现代码如下：

```
Image imageMoveQuickly = (Image)findComponentById(ResourceTable.Id_image_move_quickly);
if(imageMoveQuickly != null) {
imageMoveQuickly.setTouchEventListener(new Component.TouchEventListener() {
   @Override
   public boolean onTouchEvent(Component component, TouchEvent touchEvent) {
      if(touchEvent.getAction() == TouchEvent.PRIMARY_POINT_DOWN) {
         // 手指按下，快速下落
         tetris.moveQuickly();
      } else if(touchEvent.getAction() == TouchEvent.PRIMARY_POINT_UP) {
         // 手指抬起，恢复正常的下落速度
         tetris.moveNormally();
      }
      return true;
   }
});
```

图 12-15 快速下落按钮

运行程序，然后开始游戏，用手指按下（不要抬起来）图 12-15 所示的快速下落按钮，会发现方块会快速下落，然后抬起手指，方块的下落速度又恢复正常了。

12.5.5 处理积分

Tetris 组件提供了消除行监听器，本游戏规定，每消除一行，积分会增加 10，所以可以在这个监听器的 rowRemoved 方法中更新积分，实现代码如下：

```
tetris.setOnTetrisRowRemovedListener(new TetrisRowRemovedListener() {
   @Override
   public void rowRemoved(int rowCount) {
      // 获取当前积分
      int currentScore = Integer.parseInt(textScore.getText());
      // 积分累加
      currentScore += rowCount * 10;
      // 更新积分
      textScore.setText(String.valueOf(currentScore));
   }
});
```

运行程序，并开始游戏，玩一会儿游戏，让某些行可以消除，就会发现右侧的积分变化了，如图 12-16 所示。

积分
40

图 12-16 显示积分

12.5.6 游戏结束

Tetris 组件提供了可以监听游戏结束的事件，如果要提示游戏结束，就需要使用这个事件，代码如下：

```
tetris = new Tetris(this,layoutTetris);
tetris.setOnTetrisGameoverListener(new TetrisGameoverListener() {
   @Override
   public void onGameover() {
       Tools.showTip(MainAbility.this,"游戏结束");
   }
});
```

在这段代码的 onGameover 事件中只显示了 Toast 信息框用于提示游戏结束，读者也可以使用对话框或其他形式向玩家提示游戏结束。

12.6 总结与回顾

本章通过项目案例讲解了一个游戏类 App 的实现，由于目前 HarmonyOS 还不支持 OpenGL，也不支持 Vulkan，无法移植游戏引擎，很难实现复杂的游戏，因此本章的俄罗斯方块游戏采用画布（Canvas）绘制所有的游戏元素。因为俄罗斯方块在 UI 上并不复杂，所以采用画布完全可以实现预期的效果。

本章讲解的游戏项目是将这款俄罗斯方块游戏做成组件形式，这也符合目前大多数游戏开发团队的需求。现在开发的 App 越来越复杂，一个 App 可能由多人同时开发，如果多人共同维护同一个工程，可能会造成不必要的麻烦，例如，多人同时修改一个源代码文件，就有可能造成冲突。因此，最好的方式是将一些核心的功能提炼出来，做成组件，单独调试组件，然后再进行集成和联调，这样就可以在最大限度上避免冲突。同时核心功能以组件的形式提供，还可以复用，大幅提高开发效率。例如，如果使用俄罗斯方块游戏组件（Tetris 组件）开发游戏，调用组件的核心代码只不过十几行，完全可以在几分钟内搞定，这也是本章章名为“5 分钟搞定俄罗斯方块”的原因。

俄罗斯方块游戏主要使用了图形绘制技术。要实现这款游戏，还需要了解这款游戏的规则和算法。基本规则就是行消除。算法的核心就是如何判断一个俄罗斯方块在落下时何时该停止，以及停止后什么情况下应该消除行。剩下的就是具体的业务逻辑了，例如，积分如何计算、是否产生下一个方块、顺时针旋转方块等。

本章中实现的俄罗斯方块游戏已经具备了基本的功能，但还可以更进一步完善，读者可以尝试在如下几个方面完善这款游戏。

- 为 Tetris 组件增加更多的属性和事件，让 Tetris 组件变得更灵活、更强大。
- 增加更多的方块。
- 支持用按钮控制方块的旋转。
- 支持顺时针和逆时针两种旋转方式。
- 设计更复杂的积分规则。
- 可以利用远程调用 Service Ability 的特性增加双人对战，其基本规则是 A 和 B 一起玩，如果 A 消除了 n 行，B 就会增加 n 行，反之亦然。
- 增加多人观战模式，允许更多人加入，观看实时对战。
- 增加 TV 对战模式，也就是说，两个人同时连接 TV，这时两部手机会变成两个遥控器，TV 屏幕上会显示左右两个游戏界面，分别代表两个玩家。这种双人对战模式，双方可以互相看到对方的屏幕，用手机遥控各自的游戏屏幕。

读者可以尝试在上述几个方面对俄罗斯方块游戏进行完善，也可以加入更多更酷的功能，并将其发布到码云（Gitee）或 GitHub 上，让更多人学习使用。